ANNALS OF
THE NEW YORK ACADEMY
OF SCIENCES

Volume 1045

EDITORIAL STAFF

Director, Publishing and New Media
SARAH GREENE

Managing Editor
JUSTINE CULLINAN

Associate Editor
JOHN W. KENNEDY

The New York Academy of Sciences
2 East 63rd Street
New York, New York 10021

THE NEW YORK ACADEMY OF SCIENCES
(Founded in 1817)
BOARD OF GOVERNORS, September 2004–September 2005
TORSTEN N. WIESEL, *Chairman of the Board*
GERALD D. FISCHBACH, *Vice Chairman*
MICHAEL SCHMERTZLER, *Treasurer*
ELLIS RUBINSTEIN, *Chief Executive Officer* [ex officio]

Honorary Life Governors
WILLIAM T. GOLDEN   JOSHUA LEDERBERG

Governors

KAREN E. BURKE	VIRGINIA W. CORNISH	PETER B. CORR
R. BRIAN FERGUSON	RONALD L. GRAHAM	MARNIE IMHOFF
WENDY EVANS JOSEPH	JACQUELINE LEO	ROBERT W. LUCKY
PAUL MARKS	BRUCE McEWEN	RONAY MENSCHEL
JOHN T. MORGAN	JOHN F. NIBLACK	SANDRA PANEM
PETER RINGROSE	DAVID D. SABATINI	JOHN SEXTON
	DEBORAH WILEY	

VICTORIA BJORKLUND, *Counsel* [ex officio]   LARRY R. SMITH, *Secretary* [ex officio]

NONLINEAR DYNAMICS IN ASTRONOMY AND PHYSICS
In Memory of Henry E. Kandrup

ANNALS OF THE NEW YORK ACADEMY OF SCIENCES
Volume 1045

NONLINEAR DYNAMICS IN ASTRONOMY AND PHYSICS
In Memory of Henry E. Kandrup

Edited by
Stephen T. Gottesman, J.-R. Buchler, and M.E. Mahon

The New York Academy of Sciences
New York, New York
2005

Copyright © 2005 by the New York Academy of Sciences. All rights reserved. *Under the provisions of the United States Copyright Act of 1976, individual readers of the* Annals *are permitted to make fair use of the material in them for teaching and research. Permission is granted to quote from the* Annals *provided that the customary acknowledgment is made of the source. Material in the* Annals *may be republished only by permission of the Academy. Address inquiries to the Permissions Department (permissions@nyas.org) at the New York Academy of Sciences.*

Copying fees: *For each copy of an article made beyond the free copying permitted under Section 107 or 108 of the 1976 Copyright Act, a fee should be paid through the Copyright Clearance Center, Inc., 222 Rosewood Drive, Danvers, MA 01923 (www.copyright.com).*

♾ *The paper used in this publication meets the minimum requirements of American National Standard for Information Sciences—Permanence of Paper for Printed Library Materials. ANSI Z39.48-1984.*

Library of Congress Cataloging-in-Publication Data

Florida Workshops in Nonlinear Astronomy and Physics (16th : 2004 : Gainesville, Fla.)
  Nonlinear dynamics in astronomy and astrophysics : in memory of Henry E. Kandrup / edited by Stephen T. Gottesman, J.-R. Buchler, and M.E. Mahon
    p. cm. — (Annals of the New York Academy of Sciences ; v. 1045)
  Includes bibliographical references and index.
  ISBN 1-57331-590-7 (alk. paper) — ISBN 1-57331-591-5 (pbk. : alk. paper)
  1. Stellar dynamics—Congresses. 2. Nonlinear theories—Congresses. 3. Astrophysics—Congresses. I. Kandrup, Henry E. II. Gottesman, S. T. (Stephen T.) III. Buchler, J. R. (J. Robert) IV. Mahon, M. E. V. Title. VI. Series.
  Q11.N5 vol. 1045
  [QB810]
  500 s—dc22
  [523
                                                                                            2005014059
                                                                                                CIP

K-M Research/CCP
Printed in the United States of America
ISBN 1-57331-590-7 (cloth)
ISBN 1-57331-591-5 (paper)
ISSN 0077-8923

ANNALS OF THE NEW YORK ACADEMY OF SCIENCES

Volume 1045
June 2005

NONLINEAR DYNAMICS IN ASTRONOMY AND PHYSICS
In Memory of Henry E. Kandrup

Editors
STEPHEN T. GOTTESMAN, J.-R. BUCHLER, AND M.E. MAHON

This volume is the result of a workshop entitled **Nonlinear Dynamics in Astronomy and Physics**, dedicated to the memory of Professor Henry E. Kandrup, held November 4-6, 2004, at the University of Florida, Gainesville, Florida.

CONTENTS

Preface. *By* S.T. GOTTESMAN, J.-R. BUCHLER, AND M.E. MAHON	ix
In Memory of Henry Emil Kandrup. *By* JAMES R. IPSER	1
Henry Kandrup's Ideas About Relaxation of Stellar Systems. *By* DAVID MERRITT	3
Chaotic Collisionless Evolution in Galaxies and Charged-Particle Beams. *By* HENRY E. KANDRUP, COURTLANDT L. BOHN, RAMI A. KISHEK, PATRICK G. O'SHEA, MARTIN REISER, AND IOANNIS V. SIDERIS	12
Chaotic Dynamics in Charged-Particle Beams: Possible Analogs of Galactic Evolution. *By* COURTLANDT L. BOHN	34
The University of Maryland Electron Ring: A Platform For Study of Galactic Dynamics on a Laboratory Scale. *By* R.A. KISHEK, P.G. O'SHEA, S. BERNAL, I. HABER, J.R. HARRIS, Y. HUO, H. LI, AND M. REISER	45
Wavelet-Based Poisson Solver for Use in Particle-in-Cell Simulations. *By* BALSA TERZIĆ AND ILYA V. POGORELOV	55
Energy Trapping in Loaded String Models with Long- and Short-Range Couplings. *By* ILYA V. POGORELOV AND HENRY E. KANDRUP	68
Characterization of Chaos: A New, Fast, and Effective Measure. *By* IOANNIS V. SIDERIS	79
Hard Sphere Dynamics for Normal and Granular Fluids. *By* JAMES W. DUFTY AND APARNA BASKARAN	93
Nonlinear Stability of Newtonian Galaxies and Stars from a Mathematical Perspective. *By* GERHARD REIN	103

Chaos in Orbits Due to Disk Crossings. *By* C. HUNTER	120
Systems with Escapes. *By* G. CONTOPOULOS AND M. HARSOULA	139
On Bars and Haloes: Their Interaction and Their Orbital Structure. *By* E. ATHANASSOULA	168
The Basic Dynamical Mechanism in Spiral Galaxies. *By* DANIEL PFENNIGER AND YVES REVAZ	193
The Two Pattern Speeds of NGC 3359. *By* VEERA BOONYASAIT, P.A. PATSIS, AND S.T. GOTTESMAN	203
Evolution of Binary Supermassive Black Holes via Chain Regularization. *By* ANDRAS SZELL, DAVID MERRITT, AND SEPPO MIKKOLA	225
Gravitomagnetic Field and Penrose Scattering Processes. *By* REVA KAY WILLIAMS	232
Self-Gravity Driven Instabilities at Accelerated Interfaces. *By* ROBERT M. HUECKSTAEDT, JAMES H. HUNTER, JR., AND RICHARD V.E. LOVELACE	246
Dynamics of Intracluster Gas and Bulk Motions in Clusters. *By* RENATO DUPKE	260
Resonance Bands and Binary-Star Formation. *By* NORMAN R. LEBOVITZ	276
The Symplectic Group and Classical Mechanics. *By* ALEX J. DRAGT	291
Chaos and Quantum Mechanics. *By* SALMAN HABIB, TANMOY BHATTACHARYA, BENJAMIN GREENBAUM, KURT JACOBS, KOSUKE SHIZUME, AND BALA SUNDARAM	308
Afterword. *By* Chris Hunter	333
Index of Contributors	335

Financial assistance was received from:

- THE FLORIDA SPACE GRANT CONSORTIUM
- THE DEPARTMENT OF ASTRONOMY, UNIVERSITY OF FLORIDA
- THE DEPARTMENT OF PHYSICS, UNIVERSITY OF FLORIDA
- THE COLLEGE OF LIBERAL ARTS AND SCIENCES, UNIVERSITY OF FLORIDA
- THE OFFICE OF THE VICE PRESIDENT OF RESEARCH, UNIVERSITY OF FLORIDA
- THE INSTITUTE FOR FUNDAMENTAL THEORY, UNIVERSITY OF FLORIDA

The New York Academy of Sciences believes it has a responsibility to provide an open forum for discussion of scientific questions. The positions taken by the participants in the reported conferences are their own and not necessarily those of the Academy. The Academy has no intent to influence legislation by providing such forums.

Professor Henry E. Kandrup (1955 – 2003)

(Photograph reproduced courtesy of the Kandrup family.)

Preface

The Departments of Astronomy and Physics at the University of Florida have jointly sponsored sixteen annual workshops on aspects of nonlinear astronomy and physics. The most recent to appear in print was *The Onset of Nonlinearity*, published in 2001 as Volume 927 of the *Annals of the New York Academy of Sciences*. A common thread through almost all of these publications was Professor Henry E. Kandrup, who acted as a coeditor. In October 2003 Professor Kandrup died. In November 2004, a workshop in his memory was held at the University of Florida, sponsored again by the two departments. The title of this meeting was "Nonlinear Dynamics in Astronomy and Physics." Currently, this is a most important sphere of research in the physics and astrophysics communities. In addition, this topic encompasses the broad area of Henry Kandrup's interests.

The objective of the workshop was to bring together for three days an eclectic mix of experts from applied mathematics, astrophysics, galactic dynamics, plasma physics, cosmology, accelerator design, and statistical mechanics to discuss a variety of rapidly developing and overlapping subjects in nonlinear dynamics. Our emphasis was on the cross-disciplinary nature of much current research in dynamics within the astronomy and physics communities. It is clear from many examples that developments in one field can advance more quickly when there is information available concerning parallel developments in other fields. This occurs not only by the exchange of methods and techniques, but also from conceptual refinement resulting from different perspectives on similar problems. We expected that the interdisciplinary nature of the presentations and discussions at the workshop would fertilize further developments. Publication of this volume in the *Annals of the New York Academy of Sciences* series offers an important record of these exchanges.

The topic of the 2004 Workshop and this volume of the *Annals*, "Nonlinear Dynamics in Astronomy and Physics," pervades dynamical problems on all astrophysical scales from those of the Sun and the Solar System, exo-planetary systems, stellar and galactic systems, to the largest cosmological scales. Now is a particularly appropriate time for these ideas because observations have provided compelling evidence that many galaxies are genuinely three-dimensional, that is, they are neither spherical nor axisymmetric, that they typically have a central density cusp, and that the center of the galaxy often contains a super massive black hole. Furthermore, the Hubble Space Telescope has produced exciting data about the spatial and temporal structure of cosmological space–time and especially about the epoch of galaxy formation. The research specialties of the papers here reflect this diversity and we hope this volume will appeal to a broad audience.

We cannot end these introductory comments without saying something about Henry Kandrup and his contributions. Most of Henry's research was centered on stochastic problems in stellar dynamics. His work was fertile and seminal. He knew all of the contributors and collaborated directly with many. The presentations at the workshop and papers in these proceedings reflect the mutual stimuli of these connections.

Professor Bohn has most graciously and generously put Henry Kandrup's name as lead author of the third paper in this volume. This reflects not only the respect of the community for Henry but also the stature of the man and his contributions, especially to this new field of what he liked to call "laboratory galactic dynamics." Henry was an essential part of the earlier workshops sponsored by the Departments of Physics and Astronomy and it is altogether fitting that the meeting and these published proceedings be dedicated to his memory. To paraphrase Salman Habib (in this volume), although we do not know what Henry's opinions would have been on this workshop, we can be sure he would not have been quiet.

The workshop and this volume were sponsored and supported by the Florida Space Grant Consortium, and by the following: the departments of Astronomy and Physics; the College of Liberal Arts and Sciences; the Office of the Vice President for Research; and the Institute for Fundamental Theory, all of the University of Florida. In addition, we recognize the ever helpful assistance of the editorial offices of the New York Academy of Sciences, especially Ms. Justine Cullinan, Ms. Linda Mehta, and especially Dr. John W. Kennedy, who so skillfully saw this publication through the press.

S.T. GOTTESMAN, J.-R. BUCHLER, AND M.E. MAHON
Editors and Local Organizers

In Memory of Henry Emil Kandrup

I met Henry Kandrup in 1978 at the University of Chicago, when he came into the Fermi Institute as a graduate student. He expressed an interest in working on problems in stellar dynamics. That area was one of my interests, and we entered into a dialogue. Within a year Henry had submitted his first paper for publication, on the gravothermal catastrophe in stellar systems. I felt privileged to be coauthor of that paper.

Henry progressed rapidly (he was always very "fast"). By the Spring of 1979 he had developed, largely on his own, a thesis topic on the stellar-dynamics problem of distribution functions in gravitation systems. Indeed, Henry's rapid development and mastery of his thesis-topic area were confirmed by the request, to which he acquiesced, that he write a review for *Physics Reports* on gravitational fluctuations and stellar distributions in gravitational systems. This occurred while he was still a graduate student!

Henry defended his thesis in early 1980, an event I will never forget. As soon as I walked into the examination room on the morning of Henry's defense, another member of Henry's committee came up to me and remarked not only that Henry had done a marvelous job in his research, but also that Henry's clarity, coherence, and overall writing style were the best this faculty member could remember seeing in a thesis. Indeed, throughout his career the quality of Henry's exposition, both orally and in writing, was of the highest standard.

Henry made the usual journey through a variety of institutions before joining the faculty of the Astronomy Department at the University of Florida in 1990, where he remained until his untimely death. His contributions to stellar dynamics. and theoretical astrophysics in general, span a wide range of topics, including the theory of stellar distributions, violent relaxation, dynamical and secular instabilities, entropy generation, Landau damping and phase mixing, Hamiltonian formulations, chaos and Lyapunov exponents, discreteness effects in the N-body problem, and problems in relativistic astrophysics.

Although he was a theoretician, Henry always was on the lookout for possible connections of his work with observational and experimental results. In fact, at the time of his death, Henry and his coworkers were very excited about the possibility of developing experiments to test for the presence of phenomena in charged particle beams that are related to his ideas on chaos and chaotic mixing in N-body systems.

Henry was devoted to his students, and that devotion was returned with warm and deep affection. Tragically, we have lost Henry, but I remain confident that his spirit and science will live on, in particular through the work of the students he so conscientiously nurtured.

JAMES R. IPSER

Henry Kandrup's Ideas About Relaxation of Stellar Systems

DAVID MERRITT

Rochester Institute of Technology
Rochester, New York, USA

ABSTRACT: Henry Kandrup wrote prolifically on the problem of relaxation of stellar systems. His picture of relaxation was significantly more refined than the standard description in terms of phase mixing and violent relaxation. In this article, I summarize Henry's work in this and related areas.

KEYWORDS: Henry Kandrup; relaxation; stellar systems

Henry Kandrup was a leading figure in galactic dynamics, distinguishing himself both as an original thinker and as an educator. His published contributions were prolific and wide-ranging. Although I never formally collaborated with Henry, we often discussed dynamics during my visits to the University of Florida or at scientific meetings, and both of us were conscientious readers of the other's papers. Henry's ideas often influenced my own work, and I believe that I can see evidence of the reverse influence in some of Henry's published papers.

Here I focus on a topic that occupied Henry's attention throughout his career: dynamical chaos and its connection with relaxation and equilibrium of stellar systems. Henry first addressed this topic in his Ph.D. Thesis, *Stochastic Processes in Stellar Dynamics*,[1] and returned to it in one of his last papers, "Chaos and Chaotic Phase Mixing in Galaxy Evolution and Charged Particle Beams".[2]

Henry's unique contribution was to associate relaxation in stellar systems with the exponential instability of orbits. Prior to Henry's work, discussions in the astronomy literature of dynamical relaxation rarely addressed the mixing properties of flow, and discussions of stochasticity rarely drew conclusions about relaxation. Henry argued that there was a fundamental connection between the two. He pointed out that the trajectories of stars even in "collisionless" systems like galaxies, could often mimic the exponentially unstable trajectories of molecules in a collisional fluid, due either to non-integrability of the steady-state potential, or to time variations in the potential associated with external perturbations or departures from a steady state. Henry demonstrated that this chaos or near-chaos could be very effective at inducing evolution to a steady state, in much the same way that collisions between molecules in a gas erase memory of the initial conditions. In this way, Henry established a new and important link between dynamical chaos, statistical mechanics, and the structure and evolution of stellar systems.

Many stellar and galactic systems are smooth and symmetric in appearance. This is surprising at first sight, since the time required for gravitational encounters to

---

Address for correspondence: David Merritt, Rochester Institute of Technology, 54 Lomb Memorial Drive, Rochester, NY 14623-5603, USA.
merritt@astro.rit.edu

"smear out" trajectories is very long in such low-density systems, often much longer than the age of the universe. One of the first to note this puzzle was Fritz Zwicky.[3] Zwicky was so impressed by the regular appearance of the Coma galaxy cluster that he argued against Hubble's expanding-universe model—on the grounds that the age of the universe in the Hubble model was too short for encounters between galaxies to remove irregularities in their spatial distribution.

At the time that Henry began his career, the smooth appearance of galaxies and galaxy clusters was generally understood to be due to a combination of phase mixing and "violent relaxation". Phase mixing is the gradual shearing of points in a fixed, integrable potential; after many orbital periods, phase mixing results in a coarse-grained density that is constant with respect to angle over the invariant torus. "Violent relaxation" was defined as the more extreme redistribution of particles that occurs when the gravitational potential is rapidly varying. King,[4] Hénon,[5] Lynden-Bell,[6] and others realized that relaxation under conditions of a rapidly varying potential might be very efficient, and the last-named author identified the relaxation rate directly with the rate of change of the potential. Support for this hypothesis was seen in numerical simulations of the collapse of a cold cloud of stars, where a nearly steady state is reached after just a few crossing times.

Henry and I often discussed violent relaxation during my visits to the University of Florida. During one of these conversations, Henry told me that he had been very impressed by a letter written by Richard Miller to Donald Lynden-Bell on July 21, 1966. In the letter, Miller argued that the identification of relaxation with changes in the potential was problematic:

> "A counter example is furnished by a potential that depends only upon the time. Consider a stellar dynamical system described by a Hamiltonian H_0 and another described by $H = H_0 + V(t)$, where V depends only upon the time. All motions in the two systems are identical; no relaxation is induced by $V(t)$, contrary to the assertion of your equation 1."

Miller went on to note:

> "I think there is a germ of an idea in your assertion [that relaxation can be identified with changes in the potential], but it wants more complete working-out. Essentially, I think that the kind of term that might replace $\langle \dot{\Phi}^2 \rangle$ in your equations is $\langle (G\rho)^{-1/2} \dot{\Phi}(a \cdot \nabla\Phi) \rangle$ or some rather fancy term of that character—displaying both time and space derivatives, but averaged over the cluster and measured in time units characteristic of the cluster (a is some length characterizing the cluster)."

Henry told me that he shared Miller's reservations about equating relaxation with $\dot{\Phi}$. As Henry put it:[7]

> "More pragmatically, one infers from N-body simulations that a strongly convulsing mean field potential is not necessary. One observes a comparably efficient approach towards a metaequilibrium on a time scale [of order] t_{cr} [the crossing time] both for 'violent' evolution, where Φ exhibits huge changes on a very short time scale, and for 'nonviolent' evolution, where Φ exhibits only relatively small changes. Nonviolent relaxation can be just as efficient as violent relaxation."

In this passage, Henry is referring to what he elsewhere called "chaotic mixing"—the exponential spreading of an ensemble of initially localized, stochastic trajectories. He continued:

> "This means that phase mixing can proceed *much* more efficiently for chaotic flows than for regular flows, where any approach towards a (near-)equilibrium typically proceeds as a power law in time. Chaotic flows should relax much more efficiently than do regular flows. It would thus seem that the phase mixing of chaotic flows ... could serve to provide an explanation of why various systems in nature seem to approach an equilibrium or near-equilibrium as fast as they do. In particular, chaotic mixing could help explain the remarkable efficacy of violent relaxation: Why do galaxies look 'so relaxed' when the nominal relaxation time t_R is typically much longer than t_H, the age of the Universe?

Rather than identify relaxation with either phase mixing in a fixed potential, or "violent relaxation" in a time-varying potential, Henry proposed that the proper distinction was between phase mixing and chaotic mixing, and that the time dependence of the potential was secondary. The efficient relaxation observed in simulations of collapse was due, he argued, to the more chaotic nature of the phase-space flow when the potential was rapidly varying, and not simply due to the redistribution of energies that takes place when the potential has a time-dependent component.

Although the existence of stochastic orbits in galactic potentials had been appreciated since the work of Hénon, Contopoulos, Miller, and others in the 1960s, Henry was one of the first to ask what the *consequences* of the chaos might be for the evolution of an ensemble of orbits toward a statistical steady state. Henry began a systematic investigation of this question by looking at the effects of chaos in time-independent potentials; by definition, violent relaxation cannot occur if the potential is fixed. In a series of papers with M.E. Mahon and other collaborators,[8–11] Henry investigated the relation between stochasticity of individual trajectories—as measured, for instance, by Liapunov exponents—and the rate at which an initially compact *ensemble* of stars evolves toward a uniform distribution over the accessible phase space. These papers showed that the coarse-grained distribution function typically exhibits an exponential approach toward equilibrium at a rate that correlates well with the mean Liapunov exponent for the ensemble. When evolved for much longer times, the phase space density slowly changes as orbits diffuse into regions that, although accessible, are avoided over the shorter time interval. Henry coined the term "near-invariant distribution" to describe the end-point of chaotic mixing of an isoenergetic ensemble of points.

In another paper,[12] Henry compared the efficiency of phase mixing and chaotic mixing. He noted that—for initially very localized ensembles—the two processes occur at very different rates: chaotic mixing takes place on the Liapunov, or exponential divergence, time scale, whereas the phase mixing rate falls to zero. However, phase mixing of a group of points with a finite extent can be much more rapid. Furthermore, the mixing rate of chaotic ensembles typically falls below the Liapunov rate once the trajectories separate; this is especially true for those stochastic orbits that are confined over long periods of time to restricted parts of phase space. The effective rates of phase mixing and chaotic mixing might, therefore, be comparable in real galaxies. Henry noted also that chaotic mixing in three-dimensional potentials can occur at substantially different rates in different directions.

Henry recognized that the existence of invariant or near-invariant distributions, in regions of phase space that were not characterized by three isolating integrals of motion, implied the existence of a new class of equilibrium or near-equilibrium states for galaxies. The classical Jeans theorem requires that the phase-space density of a stationary stellar system be expressed solely in terms of the isolating integrals in that potential. Henry pointed out that there existed a far larger class of systems that could be in a stationary state. In the abstract of "Invariant Distributions and Collisionless Equilibria",[13] he wrote:

> "This paper discusses the possibility of constructing time-independent solutions to the collisionless Boltzmann equation which depend on quantities other than global isolating integrals such as energy and angular momentum. The key point is that, at least in principle, a self-consistent equilibrium can be constructed from *any* set of time-independent phase-space building blocks which, when combined, generate the mass distribution associated with an assumed time-independent potential."

Noting that strictly "time-independent" phase-space building blocks were mathematical idealizations, Henry pointed out that his "near-invariant distributions" were effectively time independent, and could in principle be used as building blocks in the construction of stationary galaxies. Indeed he argued that chaotic orbits were in a sense more natural components of steady-state galaxies than regular orbits, since an efficient mechanism (chaotic mixing) exists that can convert a generic distribution of points in chaotic phase space into a time-independent one. By contrast, an ensemble of points on a regular torus does not evolve toward a coarse-grained steady state: it simply translates, unchanged, around the torus. The Jeans theorem *postulates* a uniform distribution over each torus but says nothing about how this unlikely distribution is to be achieved.

Henry's insight constituted the most significant update of the Jeans theorem since the 1960s, when various authors pointed out the distinction between isolating and non-isolating integrals. I propose that a *generalized Jeans theorem* be attributed to Henry:

> **Generalized Jeans Theorem:** The phase-space density of a stationary stellar system is constant within every connected region.

A connected region is one that can not be decomposed into two finite regions, such that all trajectories remain for all time in either one or the other. Invariant tori are such regions, but so are the more complex parts of phase space associated with stochastic orbits.

As Henry once pointed out to me (with some amusement), people have actually been invoking the generalized version of the Jeans theorem for years without realizing it. A textbook example of a system satisfying the classical Jeans theorem (and one that was discussed by Jeans himself[14]) is an axisymmetric galaxy in which f is a function of the two classical integrals of motion, the energy E and the angular momentum J_z. Writing $f = f(E, J_z)$ implies that the phase space density is constant on hypersurfaces of constant E and J_z. However, not all orbits in axisymmetric potentials are characterized by a third isolating integral, hence parts of these hypersurfaces are associated with chaotic trajectories. The two-integral approach to axisymmetric modelling—which assigns a constant density to these regions—thus depends on the generalized form of the Jeans theorem for its justification. Henry went on to note[13] that one could in principle construct novel axisymmetric models, in which the surfaces of constant E and J_z are *not* sampled uniformly; for instance, one could exclude all the chaotic orbits, or assign different densities to different chaotic regions on the same (E, J_z) hypersurface. As far as I know, no one has yet attempted to construct models of this form, although it would be relatively straightforward to do so via orbital superposition. Nevertheless, it has become clear in the last few years that chaotic orbits can be major components of steady-state galaxies, demonstrating that Henry's generalized theorem is potentially very significant for our understanding galactic structure. For instance, one recent study[15] found that 50% or more of the mass in steady-state triaxial nuclei could e associated with chaotic orbits.

Smooth potentials are idealizations of real galaxies. As seen by a single star, any lumpiness or distortions in the stellar density would add small-amplitude perturbing forces to the mean field. Such perturbations would not be expected to have much consequence for either strongly chaotic or precisely regular orbits, but Henry realized that they might have an appreciable effect on the evolution of weakly stochastic orbits, by scattering trajectories away from a trapped region into a region where the mixing is more rapid. Henry, S. Habib, and M.E. Mahon[16–18] investigated this idea, adding random noise to otherwise smooth potentials and observing the effects on the mixing rate. They found that even very weak noise, with a characteristic time scale

$$\left| \frac{1}{v} \frac{\delta v}{\delta t} \right|^{-1}$$

of order 10^6 crossing times, could induce substantial changes in the motion of trapped stochastic orbits after only 10^2 orbital periods.

In a strongly time-dependent potential, chaos should be even more prevalent, if only because time-dependent potentials lack at least one isolating integral of the motion (the energy) that is always present in stationary potentials. In three studies,[19–21] Henry and collaborators considered the effects of two sorts of strongly time-dependent perturbations on the structure of orbits in two-dimensional and three-dimensional potentials: a time-dependent scale factor $R(t) = t^p$ mimicking expansion or collapse; and strictly periodic driving, a crude model of the oscillations that accompany the final stages of relaxation. By computing the values of short-time Liapunov exponents, Henry and coworkers found that trajectories in the expanding or contracting potential could mimic regular orbits part of the time and stochastic orbits at other times. Contraction tended to make the effects of chaos stronger, and expansion tended to make the chaos weaker. In the oscillatory model, orbits appeared to remain either regular or chaotic for all times, although the periodicity in the global potential seemed to induce some orbits to become what Henry termed "wildly chaotic," exhibiting substantial changes in energy as they chaotically diffused.

Having demonstrated the importance of chaotic mixing in stationary, weakly time dependent, and strongly time dependent potentials, it remained only for Henry to make explicit the link between chaotic mixing and violent relaxation. Henry did not get quite this far before his death, but he had a clear idea of how to proceed. In this passage, he describes a simple and beautiful scheme for establishing the connection:

> "At least crudely, one can visualize an evolution described by the collisionless Boltzmann equation as involving a collection of characteristics corresponding to orbits evolved in a specified time-dependent potential, ignoring the fact that the potential is generated self consistently ... to the extent that this picture is valid, one might then anticipate that the efficacy with which an initial ensemble approaches an equilibrium or near-equilibrium ... will depend on the degree to which the flow in the specified potential is chaotic. In particular, to the extent that the flow is chaotic, one would expect a rapid and efficient approach towards a near-equilibrium."

In Henry's scheme, one would first carry out a fully self-consistent simulation of collapse and virialization, via an N-body code say, recording the gravitational potential on a grid in both space and time. Returning to the initial conditions, one would then select out initially localized ensembles of phase-space points and evolve them forward in the previously-recorded potential, this time ignoring the self-gravity of the ensemble. The sum total of all such integrations would be a reproduction of the self-consistent collapse, and by analyzing the rate of approach of each ensemble to its near-invariant distribution, one could generate a potentially complete picture of the way in which the properties of the phase-space flow were related to the rate of violent relaxation. Shortly before his death, Henry told me that he was hoping to carry out this program in collaboration with a student, but apparently this wish never came to fruition.

Like most dynamicists, Henry was fascinated by the concept of entropy. Henry's ideas about entropy were complex, and I am not sure that I understood them completely, but overall I felt that Henry was skeptical about the relevance of entropy arguments to galactic dynamics. Here, I can not resist quoting from V.A. Antonov whose skeptical view of entropy was similar, I think, to Henry's:

> "True diffusion is well described by differential equations. On the contrary, mixing is not represented in terms of differential operations. There is the phase density before the mixing, and the phase density after the mixing, but it is difficult to define when and where the transmutation occurs."

Henry understood that chaotic mixing is irreversible, in the sense that an infinitely precise fine-tuning of the velocities would be required in order to undo its effects. This is a sort of entropy increase and it implies an evolution toward a state whose properties are in some ways predictable. However, unlike Antonov, Henry realized that it is difficult to establish very general rules that link the initial and final states of a stellar system that evolves via collisionless relaxation, in particular, rules that would allow one to make statements about which final states are preferred. Henry was critical,[23] for instance, about a purported demonstration[24] of an "H-theorem" for collisionless systems:

> "It is, moreover, clear physically that there exist 'reasonable' choices of initial data, such as those leading to nearly homologous collapse, which exhibit nearly periodic motion; and for such data, one might anticipate that, after one approximate period, H will have returned very nearly to its initial value ... Because the H-functions need not increase monotonically, they cannot be used to provide a useful characterization of the continuous dynamics."

Henry also made fundamental contributions to our understanding of another sort of chaos characteristic of stellar systems. Already in 1964,[25] Richard Miller had shown that the trajectories of stars in small N-body integrations were generically chaotic, in the sense that the $6N$-dimensional phase path was exponentially unstable to small changes in the initial conditions. In a series of papers,[26–29] Henry and collaborators carried out a systematic numerical study of this instability. They found that the time scale for growth of perturbations tended to *decrease* with increasing N, remaining of order the crossing time for values of N as large as 4,000. This result was consistent with Henry's earlier theoretical arguments[30–32] and in contradiction with a prediction of Gurzadyan and Savvidy[33] that the growth rate should fall as $1/N^{1/3}$. (Henry's prediction that the grow rate should increase with increasing N has recently been verified for values of N as large as 10^{5}.[34]) Henry recognized that this generic instability of the N-body equations did not necessary imply that mixing or relaxation would be efficient, since the exponential instability often seemed to *saturate* on scales much smaller than the scale of the system. However, he made the interesting point[7] that the existence of the instability made it difficult to reconcile the N-body equations of motion with the collisionless Boltzmann equation:

> "This leads, however, to an important question of principle. The N-body problem appears to be chaotic on a time scale about t_{cr} [the crossing time], but the flow associated with the CBE is often integrable or near-integrable in the sense that many or all of the characteristics are regular, i.e., nonchaotic. So what do the (often near-integrable) CBE characteristics have to do with the true (chaotic) N-body problem?"

Henry asked: In what sense do the N-body equations of motion "go over" to the collisionless Boltzmann equation as $N \to \infty$? Henry considered several possible ways in which this might happen, and concluded

> "Given the fact that the N-body problem is chaotic on a time scale about t_{cr}, it would seem reasonable to conjecture that the orbits generated in two different N-body realizations will diverge exponentially on a time scale about t_{cr} ... However, one might nevertheless expect that, for sufficiently large N, the ensemble average of the different N-body orbits generated from the same (x_0, v_0) will closely track the CBE characteristic for some finite time. In particular, one might conjecture that the rms configuration space deviation between the N-body orbits and the CBE characteristics will scale as
>
> $$\delta r_{\text{rms}}(t) \approx F(N) e^{t/\tau},$$
>
> where $\tau \approx t_{cr}$, roughly independent of the total particle number N, and where the prefactor $F(N) \to 0$ for $N \to \infty$."

Henry was proposing here a "weak" correspondence between the N-body and CBE descriptions, in the sense that an ensemble average of the N-body trajectories might mimic the orbit in the smoothed-out potential. This suggestion was characteristically cautious, and soon after, Henry and I. Sideris[35] numerically demonstrated a stronger sense in which the N-body trajectories approach the smooth-potential orbits:

> "... there is a clear, quantifiable sense in which, as N increases, chaotic orbits in the frozen-N systems remain 'close to' integrable characteristics in the smooth potential for progressively longer times. When viewed in configura-

tion or velocity space, or as probed by collisionless invariants like angular momentum, frozen-N orbits typically diverge from smooth potential characteristics as a power law in time, rather than exponentially, on a time scale about $N^p t_D$, with $p \sim 1/2$ and t_D a characteristic dynamical, or crossing, time."

By "frozen-N" orbits, Henry meant trajectories computed in a potential where the smooth density had been replaced by a matrix of fixed point masses, the gravitational analog of the Lorentz gas.[36] For large N, these experiments showed that almost all trajectories, and not just their ensemble averages, tended in their finite behavior toward the behavior of regular orbits, even though the growth of infinitesimal perturbations (as measured by Liapunov exponents) remained large. In one of his last papers,[37] Henry and Sideris showed that the chaos associated with a smooth potential could often be distinguished from the N-body chaos, despite the latter having a much shorter growth time, by comparing initially nearby orbits in a single N-body system, or by tracking orbits with the same initial conditions evolved in two different N-body realizations of the same smooth density.

ACKNOWLEDGMENT

My thanks to Richard Miller and Haywood Smith for their careful reading of the manuscript.

REFERENCES

1. KANDRUP, H.E. 1989. Stochastic Problems in Stellar Dynamics. Ph.D. Thesis, University of Chicago.
2. KANDRUP, H.E. 2003. Chaos and chaotic phase mixing in galaxy evolution charged particle beams. *In* Galaxies and Chaos. G. Contopoulos & N. Voglis, Eds. Lecture Notes in Physics **626**: 154–168.
3. ZWICKY, F. 1939. On the formation of clusters of nebulæ and the cosmological time scale. Proc. Natl. Acad. Sci. **25**: 604–609.
4. KING, I. 1962. The structure of star clusters. I. An empirical density law. Astron. J. **67**: 471–485.
5. HÉNON, M. 1964. L'Evolution initiale d'un amas sphérique. Ann. d'Astrophys. **27**: 83–91.
6. LYNDEN-BELL, D. 1967. Statistical mechanics of violent relaxation in stellar systems. Mon. Not. R. Astron. Soc. **136**: 101–121.
7. KANDRUP, H.E. 1998. Collisionless relaxation in galactic dynamics and the evolution of long-range order. *In* Long-Range Correlations in Astrophysical Systems. J.R. Buchler, J.W. Dufty & H.E. Kandrup, Eds. Ann. N.Y. Acad. Sci. **848**: 28–47.
8. KANDRP, H.E. & M.E. MAHON. 1994. Relaxation and stochasticity in a truncated Toda lattice. Phys. Rev. E **49**: 3735–3747.
9. KANDRUP, H.E. & M.E. MAHON. 1994. Short times characterisations of stochasticity in nonintegrable galactic potentials. Astron. Astrophys. **290**: 762–770.
10. MAHON, M.E., R.A. ABERNATHY, B.O. BRADLEY & H.E. KANDRUP. 1995. Transient ensemble dynamics in time-independent galactic potentials. Mon. Not. R. Astron. Soc. **275**: 443–453.
11. KANDRUP, H.E. & C. SIOPIS. 2003. Chaos and chaotic phase mixing in cuspy triaxial potentials. Mon. Not. R. Astron. Soc. **345**: 727–742.
12. KANDRUP, H.E. 1998. Phase mixing in time-independent Hamiltonian systems. Mon. Not. R. Astron. Soc. **301**: 960–974.
13. KANDRUP, H.E. 1999. Invariant distributions and collisionless equilibria. Mon. Not. R. Astron. Soc. **299**: 1139–1145.

14. JEANS, J.H. 1915. On the theory of star-streaming and the structure of the universe. Mon. Not. R. Astron. Soc. **76:** 70–84.
15. POON, M.Y. & D. MERRITT. A self-consistent study of triaxial black hole nuclei. Astrophys. J. **606:** 774–787.
16. HABIB, S., H.E. KANDRUP & E.M. MAHON. 1996. Chaos and noise in a truncated Toda potential. Phys. Rev. E **53:** 5473–5476.
17. HABIB, S., H.E. KANDRUP & E.M. MAHON. 1997. Chaos and noise in galactic potentials. Astrophys. J. **480:** 155–166.
18. KANDRUP, H.E., I.V. POGORELOV & I.V. SIDERIS. 2000. Chaotic mixing in noisy Hamiltonian systems. Mon. Not. R. Astron. Soc. **311:** 719–732.
19. KANDRUP, H.E. & J. DRURY. 1998. Chaos in cosmological Hamiltonians. Ann. N.Y. Acad. Sci. **867:** 306–000.
20. KANDRUP, H.E., I.M. VASS & I.V. SIDERIS. 2003. Transient chaos and resonant phase mixing in violent relaxation. Mon. Not. R. Astron. Soc. **341:** 927–936.
21. TERZIÓ, B. & H.E. KANDRUP. 2004. Orbital structure in oscillating galactic potentials. Mon. Not. R. Astron. Soc. **347:** 957–967.
22. ANTONOV, V.A. 2005. Brouwer Award Lecture (unpublished).
23. KANDRUP, H.E. 1987, An H-theorem for violent relaxation? Mon. Not. R. Astron. Soc. **225:** 995–998.
24. TREMAINE, S., M. HÉNON & D. LYNDEN-BELL. 1996. H-functions and mixing in violent relaxation. Mon. Not. R. Astron. Soc. **219:** 285–297.
25. MILLER, R.H. 1964. Irreversibility in small stellar dynamical systems. Astrophys. J. **140:** 250–256.
26. KANDRUP, H.E. & H. SMITH. 1991. On the sensitivity of the N-body problem to small changes in initial conditions. Astrophys. J. **374:** 255–265.
27. KANDRUP, H.E. & H. SMITH. 1992. On the sensitivity of the N-body problem to small changes in initial conditions. II. Astrophys. J. **386:** 635–645.
28. KANDRUP, H.E., H. SMITH & D.E. WILLMES. 1992. On the sensitivity of the N-body problem to small changes in initial conditions. III. Astrophys. J. **399:** 627–633.
29. KANDRUP, H.E., M.E. MAHON & H. SMITH. 1994. On the sensitivity of the N-body problem to small changes in initial conditions. IV. Astrophys. J. **428:** 458–465.
30. KANDRUP, H.E. 1989. The time scale for "mixing" in a stellar dynamical system. Phys. Lett. A. **140:** 97–100.
31. KANDRUP, H.E. 1990. How fast can a galaxy "mix"? Physica A **169:** 73–94.
32. KANDRUP, H.E. 1990. Divergence of nearby trajectories for the gravitational N-body problem. Astrophys. J. **364:** 420–425.
33. GURZADIAN, V.G. & G.K. SAVVIDY. 1986. Collective relaxation of stellar systems. Astron. Astrophys. **160:** 203–210.
34. HEMSENDORF, M. & D. MERRITT. 2002. Instability of the gravitational N-body problem in the large-N limit. Astrophys. J. **580:** 606–609.
35. KANDRUP, H.E. & I.V. SIDERIS. 2001. Chaos and the continuum limit in the gravitational N-body problem: integrable potentials. Phys. Rev. E. **64:** 056209-1-11.
36. LORENTZ, H.A. 1905. The motion of electrons in metallic bodies. Proc. Amst. Acad. **7:** 438–453.
37. KANDRUP, H.E. & I.V. SIDERIS. 2003. Smooth potential chaos and N-body simulations. Astrophys. J. **585:** 244–249.

Chaotic Collisionless Evolution in Galaxies and Charged-Particle Beams

HENRY E. KANDRUP,[a] COURTLANDT L. BOHN,[b,c] RAMI A. KISHEK,[d] PATRICK G. O'SHEA,[d] MARTIN REISER,[d] AND IOANNIS V. SIDERIS[b]

[a]*Departments of Astronomy and Physics and Institute for Fundamental Theory, University of Florida, Gainesville, Florida, USA (deceased)*

[b]*Department of Physics, Northern Illinois University, DeKalb, Illinois, USA*

[c]*Fermi National Accelerator Laboratory, Batavia, Illinois, USA*

[d]*Institute for Research in Electronics and Applied Physics, University of Maryland, College Park, Maryland, USA*

> ABSTRACT: Both galaxies and charged particle beams can exhibit collisionless evolution on surprisingly short time scales. This can be attributed to the dynamics of chaotic orbits. The chaos is often triggered by resonance caused by time dependence in the bulk potential, which acts almost identically for attractive gravitational forces and repulsive electrostatic forces. The similarity suggests that many physical processes at work in galaxies, although inaccessible to direct controlled experiments, can be tested indirectly via controlled experiments with charged particle beams, such as those envisioned for the University of Maryland electron ring currently nearing completion.
>
> KEYWORDS: chaos; N-body problem; nonlinear dynamics; collisionless

PREAMBLE

Henry Kandrup and Court Bohn had independently realized that there were important parallels between the collisionless evolution of charged-particle beams and large stellar systems. Both desired to pursue this matter explicitly by way of direct experiments with beams. Independently, Martin Reiser obtained funding to build the University of Maryland electron ring (UMER) for the expressed purpose of doing controlled experiments to measure the dynamical consequences and evolutionary time scales associated with internal Coulomb forces; that is, space charge. All of these circumstances led to a strong collaboration. Henry had been eagerly anticipating the completion of UMER and experiments that the collaboration was planning.

We all endeavored to introduce the notion of an analogy between the dynamics of beams and galaxies to a broad spectrum of investigators. Before Henry passed away, we had completed a paper, one that excited Henry immensely, to review the pertinent literature and introduce this idea. Feedback from referees was generally negative toward publication but positive toward pursuit of the idea. Loosely translated, the

Address for correspondence: Courtlandt L. Bohn, Department of Physics, Northern Illinois University, DeKalb, Illinois, USA. Voice: 815-753-6473; fax: 815-753-8565.
clbohn@niu.edu

referee reports stated that we have a nice proposal, for example, to submit to a funding agency, but we should complete new experiments prior to journal publication.

The paper has evolved considerably since Henry's passing, but it retains much of his language, particularly with reference to galactic dynamics. We, his colleagues, hereby offer this paper as part of the Symposium that honors Henry. What follows is a version that incorporates all referee comments and that is edited to mesh with other related Symposium contributions, but that retains the original flavor and Henry's unique touch. It would surely have his imprimatur.

INTRODUCTION

Many-body systems whose constituents interact via long-range inverse-square-law Coulomb forces, both gravitational and electrostatic, can exhibit macroscopic relaxation and loss of coherence on time scales much shorter than might be expected on dimensional grounds. This process moves the system toward a long-lived *metaequilibrium* state, a state that differs from true thermal equilibrium (which, in the case of galaxies, cannot be accessed dynamically). When a galaxy has a sizeable gaseous component, the gas will interact with the stellar component and thereby enhance its relaxation. However, observations and simulations agree that even a relatively gas-poor (and thus, presumably nearly dissipation-free) elliptical galaxy displaced from a metaequilibrium state as a result of an encounter with another galaxy can readjust itself toward a new metaequilibrium state within a few hundred million years (i.e., within about 10% of the age of the galaxy) although the nominal relaxation time t_R associated with *collisions* is orders of magnitude longer than the age of the Universe. Similarly, charged-particle beams, which would be expected to maintain coherence while traveling some 100km or more through an accelerator, can lose coherence and disperse significantly within distances as short as 10m.

Because collisions cause relatively slow relaxation, any rapid relaxation must be due to collisionless, that is, collective, processes. More specifically, the collective behavior must be connected with mixing, that is, the tendency of initially localized clumps of orbits to disperse. Mixing is *much* more efficient in a chaotic system than in a system in which the bulk coarse-grained potential is integrable or near-integrable. An initially localized clump of regular, that is, non-chaotic, orbits will typically disperse secularly, that is, as a power law in time; a clump of chaotic orbits will instead disperse exponentially.

Allowing for a bulk potential that is strongly chaotic, thereby supporting *chaotic mixing*, enables one to understand how a galaxy can *relax* toward a metaequilibrium state on a comparatively short time scale. Such an understanding is of *practical* importance in reference to charged-particle beams. There, rapid collisionless relaxation places strong constraints on *emittance compensation*, that is, processes designed to confine the constituent particles to a compact volume of phase space, as is required for high-brightness beams.

Theoretical considerations and detailed numerical simulations suggest that, in this setting, the origin of the chaos that drives the evolution is largely irrelevant. In particular, whether the two-body forces are attractive or repulsive should not be crucial. What *is* important is that the long-range scalings of gravitational and electrostatic forces are identical and that, in both cases, the early stages of evolution should be driven by long-range, collective interactions (acting *globally*) as opposed to short-range collisional encounters (acting *locally*). All that seems to matter is whether the bulk potential associated with the many-body system admits a large measure of chaotic orbits.[1]

A complete understanding of these phenomena requires a synthesis of theory, simulations, and experiments. Performing experiments on self-gravitating systems like galaxies is impossible. However, controlled experiments *can* be performed with charged-particle beams, and combining the results of such experiments with simulations and theory should lead to a clear picture of the role of chaotic phase mixing in beams. Moreover, as we exemplify in the next section, the physics should not depend crucially on whether the force between particles is attractive or repulsive, and one would thus expect that many results about beams should translate more or less directly into detailed predictions about the structure and evolution of galaxies. Indeed, one can go one step further and argue that, in a real sense, carefully constructed experiments involving charged-particle beams can be used as semidirect probes of the physics of self-gravitating systems like galaxies.

THE BEAM-GALAXY ANALOGY: THEORETICAL CONSIDERATIONS

That collisional relaxation should be largely irrelevant in many settings involving galaxies and beams is easily seen. Viewing such effects as an incoherent sum of binary encounters, one computes, respectively, for galaxies[2] and (in Gaussian units) for charged particle beams[3,4] relaxation times

$$t_R \sim \frac{v^3}{(Gm)^2 n \log \Lambda} \quad \text{and} \quad t_R \sim \frac{v^3}{q^2 n \log \Lambda}, \qquad (1)$$

where v is a typical speed associated with random motions, G is the gravitational constant, m and q are, respectively, the typical stellar mass and particle charge, n is a characteristic number density, and $\log \Lambda$ is the so-called Coulomb logarithm, which scales as a positive power of the number of constituent particles N. (If one assumes that collisions act as a source of Brownian motion, t_R can be related to the time integral of the quantity $N \langle \mathbf{F}(0) \cdot \mathbf{F}(t) \rangle$, where $\langle \mathbf{F}(0) \cdot \mathbf{F}(t) \rangle$ is the autocorrelation function for the test particle interacting with a single field particle.[5])

In either case, assuming the bulk random kinetic and potential energies are comparable in magnitude implies that $t_R \sim (N/\log \Lambda) t_D$, with $t_D \sim R/v$ denoting the *dynamical time*, a characteristic orbital time scale defined in terms of the *size R* of the system. For large N (typically $N \sim 10^9$–10^{12} in realistic, large stellar and particle-beam systems) the relaxation time t_R is clearly orders of magnitude longer than the dynamical time t_D; collisional relaxation is slow. In contrast, mixing of chaotic orbits, that is, *chaotic mixing*, can proceed extremely fast; the *e*-folding time associated with the dispersal of an initially localized *clump* of particles, given as the inverse of the largest positive Lyapunov exponent respective to that clump,[6] is typically comparable in magnitude to t_D. This is, for example, the case for the systems illustrated in FIGURES 2 and 4 discussed below.

Presently there is no known generic algorithm permitting accurate analytic or quasianalytic estimates of the largest Lyapunov exponent in three-dimensional bulk potentials. However, recent work[7–9] has shown that, in many cases, an analytic technique developed for systems with many degrees of freedom[10] can be adapted to provide reasonable estimates for lower-dimensional systems, the breakdown of that approach reflecting typically systems in which autocorrelation functions for properties of representative orbits have long time tails.[11] It is, therefore, relevant to recall the analytic results for the largest Lyapunov exponent χ in a three-dimensional time-independent bulk potential, for this then becomes a quantitative measure of the rate of collisionless relaxation by way of chaotic mixing:

$$\chi(\xi) \approx \frac{1}{\sqrt{3}} \frac{L(\xi) - 1}{L(\xi)} \sqrt{\kappa},$$

$$L(\xi) = [T(\xi) + \sqrt{1 + T^2(\xi)}]^{1/3}, \qquad (2)$$

$$T(\xi) = \frac{3\pi\sqrt{3}}{8} \frac{\xi^2}{2\sqrt{1+\xi} + \pi\xi}.$$

The auxiliary quantities κ and ξ are determined from the potential $V(x)$,

$$\kappa = \frac{1}{2} \langle \nabla^2 V \rangle, \qquad \xi = \frac{1}{\kappa\sqrt{2}} \sqrt{\langle (\nabla^2 V)^2 \rangle - \langle \nabla^2 V \rangle^2}, \qquad (3)$$

where the averages are taken over the microcanonical ensemble in the following manner

$$\langle A \rangle \equiv \frac{\int d\mathbf{x} \int d\mathbf{p}\, A(\mathbf{x}) \delta[H(\mathbf{x}, \mathbf{p}) - E]}{\int d\mathbf{x} \int d\mathbf{p}\, \delta[H(\mathbf{x}, \mathbf{p}) - E]}, \qquad (4)$$

in which E denotes the total particle energy. Upon invoking the Poisson equation, we see immediately that the auxiliary quantities are determined from the density distribution. For a *galaxy*, we have

$$\nabla^2 V = 4\pi G \rho, \qquad \kappa = 2\pi G \rho(0) \langle \zeta \rangle, \qquad \xi = \frac{\sqrt{2}}{\langle \zeta \rangle} \sqrt{\langle \zeta^2 \rangle - \langle \zeta \rangle^2}, \qquad (5)$$

where $\zeta = \rho(\mathbf{x})/\rho(0)$, with $\rho(\mathbf{x})$ denoting the mass density. For a *beam*, we take the external focusing potential V_f to be quadratic in the coordinates $\mathbf{x}$ comoving with the bunch; that is, $V_f(\mathbf{x}) = (\boldsymbol{\omega} \cdot \mathbf{x})^2/2$, where $\boldsymbol{\omega} = (\omega_x, \omega_y, \omega_z)$ corresponds to the focusing strength; the total potential is $V = V_f + V_s$. Then we have

$$\nabla^2 V_s = -4\pi\rho, \qquad \kappa = \frac{\omega_{p0}^2}{2}\left[\left(\frac{\omega}{\omega_{p0}}\right)^2 - \langle \zeta \rangle\right], \qquad \xi = \frac{\sqrt{2}\sqrt{\langle \zeta^2 \rangle - \langle \zeta \rangle^2}}{(\omega/\omega_{p0}) - \langle \zeta \rangle}, \qquad (6)$$

where ω_{p0} is the plasma frequency at the bunch centroid and $\rho(\mathbf{x})$ now refers to charge density. The time scale for chaotic mixing is now $t_m = 1/\chi \propto f(\xi)\kappa^{-1/2}$. The analogy between chaotic mixing in beams versus galaxies becomes apparent: for both classes of systems, the dynamical time is $t_D \sim \kappa^{-1/2}$, the auxiliary quantity ξ involves a product of the dispersion in the density profile and the square of the dynamical time, and $f(\xi)$ is the same function for both systems. For beams, space charge is a repulsive collective force that acts to lengthen the dynamical time by weakening the net focusing force acting on a particle (resulting in what is called the *space-charge-depressed period*). For galaxies no such weakening appears; gravity is strictly attractive.

To do a computational test of this result, one chooses an energy E and integrates a large number (say, 2,000) of tightly localized initial conditions corresponding to an energy very close to E. These trajectories then spread, and one can calculate moments, such as $\langle x^2(t)\rangle$, of the corresponding distribution of trajectories versus time and assess whether they grow exponentially. If they do, then one can extract the e-folding time and compare it to the analytic estimate. Examples of such comparisons in galactic and beam systems appear in FIGURE 1. The galactic system is a uniform-density ellipsoid containing a supermassive black hole at its centroid.[8] The beam system is a configuration of thermal equilibrium having triaxial symmetry.[9]

The preceding analytic results follow from a geometric treatment of scleronomous Hamiltonian systems in the spirit of Pettini and his collaborators.[10] It does not apply to time-dependent systems, and thus, it is not presently possible to point to an unambiguous analogy between the dynamics of beams and galaxies involving a rheonomous Hamiltonian. A geometric treatment of the latter would be based on a Finsler metric, that is, a metric that incorporates velocities and time, but it becomes unclear how to define an invariant measure to use in place of the microcanonical ensemble for evaluating phase-space averages, particularly when one considers that resonances between orbital frequencies and the frequency spectrum of the time-dependent potential come into play. Nonetheless, a reasonable ansatz is that a successful geometric treatment of rheonomous systems should result in a connection between beams and galaxies analogous to that of time-independent systems. The underlying reason is that both systems involve an inverse-square long-range force, and this force is what drives chaotic mixing.

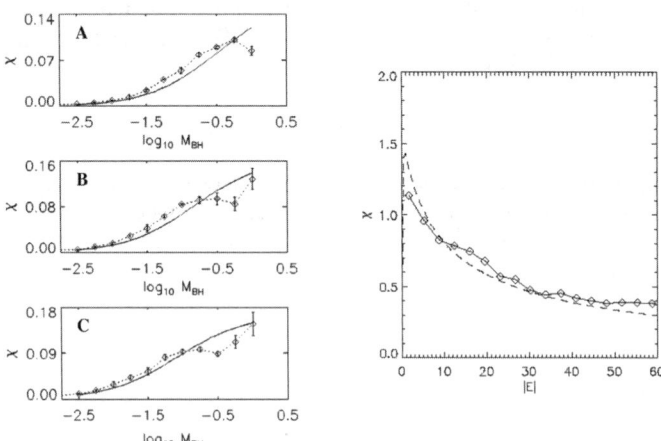

FIGURE 1. (*Left*) Numerical results (*diamonds*) and analytic (*solid line*) estimates of the mixing rate for chaotic orbits evolved in a triaxial galactic potential as a function of black-hole mass M_{BH} and for total particle energy: (**A**) $E = 1.0$, (**B**) $E = 0.6$, and (**C**) $E = 0.4$. (*Right*) Numerical (*diamonds*) and analytic estimates (*dashed line*) of the mixing rate for chaotic orbits in a triaxial thermal-equilibrium beam as a function of E. In both figures the unit of χ is t_D^{-1}. Additional details can be found in References 8 and 9, from which these figures are reprinted by permission of the American Physical Society.

REGULAR VERSUS CHAOTIC ORBITS: A TORTURED HISTORY

Chaos has been largely ignored, until comparatively recently, in both the galactic and accelerator-dynamics communities. For example, although the famous Hénon–Heiles model[12] arose originally in attempts to understand meridional motions in axisymmetric galaxies, as recently as 15 years ago the potential role of chaos in galaxy structure and evolution was almost completely neglected (with the exception of a handful of groups in Europe). Only with the advent of high-resolution photometry, facilitated in part by the Hubble space telescope, did many galactic astronomers begin to recognize that the bulk potentials associated with realistically shaped galaxies are likely to admit significant measures of chaotic orbits.

Galaxies

It has been long recognized that the dominant mechanism for relaxation in galaxies cannot be *collisional*. For example, in the 1940s Chandrasekhar[2] showed that the relaxation time scale t_R on which binary encounters between individual stars could significantly alter the trajectories of stars in the Milky Way must be about 10^{12} years or more. Shorter-time relaxation must somehow involve collective effects. Two decades later, Lynden-Bell[13] proposed a theory of *violent relaxation* which argued, *inter alia*, that regular (i.e., nonchaotic) phase mixing associated with a time-dependent potential might explain such collective effects. Substantial evidence for rapid relaxation accumulated over the next twenty years derived both from numerical simulations of many-body systems and from the interpretation of observations of galaxies that have been involved in collisions with other galaxies.[14] Despite this, however, when subjected to closer scrutiny, it seemed that, at least in its simplest guise where the orbits that phase mix are assumed to be regular, violent relaxation could not explain why relaxation was as fast as it appears to be. An ingredient seemed to be missing. Today there is good reason to think that the missing ingredient is *chaos*.

In the early 1990s Kandrup and Mahon[15] recognized that, because of their exponentially sensitive dependence on initial conditions, chaotic orbits should mix far more rapidly than regular orbits, in fact exponentially fast. In the astronomy community this phenomenon, now termed *chaotic mixing*,[16] led to speculations that chaos could play a critical role in violent relaxation. However, chaotic mixing in itself does not constitute a complete and satisfactory explanation. It cannot drive collective relaxation unless many/most of the orbits are chaotic, a prerequisite whose fulfillment is far from obvious. A few years later, motivated in part by the work of accelerator physicists,[17–19] astronomers[20] recognized that time-dependent pulsations in the bulk potential of a galaxy readjusting toward a metaequilibrium state could, via resonant couplings, make many/most of the orbits in a galaxy chaotic with large finite-time Lyapunov exponents,[21,22] and that the resulting *resonant phase mixing* might be sufficiently strong and ubiquitous to explain violent relaxation.

Finite-time Lyapunov exponents probe the average exponential instability of orbit segments over finite time intervals. Formally, they satisfy

$$\chi(t) = \frac{1}{t} \lim_{\delta Z(0) \to 0} \left\{ \ln \frac{|\delta Z(t)|}{|\delta Z(0)|} \right\},$$

where $|\delta Z|^2 = (|\delta r|^2 + |\delta v|^2)$. The largest such exponent can be estimated numerically using an algorithm introduced by Benettin, Galgani, and Strelcyn.[23]

Charged-Particle Beams

Concerns about collisionless relaxation in charged-particle beams have arisen with recent advances in technology for the production of high-brightness beams, wherein the collective Coulomb self-force, that is, the space-charge force, becomes important. In the laboratory frame this force decreases inversely as the square of the beam energy.[24] For the transverse component, this is due to the partial cancellation between the self-magnetic and self-electrostatic forces; for the longitudinal component, it is due to Lorentz contraction. Nonetheless, there are still many situations involving high-brightness beams where space charge is important. Examples include both low-to-medium-energy hadron accelerators, such as those envisioned to drive spallation-neutron sources or heavy-ion fusion or that serve as boosters for high-energy machines, as well as low-energy lepton, (e.g., electron) accelerators, such as photoinjectors.[25]

One of the earliest papers to treat space charge in beams was by Kapchinskij and Vladimirskij,[26] who considered a direct-current beam with uniform charge density and elliptical cross section confined by linear external focusing forces, and derived the equations governing the motion of the beam envelope. The corresponding distribution function in the four-dimensional transverse phase space of a single charge, commonly called the *KV distribution*, is a hyperellipsoidal shell. A decade later, Sacherer[27] noted that these results can readily be generalized to three-dimensional bunched beams (i.e., to six-dimensional phase space) so as to include the influence of space charge on bunch length and energy spread. These two papers, regarded as classics by the accelerator community, set the stage for many of the subsequent investigations concerning space charge, from which evolved now-conventional design strategies for high-brightness accelerators, strategies based on controlling root-mean-square (rms) properties of the beam.

However, the past decade has brought the realization that, albeit necessary, controlling the rms properties of a beam is not sufficient. Perhaps the most prominent example concerns beam halos, that is, particles that reach large orbital amplitudes due to a time-dependent space-charge potential arising because irregularities in the beamline prevent the beam from reaching a long-lived equilibrium state.[28] The concern is that a tiny impingement of particles on the beamline, about 1 W/m for beam energies exceeding about 20 MeV, can generate sufficient radioactivation to preclude routine, hands-on maintenance. Efforts to push our understanding of space charge beyond that required for computing bulk moments of the beam brought, as a spin-off, the realization that early-time dynamics in fully self-consistent charged-particle systems resembled that of violent relaxation in stellar systems.[18] However, that resemblance was explored no further—until now.

EVIDENCE FOR CHAOS AND CHAOTIC MIXING

Chaos in Galaxies

The Inevitability of Chaos

High-resolution observations of galaxies over the past decade or so have provided compelling evidence that many galaxies are more irregularly shaped than had been assumed as recently as 15 years ago; and attempts to model such irregularly shaped objects have led many galactic dynamicists to conclude that the bulk potentials associated with realistic galaxies admit large measures of chaotic orbits. It has been argued[29] that nonaxisymmetric elliptical galaxies containing central density cusps of the form inferred from observations[30] are very likely to admit large numbers of chaotic orbits. Similarly, models of rotating barred spiral galaxies suggest[31–33] that breaking axisymmetry with even a comparatively weak bar can trigger large numbers of chaotic orbits, especially near certain resonances. More generally, as first stressed by Udry and Pfenniger,[34] making a galaxy less symmetric, for example, by deforming it from axisymmetric to triaxial or by introducing *local* asymmetries, tends generically to increase both the relative measure of chaotic orbits and the size of the largest Lyapunov exponents. Although it is possible to contrive models of cuspy, nonaxisymmetric galaxies that are integrable or near-integrable,[35–37] they are not generic. Instead, there has emerged a general sense in much of the galactic dynamics community that *generic* irregularly shaped galaxies might be expected to contain large numbers of strongly chaotic orbits.

Are Galaxies Really in Equilibrium?

One intriguing possibility is that, perhaps because of the presence of chaos, evolving galaxies find it difficult, if not impossible, to approach a true equilibrium. It may well be that, at the time of formation, a galaxy settles down toward a long-lived *metaequilibrium* rather than a true equilibrium; and subsequently, in response to, for example, external irregularities associated with a densely populated galactic cluster, exhibits a slow, secular evolution.[29,38] To the extent that this is true, a basic question is if a galaxy originally in a nonaxisymmetric metaequilibrium will evolve toward a more nearly axisymmetric state;[29] or if, instead, a galaxy originally containing large numbers of strongly chaotic orbits might evolve toward other metaequilibria, not necessarily more nearly axisymmetric, that contain smaller numbers of chaotic orbits.[39] In any event, it is generally accepted that a robust, stable metaequilibrium must contain large measures of regular[40] and/or sticky chaotic[41] orbits to provide the *skeleton* (i.e., foundation) of the interesting configurations that support chaotic orbits in the first place.

The Role of a Time-Dependent Bulk Potential

There is also emerging evidence that chaos should be even more ubiquitous in systems that feel a strongly time-dependent bulk potential, especially a time dependence involving roughly periodic oscillations. Nonlinear dynamicists argue that chaos typically arises via resonance overlaps,[6] and this time-dependent chaos is simply another example thereof. When the time dependence influencing stellar orbits in a galaxy has power at frequencies sufficiently close to (multiples of) the frequencies at which the orbits themselves have power, the orbits and the time dependence can resonate with the result that the orbits become strongly chaotic. If the time dependence is weak, such resonances may require a near-perfect frequency match, but for stronger time dependence it often suffices for the pulsation and orbital time scales to agree within an order of magnitude.[20] However, in a nearly collisionless system like a galaxy, dimensionally there is only one natural time scale, the dynamical time $t_D \sim (G\rho)^{-1/2}$, with G the gravitational constant and a characteristic density. (Assuming the bulk kinetic and potential energies are comparable in magnitude, then $(G\rho)^{-1/2} \sim R/v$, which is the time scale of a typical orbit.) Consequently the pulsation and orbital times are likely to be comparable in magnitude throughout much of the galaxy, thus rendering chaos extremely common. Simple models suggest that galaxies subjected to damped oscillations could (1) become almost completely mixed *and* (2) settle down towards a nearly integrable metaequilibrium within a time as short as about $10 t_D$. Analogous effects can also be triggered by other nearly periodic phenomena, such as localized, nonstationary collective modes, or a supermassive black hole binary orbiting near the center of a galaxy.[42] Indeed, such a binary could produce anomalous dips observed in the surface-brightness profiles of galaxies like NGC 3706 or NGC 4406, which suggest their respective mass densities do not decrease monotonically with distance from the center.[43]

An example of such resonant phase mixing is illustrated in FIGURE 2. It tracks three initially localized clumps of test stars evolved in a galactic potential with periodic driving that damps as a power law in time. The left and center panels exhibit the (x,y)-coordinates at several different times; the right panel exhibits the exponential growth of components of an *emittance*-like quantity ε_i ($i = x, y, z$), which measures the area of the occupied phase-space planes corresponding to the coordinate r_i. Here, for example,

$$\varepsilon_x = \sqrt{\langle x^2 \rangle \langle v_x^2 \rangle - \langle xv_x \rangle^2}, \qquad (7)$$

where $\langle \ldots \rangle$ denotes an average over the clump. (In the context of charged-particle beams, emittance is given a more precise definition, described later in CHAOS IN CHARGED-PARTICLE BEAMS.) As was argued previously, initially localized clumps of regular orbits typically diverge secularly, whereas clumps of chaotic orbits diverge exponentially at a rate set by the typical value of the largest finite-time Lyapunov exponent χ,

$$\varepsilon_i \propto \left(\frac{t}{t_D}\right)^p \text{ (regular orbits)} \qquad \text{and} \qquad \varepsilon_i \propto e^{\chi t} \text{ (chaotic orbits)}, \qquad (8)$$

with p a constant of order unity.

Experimental Evidence for Chaos in Galaxies

There can, of course, be no direct experimental evidence for chaos in galaxies. However, careful analysis of observable velocity fields in suitably oriented galaxies provides compelling evidence that the gas flows in such spirals as NGC 3632 could be chaotic, especially near various resonances.[44]

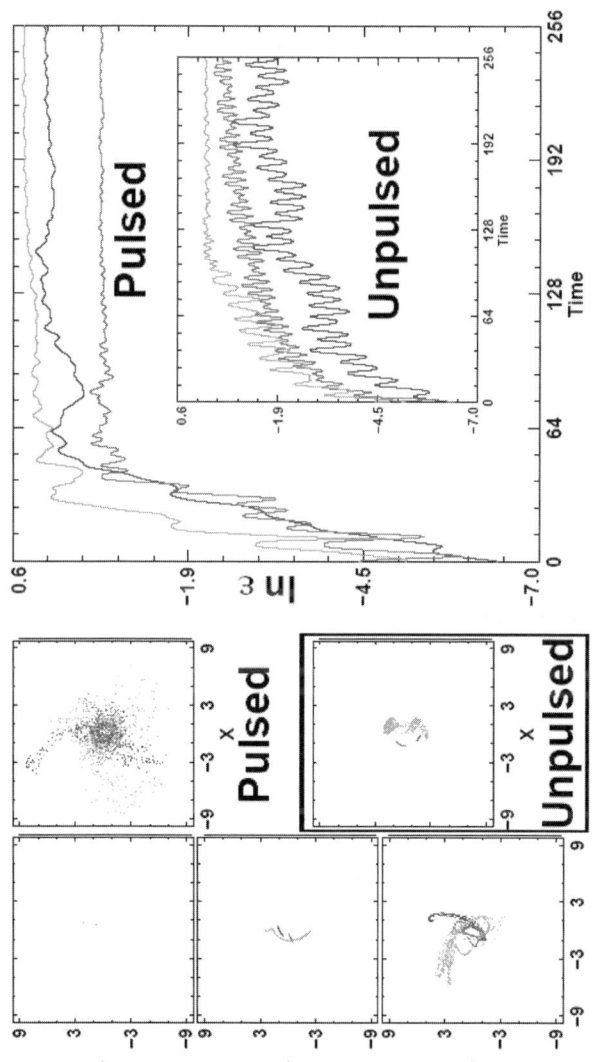

FIGURE 2. Left and **top middle**: x and y coordinates for three clumps of 1,600 test stars evolved in the time-dependent galactic potential $V(t) = -A(t)/(1 + x^2 + y^2 + z^2)^{1/2}$, with $A(t) = 1 + (a\sin\omega t)/(1 + t/t_0)^2$, for $a = 0.5$, $\omega = 1.25$, and $t_0 = 100$. From **top** to **bottom**, the snapshots are at times $t = 0, 32, 64$, and for the **top middle panel**, $t = 128$. The dynamical time $t_D \sim 20$. The clumps had initial size $\delta x = \delta y = 0.04$. **Bottom middle panel**: snapshot at $t = 128$ for the same clumps evolved in a time-independent potential with $A = 1$. **Right panel**: the quantity $\varepsilon = (\varepsilon_x^2 + \varepsilon_y^2)^{1/2}$ for the time-dependent clumps, where, in terms of velocity components, the emittances ε_x and ε_y satisfy, for example, $\varepsilon_x^2 = \langle x^2 \rangle \langle v_x^2 \rangle - \langle x v_x \rangle^2$. In this case the angular brackets represent an average over the 1,600 stars. **Right panel (inset)**: same for $A = 1$.

Chaos in Charged-Particle Beams

Intense charged-particle beams are, like galaxies, typically collisionless Hamiltonian systems wherein the density distribution self-consistently governs the dynamics via the Poisson equation. Transients in the beam distribution often arise as the accelerator manipulates the beam, whereby questions of equilibration, damping, and reversibility become fundamentally important to establishing and preserving the desired phase-space properties of the beam. For example, equipartitioning of anisotropic beams involves nonlinear energy transfer and evolution toward an isotropic metaequilibrium state.[45] As we show here, this is a consequence of chaotic mixing. Strictly speaking, chaotic mixing is a reversible process in that it is governed by Vlasov's equation. However, an essential question for the accelerator designer is whether this process is *operationally* reversible. Although it may be possible, in principle, to compensate operationally against phase-space dilution,[46] this compensation must be completed before any mixing has smeared a significant number of particles through global regions of phase space.[9,47] The question then becomes one of time scales. It arises in relation to any process for manipulating a beam with space charge, be this changing the transverse geometry of the beam (flat-to-round or round-to-flat transformations[48]), its longitudinal geometry (bunch compression[49]), or controlling the beam through sudden changes or imperfections in the beamline.[50]

Emittance and Space Charge

Consider, for simplicity, an infinitely long, that is, direct-current, beam that coasts without acceleration in the z-direction while confined by an external transverse focusing force. It is then natural to compute particle dynamics in a reference plane that comoves with the beam and is oriented transversely with respect to the beam motion. The particle velocities may in general exhibit both a systematic and a random component. For the former, the (x,y)-coordinates are measured from the beam centroid. For the latter, an average kinetic temperature can be defined for each transverse (x,y)-axis. Roughly speaking, the product of this temperature and the rms beam size is defined as the *rms emittance* of the beam, and this quantity is conserved for the special case that the (x,y)-components of the total transverse force (focusing plus space charge) are decoupled, linear, and time-independent in the reference frame comoving with the beam. More precisely, the rms emittance for the x-direction is calculated as

$$\varepsilon_x = \frac{1}{\beta\gamma mc}\sqrt{\langle x^2\rangle\langle p_x^2\rangle - \langle xp_x\rangle^2}, \tag{9}$$

where the averages involve moments defined with respect to the single-particle distribution function $f(x,p_x,y,p_y)$. Here (p_x,p_y) are the components of the transverse particle momentum with respect to the reference trajectory, $\beta = v_z/c$, $\gamma = (1-\beta^2)^{-1/2}$, m is the particle rest mass, and c is the speed of light. The *effective emittance*, often called simply *emittance*, is $4\varepsilon_x$, a quantity that corresponds to the area of the (x,p_x) phase space subtended by the beam and which has units of length. The respective emittances in the y-direction are calculated analogously.

Suppose the external transverse focusing force is linear, axisymmetric, and time-independent. For a beam with a small number of particles, the individual particles will oscillate harmonically (they execute what accelerator physicists call "betatron oscillations") at the *undepressed* betatron frequency, ω_{β_0}, determined solely by the external force. The amplitude of the oscillation for each particle depends on its total energy, and this is determined completely by the initial position and velocity of that particle. Now, as the beam current is increased, the superimposed electric field generated by the particles themselves becomes non-negligible, a phenomenon known as space charge. Space charge alters the net force seen by the individual particles in a way that is generally nonlinear and dependent on the density distribution of the beam. One can quantify space charge using a single parameter: the dimensionless intensity parameter μ defined as the ratio of the average space charge force to the external focusing force at the beam edge.

Since space charge is repulsive, it lowers the frequency of the betatron oscillations, resulting in a *depressed* betatron frequency $\omega_\beta < \omega_{\beta_0}$. The average *tune depression*, defined by $\eta \equiv \omega_\beta/\omega_{\beta_0}$ is related to the intensity parameter by $\eta = (1-\mu)^{1/2}$. Another important effect of space charge is the tendency to induce waves in the beam, a collective effect. These waves are characterized by the plasma frequency ω_p, which in turn relates to the intensity parameter μ as $\omega_p/\omega_{\beta_0} = (2\mu)^{1/2}$. Thus, in the limit of zero space charge ($\mu = 0$), the plasma frequency is zero, the tune depression is unity, and the particles behave as individual particles that only see the external focusing force. At the opposite end, the space charge limit ($\mu = 1$), the tune depression is zero, and the plasma frequency reaches its maximum value, meaning that collective oscillations dominate over the betatron motion of individual particles. At intermediate values of μ, excepting certain contrived theoretical distributions, for example, the KV distribution,[26] the net force acting on individual particles is typically nonlinear. Inasmuch as real beams are commonly out of equilibrium and subject to time-dependent focusing, the net force is often time-dependent as well.

Chaos and Equipartitioning

Anisotropy in a beam can be caused by essentially any anisotropic external influence, such as anisotropic focusing. In addition, a recent computational study provided strong evidence that chaotic mixing due to nonlinear forces from space-charge waves is intimately connected with equipartitioning, that is, the tendency of the velocity ellipsoid (or equivalently, the temperature) to isotropize rapidly.[51] These computations were done using the (2+1/2)-dimensional (distance down the accelerator is viewed as a *time* coordinate; hence the appellation (2+1/2)-dimensional) version of the particle-in-cell code WARP,[52] which tracks macroparticles with prespecified initial conditions through external electric and magnetic fields while including the self-consistently computed self-fields. The work concerned a highly space-charge-dominated, direct-current, cylindrical beam in which the initial momentum space reflected a temperature anisotropy. Accordingly, the initial rms emittances ε_x and ε_y were unequal, but the external focusing was axisymmetric. As the beam evolved, the temperature isotropized rapidly. Full equipartitioning occurred within about 5 m, after which the temperature exhibited anisotropic oscillations that largely damped by about 50 m. The equipartitioning time scales were found to correlate with the evolution of initially localized clumps of globally chaotic particles. These clumps dispersed exponentially with an *e*-folding *time* of about 2 m (roughly two plasma periods) and filled their accessible phase spaces in about 50 m.

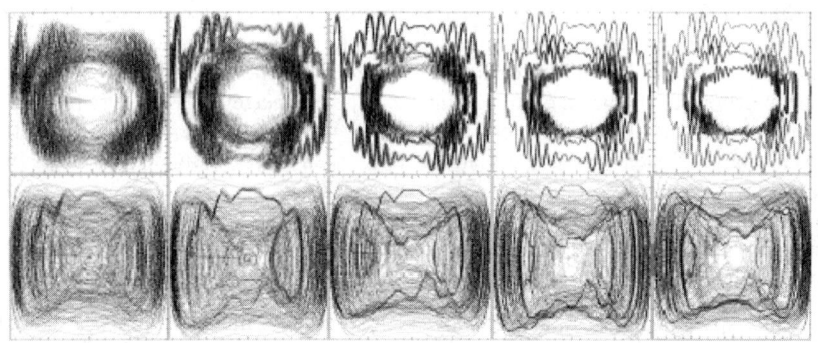

FIGURE 3. Trajectories of 20 test particles in (x,y)-space for the isotropic beam (**top**); and the anisotropic beam (**bottom**). The initial clump *emittance* decreases progressively by factors of 10 from *left* to *right*, thereby exposing the chaotic behavior of orbits in the anisotropic beam.

This first study concerned a form of *symmetry breaking*, with the broken symmetry appearing in momentum space rather than configuration space. The beam began in a nonequilibrium state and evolved toward a metaequilibrium in which the particle orbits filled an invariant measure of phase space. The transient dynamics reflected an intricate, evolving network of space-charge waves that set up a complicated time-dependent potential in which a substantial population of particle orbits became globally chaotic. By contrast, an analogously evolved symmetric, isotropic system exhibited a near-static potential that was essentially integrable, so that the orbits were essentially regular. The character of the orbits is evident in FIGURE 3, which shows trajectories of 20 test particles randomly selected from a given initially localized clump in both isotropic and anisotropic systems. Progressively reducing the area of the phase space initially spanned by the clump, as would be done in a calculation of finite-time Lyapunov exponents, reveals that the test-particle orbits are regular in the isotropic beam. However, the orbits are clearly chaotic in the anisotropic beam, this reflecting the complicated network of space-charge waves that arise in the presence of anisotropy. Equipartitioning did not lead to a significant halo because the rms properties of the beam were *matched* to the strength of the focusing forces, thereby minimizing large-scale time-dependent oscillations. A beam is said to be *matched* to the transport channel if its transverse density profile is stationary over the length of the channel. Otherwise, the density profile evolves. Consider the rms transverse radius of an evolving density profile. If the rms radius is stationary (equal to that of the matched beam), then the beam is *rms-matched*; otherwise, it is *rms-mismatched*. The density profile, hence space-charge potential, is normally more weakly time-dependent in a rms-matched beam than in a rms-mismatched beam. Only in the case of the strictly matched beam will the space-charge potential be stationary.

Merging Beamlets: Analysis of a Real Laboratory Experiment

A unique laboratory experiment concerning violent relaxation in charged-particle beams, conducted in the early 1990s, involved the propagation and merging of five beamlets in a periodic solenoidal (hence, axisymmetric) transport channel of length slightly more than 5 m.[53–55] The beamlets were initially oriented in a quincunx pattern and were close enough to each other that mutual interactions were important. The beam was nonrelativistic and subject to considerable time- dependent space charge. Given such a highly anisotropic initial density distribution and isotropic focusing, and considering that the time scale for collisional relaxation is orders of magnitude longer than the transport channel, one might naïvely expect the beamlets to merge (hence, "disappear") and reappear periodically. However, the beamlets were observed to reappear *only once*, at a point about 1 m from the source, regardless how well (or poorly) the rms beam properties were matched to the transport channel. Moreover, rms-mismatched beams led to the formation of an extended halo, with the density of the halo increasing with the degree of mismatch. Detailed simulations with a particle-in-cell code successfully reproduced the measurements.[55] The failure of the beamlets to reappear again would seem to reflect a collisionless process that, in effect, causes the particle orbits to lose memory of their initial conditions.

To explore to what extent chaotic mixing influences the dynamics of such a manifestly nonequilibrium beam, we redid the simulations using WARP. Our new simulations differ slightly from the experiment in that we considered a simpler transport channel, one that imparts a constant, linear external focusing force, whereas in the experiment the channel constituted a periodic solenoidal focusing lattice. We used a total of 4×10^6 particles distributed equally between each of the five beamlets. The idea was to generate a reasonably smooth potential. Our results correlated well with measurements of the density profile versus position along the beamline.

One might expect the strongly time-dependent space-charge potential to drive a large population of globally chaotic orbits. That this is the case is illustrated in FIGURE 4, which shows that clumps of representative test particles initially localized in phase space grew exponentially to fill much of their accessible phase-space regions. In each case, an initial extremely fast growth rate subsequently gave way to a slower rate, the transition occurring after a distance of about 5 m, at which time the beamlets had lost their identity and the phase-space distribution had become rounder. This computed finding is completely in keeping with the experimental findings.

Halo Formation

Los Alamos recently completed a laboratory experiment[56] involving the production and measurement of halo generated in a proton beam that was intentionally mismatched to a periodic focusing channel comprised of quadrupole magnets. The beam energy and current were 6.7 MeV and 75 mA, respectively, meaning the beam was nonrelativistic and space charge was strong. The length of the focusing channel spanned about 10 mismatch oscillations. The principal inferences from this experiment and corresponding simulations[57] were that (1) the phase-space volume of the beam grew in conjunction with the conversion of free energy from mismatch into "thermal" energy of the beam, and (2) parametric resonance was the principal mechanism driving halo formation. These inferences correspond to expectations from idealized theoretical models.[58,59] However, the quantitative data appeared to be sensitive to the exact phase-space distribution of the input beam, which could not be measured with precision, and the finite sensitivity of the diagnostics precluded characterization of the tenuous outermost wings of the halo profile. Moreover, the

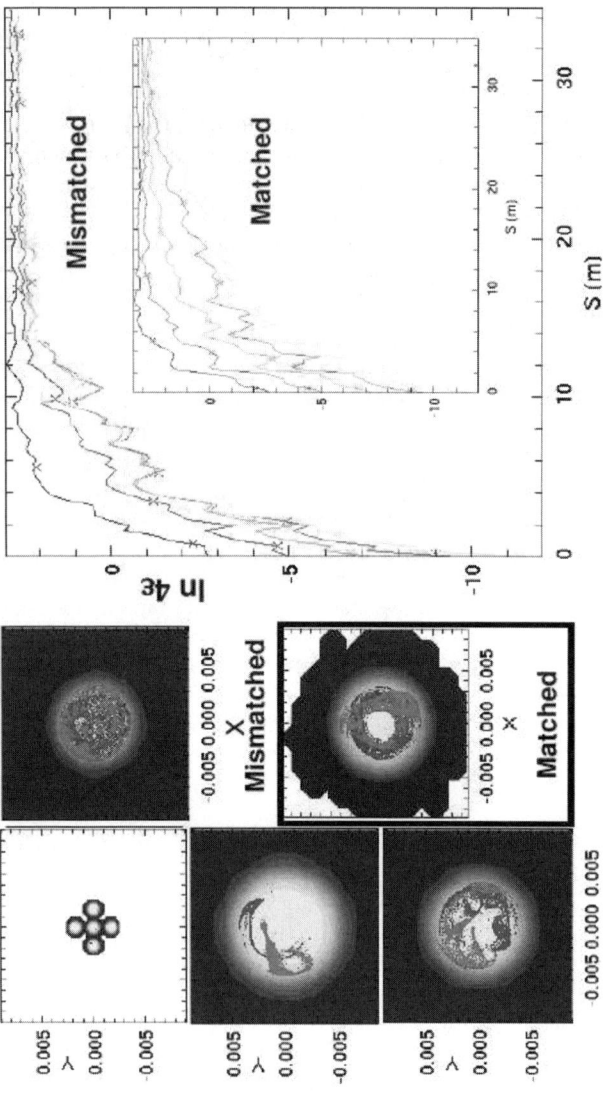

FIGURE 4. Left and **middle panels**: (x, y) plots (unit, m) for two different initially localized clumps of 20,000 test particles evolved in the total potential self-consistently computed using WARP. The snapshots are taken from a simulation of an rms-mismatched beam, at locations $s = 0.0, 10.08, 14.4,$ and 20.16 m along the beamline, with the exception of the **bottom middle** snapshot, which is at $s = 31.68$ m and pertains to a simulation with the rms-matched beam. The plasma wavelength is 0.47 m and the betatron wavelength is 2.0 m. The initial emittance of each clump is $\varepsilon_x = \varepsilon_y = 6.48 \times 10^{-10}$ m, which is 10^{-5} times the full beam emittance. **Right panel**: natural logarithm of the emittance $4\varepsilon_x$ for five clumps, each sampling a progressively smaller portion of the *red* clump on the left, hence progressively smaller initial emittances. **Right panel (inset)**: same for the rms-matched beam.

theoretical models provide no prediction of growth rates; the simulations were used to extract this information for comparison with the experiment.

As is documented in a companion paper in this volume of the *Annals*,[28] recent work has revealed that parametric resonance is not the whole story respecting halo formation.[60–62] The presence of colored noise (noise with a nonzero autocorrelation time), a real phenomenon associated with hardware imperfections and/or charge-density fluctuations, in combination with parametric resonance, can lead to much larger halos and remove the hard upper bound on the halo amplitude inferred from parametric resonance acting alone. Basically, this happens because the noise can keep a statistically small number of particles more in phase with low-order oscillatory modes of the beam. Here the fact that the orbits are chaotic is extremely important. Because chaotic orbits have power over a continuous set of frequencies, their coupling to both the modes and the noise can be significantly enhanced relative to the couplings that arise for regular, multiply periodic orbits.

Recent theoretical studies indicate that precisely the same phenomenology applies to galaxies, as well. Bohn and Sideris[60] found that substantial halo appears in gravitational systems as well as in beams. Colored noise in galaxies arises from the ambient intergalactic environment through the influence of neighboring stellar systems and/or clumpy dark-matter halos, as well as from internal density fluctuations within the subject galaxy. Subsequent work indicated that galactic halo formation is insensitive to the details of the bulk potential; generally all that is required are collective modes and noise.[63] This finding raises interesting questions: Are the observed light profiles of real galaxies primarily the product of violent relaxation at early epochs? Can remnant oscillations act over a Hubble time to alter substantially the product of violent relaxation alone? An effort toward answering these questions is underway, and a preliminary study indicates that long-time evolution and its associated halo formation can indeed influence observed properties of large galaxies.[64]

The Smooth-Potential Approximation

The foregoing discussion implies that, viewed "on the whole", discrete systems of stars or charged particles, if sufficiently large, can be approximated by a continuous density distribution and a smooth bulk potential. As pointed out in the aforementioned companion paper,[28] this assumption has been questioned in both the galactic[65] and accelerator[66,67] communities. To what extent is it really true that there actually *is* a smooth continuous-density limit? Assuming this limit exists, how large must the system be before discreteness effects (i.e., granularity associated with finite particle number) can safely be ignored? Can one, for example, treat a realistic beam bunch comprised of 10^9–10^{11} particles as a continuous charge distribution?

Numerical computations performed over the last several years, for both self-gravitating[68] and self-electrostatically interacting[69] Coulomb systems, suggest strongly that, viewed macroscopically, there is a well-defined continuum limit, and that discreteness effects can be extremely well modeled, even for individual orbits over comparatively short times, by Gaussian white noise in the context of a Fokker–Planck description. Indeed, one can estimate smooth-potential Lyapunov exponents from N-body simulations.[70]

That a Fokker–Planck description can be justified is nontrivial, since the standard derivations[4] and most experimental tests focus on the long-time behavior of orbit ensembles. Even more interesting, however, is the fact that, when applied to chaotic systems, a Fokker–Planck description implies that discreteness effects can be important on time scales much shorter than the collisional relaxation time t_R. Discreteness effects can dramatically accelerate diffusion through a complex phase space, both by facilitating transport along the Arnol'd web[5] and, in some cases, by transforming regular orbits into chaotic orbits and *vice versa*. Indeed, under certain circumstances, for example, for systems with "lumps" and/or asymmetries and/or pronounced density gradients, discreteness effects can be important on relatively short time scales even for N as large as about 10^{10}.[69]

It is important to stress that, even if discreteness effects become important over comparatively short time scales, their effect is *not* to induce collisional relaxation. Suppose, for example, that, in the absence of discreteness effects, the bulk potential, albeit chaotic, is strictly time-independent and the energies of individual "particles" are thereby conserved absolutely. Discreteness effects can then act to accelerate diffusion through a complex phase space, serving as a source of what nonlinear dynamicists call "extrinsic diffusion";[5] but, over time scales much smaller than t_R, they do *not* induce significant changes in energy.

Summary

There is growing evidence that physical processes involving chaos act very similarly in galaxies and charged-particle beams. In both cases a time-dependent potential can trigger resonances that lead to large measures of strongly chaotic orbits with large Lyapunov exponents, even if, as for the model used to generate FIGURE 2, the potential becomes integrable when the time dependence is "turned off". Manifestations of chaos can also be quite similar in time-independent systems. For example, a systematic investigation[9] of how the amount of chaos in a thermal-equilibrium beam[71] varies with the geometry of the beam yields results very similar to what is found[72] in triaxial generalizations of the Dehnen potentials of galactic dynamics[73] that have been proposed to model cuspy, triaxial galaxies.[74]

PLANS FOR FUTURE EXPERIMENTS

Charged-particle beams differ from galaxies in that beams adjust themselves to screen the external focusing force. The screening distance is the Debye length, and in a configuration of thermal equilibrium, the density profile in the outer region of the beam decreases to a low-density tail over a few Debye lengths as a result of screening the external focusing force.[9] By contrast, galaxies do not exhibit any analog of this Debye shielding. (Concerning gravitation, the length scale of interest is the Jeans length over which a gravitational instability arises, thereby leading, e.g., to the formation of galaxies.) Consequently, with a beam, the bulk potential (focusing plus space charge) cannot generally be molded to match precisely that of an evolving stellar system. For example, structures mimicking the presence of central density cusps or black holes in galaxies cannot be preserved in a beam because space charge is repulsive. Nor can a beam mimic effects from space–time evolution over cosmological time scales. However, phenomenology inherent to time-dependent collective dynamics in galaxies *can* be mimicked with beams. Gravitational examples (and their beam analogs) include colliding and merging galaxies (merging beamlets), collapsing galaxies (rms-mismatched beams), or perturbed but comparatively quiescent galaxies (beams with evolving density inhomogeneities). Thus, the key dynamics underlying galactic systems can be studied in the laboratory.

As implied in the section CHAOS IN CHARGED-PARTICLE BEAMS, to date there have been no laboratory experiments designed explicitly to explore the role of chaotic phase mixing via Coulomb forces on the evolution of nonequilibrium beams. Our simulations and interpretation of the merging-beamlet experiment point, however, to the importance of chaotic dynamics in real beams. Furthermore, the preponderance of simulations that we have highlighted herein suggests strongly that the combined effects of transient chaos and resonances are the keys toward a full understanding of violent relaxation in both beams and galaxies. Accordingly, we are planning a series of experiments to study phase mixing and attendant collisionless relaxation using the University of Maryland electron ring (UMER), a facility that is just now coming on line.[75] The ring is designed to transport the beam through many turns spanning over one kilometer, a distance spanning some 500–1,000 plasma periods, and the relative strength of the collective space-charge force is adjustable over a wide range, $0.25 \leq \mu \leq 0.97$.

The evolution and mixing of initial perturbations can be tracked using the comprehensive suite of diagnostics incorporated into UMER. As a whole, the diagnostic suite permits direct measurement of mixing time scales in units of the characteristic dynamical time, and the degree of mixing in both configuration space and in energy, by enabling the evolution of macroscopic features to be observed and quantified. It should thereby be possible to distinguish observationally between chaotic (i.e., exponential, global) phase mixing versus regular (i.e., secular, more local) phase mixing. We also plan, of course, to confirm our interpretations using simulation codes. We project an added benefit, as well: establishing the phenomenology of phase mixing in time-dependent beam potentials both experimentally and numerically should likewise provide an unambiguous mechanism for validating codes and simulation techniques in both beam physics and galactic dynamics.

CONCLUSIONS

It is clear that, in principle, chaotic mixing can account for rapid macroscopic dynamics, including collective relaxation to a metaequilibrium state. Moreover, there is substantial numerical evidence that such mixing could play an important role in the evolution of both galaxies and charged-particle beams. Although a portion of this numerical evidence arose historically as part of interpreting real laboratory experiments with beams, there is need for considerably more work. Our idea is to look for evidence of chaos and chaotic phase mixing in controlled laboratory experiments involving large Coulomb systems. Unfortunately and obviously, it is impossible to perform controlled experiments on self-gravitating systems like galaxies. However, in view of the strong indications, both theoretical and numerical, that the relevant physics is virtually identical in galaxies and charged-particle beams, it seems possible—and highly desirable—to use beamlines like UMER as laboratories in which to perform indirect tests of the predictions of galactic dynamics. The key quantities to be measured in such experiments are the evolutionary time scales attendant to charged-particle beams with well-diagnosed and freely adjustable initial conditions, as well as the efficacy of mixing in both configuration space and energy.

The suite of diagnostics on UMER is capable of detailed, time-resolved measurement of the distribution function in the six-dimensional phase space of a single beam particle. These diagnostics are designed to measure the same macroscopic observables and their respective evolutionary time scales as are generated in numerical simulations. Accordingly, UMER serves as a platform for a virtually unlimited range of experiments to explore nonlinear, transient dynamics of Coulomb systems, and our overarching plan is to exploit this capability to access the physics of collisionless relaxation that large charged-particle and self-gravitating stellar systems share in common.

We conclude by positing four categories of questions relevant to evolving galaxies that UMER will likely be able to answer: (1) Do initially localized *clumps* mix as simulations predict? Specifically, do they clearly mix exponentially in regimes where one would predict a preponderance of chaotic orbits and secularly otherwise? Do the mixing time scales agree with code predictions? The process of making related computational predictions followed by experimental measurements will, incidentally, constitute an excellent process for code validation.[76] (2) How important are collisions during the early evolution of an "inverse-square-law" system? This can be answered experimentally by measuring changes in the particle-beam evolution as the particle density is gradually lowered, so that space charge becomes gradually less important, leaving collisions to drive the evolution. (3) Do nonequilibrium inverse-square-law systems evolve so as to reduce the population of chaotic orbits? In the case of beams with space charge, it would seem the answer must be yes, for in thermal equilibrium most of this beam will have nearly uniform density, hence mostly regular orbits. However, the matter deserves careful experiments; the answer hinges on the specific metaequilibrium states that the beam accesses during the course of its evolution, as will also be true for galaxies. (4) For systems initially far from equilibrium, what is the relative importance of *violent* short-time evolution versus *calm* long-time evolution? Can the latter significantly alter the macroscopic properties of the system that one measures in the laboratory in the case of beams or observes on the sky in the case of galaxies? In other words, is what one observes really the product of violent relaxation exclusively? Whatever the answers to these questions turn out to be, the process of exploring them promises to be interesting and illuminating.

ACKNOWLEDGMENTS

This work was supported by the Department of Education under Grant P-116Z010035, by the Department of Energy under Grants DEFG02-04ER41323, DEFG02-94ER40855, DEFG02-92ER54178, and DEAC02-76CH00300, and the National Science Foundation under Grant AST-0307351. We are grateful to T. Antonsen, J. Ellison, and I. Haber for providing helpful comments and recommendations.

The authors declare that they have no competing financial interests.

REFERENCES

1. KANDRUP, H.E. 2003. Chaos and chaotic phase mixing in galaxy evolution and charged particle beams. *In* Galaxies and Chaos, Theory and Observations. G. Contopoulos & N. Vogilis, Eds. Springer Lecture Notes in Physics **626**: 154–168. Springer-Verlag, Heidelberg.
2. CHANDRASEKHAR, S. 1943. Principles of Stellar Dynamics. University of Chicago, Chicago.
3. REISER, M. 1994. Theory and Design of Charged Particle Beams. Wiley, New York.
4. ROSENBLUTH, M.N., W.M. MACDONALD & D.L. JUDD. 1957. Fokker–Planck equation for an inverse-square force. Phys. Rev. **107**: 1.
5. LEE, E.D. 1968. Brownian motion in a stellar system. Astrophys. J. **151**: 687.
6. LICHTENBERG, A.J. & M.A. LIEBERMAN. 1992. Regular and Chaotic Dynamics. Springer, Berlin.
7. BOHN, C.L. 2000. Nonperturbative geometrodynamic calculation of chaotic mixing time in charged particle beams. *In* The Physics of High-Brightness Beams. J. Rosenzweig & L. Serafini, Eds.: 358–368. World Scientific, Singapore.
8. KANDRUP, H.E., I.V. SIDERIS & C.L. BOHN. 2002. Chaos, ergodicity, and the thermodynamics of lower-dimensional time-independent Hamiltonian systems. Phys. Rev. E **65**: 016214.
9. BOHN, C.L. & I.V. SIDERIS. 2003. Chaotic orbits in thermal equilibrium beams: existence and dynamical implications. Phys. Rev. ST Accel. Beams **6**: 034203.
10. CASETTI, L., C. CLEMENTI & M. PETTINI. 1996. Riemannian theory of Hamiltonian chaos and Lyapunov exponents, Phys. Rev. E **54**: 5969.
11. TERZIĆ, B. & H.E. KANDRUP. 2003. Semi-analytic estimates of Lyapunov exponents in lower-dimensional systems. Phys. Lett. A **311**: 241.
12. HÉNON, M. & C. HEILES. 1964. The applicability of the third integral of motion: some numerical tests. Astron. J. **69**: 73.
13. LYNDEN-BELL, D. 1967. Statistical mechanics of violent relaxation in stellar systems. Mon. Not. R. Astron. Soc. **136**: 101.
14. BERTIN, G. 2000. Dynamics of Galaxies. Cambridge University Press, Cambridge.
15. KANDRUP, H.E. & M.E. MAHON. 1994. Relaxation and stochasticity in a truncated Toda lattice. Phys. Rev. E **49**: 3735.
16. MERITT, D. & M. VALLURI. 1996. Chaos and mixing in triaxial stellar systems. Astrophys. J. **471**: 82.
17. BOHN, C.L. 1993. Transverse phase-space dynamics of mismatched charged-particle beams. Phys. Rev. Lett. **70**: 932.
18. BOHN, C.L. & J.R. DELAYEN, 1994. Fokker–Planck approach to the dynamics of mismatched charged-particle beams. Phys. Rev. E **50**: 1516.
19. GLUCKSTERN, R.L. 1994. Analytic model for halo formation in high current ion linacs. Phys. Rev. Lett. **73**: 1247.
20. KANDRUP, H.E., I.M. VASS & I.V. SIDERIS. 2003. Transient chaos and resonant phase mixing in violent relaxation. Mon. Not. R. Astron. Soc. **341**: 927.
21. GRASSBERGER, P., R. BADII & A. POLITI. 1988. Scaling laws for invariant measures on hyperbolic and nonhyperbolic attractors. J. Stat. Phys. **51**: 135.
22. ANTONSEN, T.M., Z. FAN, E. OTT & E. GARCIA-LOPEZ. 1996. The role of chaotic orbits in the determination of power spectra of passive scalars. Phys. Fluids **8**: 3094.
23. BENETTIN, G., L. GALGANI & J.M. STRELCYN. 1976. Kolmogorov entropy and numerical experiments. Phys. Rev. A **14**: 2338.
24. CHAO, A.W. 1993. Physics of Collective Beam Instabilities in High Energy Accelerators. Wiley, New York. 26.
25. LEE, S.Y., Ed. 1996. Space Charge Dominated Beams and Applications of High Brightness Beams, AIP Conference Proceedings 377, AIP, New York.
26. KAPCHINSKIJ, I.M. & V.V. VLADIMIRSKIJ. 1959. Limitations of proton beam current in a strong focusing linear accelerator associated with the beam space charge. *In* Proceedings of the International Conference on High-Energy Accelerators and Instrumentation, CERN, 274.

27. SACHERER, F.J. 1971. RMS envelope equations with space charge. IEEE Trans. Nucl. Sci. **NS-18**: 1105.
28. BOHN, C.L. Chaotic dynamics in charged-particle beams: possible analogs of galactic evolution. Ann. N.Y. Acad. Sci. **1045**: this volume.
29. MERRITT, D. 1999. Elliptical galaxy dynamics. Publ. Astron. Soc. Pac. **111**: 129.
30. LAUER, T., E.A. AJHAR, Y.-I. BYUN, et al. 1995. The centers of early-type galaxies with HST. I. An observational survey. Astron. J. **110**: 2622.
31. SPARKE, L.S. & J.A. SELLWOOD. 1987. Dissection of an N-body bar. Mon. Not. R. Astron. Soc. **225**: 653.
32. PFENNIGER, D. & D. FRIEDLI, 1991. Structure and dynamics of 3D N-body barred galaxies. Astron. Astrophys. **252**: 75.
33. EL-ZANT, A. & I. SHLOSMAN. 2002. Dark halo shapes and the fate of stellar bars. Astrophys. J. **577**: 626.
34. UDRY, S. & D. PFENNIGER. 1988. Stochasticity in elliptical galaxies. Astron. Astrophys. **198**: 135.
35. SRIDHAR, S. 1989. Undamped oscillations of homogeneous collisionless stellar systems. Mon. Not. R. Astron. Soc. **238**: 1159.
36. HOLLEY-BOCKELMANN, K., J.C. MIHOS, S. SIGGURDSSON & L. HERNQUIST. 2001. Models of cuspy triaxial galaxies. Astrophys. J. **549**: 862.
37. POON, M.Y. & D. MERRITT. 2002. Triaxial black hole nuclei. Astrophys. J. Lett. **568**: 89.
38. KANDRUP, H.E. 2002. Should elliptical galaxies be idealised as collisionless equilibria? Space Sci. Rev. **102**: 101.
39. SIOPIS, C. & H.E. KANDRUP. 2000. Phase-space transport in cuspy triaxial potentials: Can they be used to construct self-consistent equilibria? Mon. Not. R. Astron. Soc. **319**: 43.
40. BINNEY, J. 1978. Elliptical galaxies—prolate, oblate, or triaxial? Comments Astrophys. **8**: 27.
41. ATHENASSOULA, E., O. BIENAYMÉ, L. MARTINET & D. PFENNIGER. 1983. Orbits as building blocks of a barred galaxy model. Astron. Astrophys. **127**: 349.
42. KANDRUP, H.E., I.V. SIDERIS, B. TERZIĆ & C.L. BOHN. 2003. Supermassive black hole binaries as galactic blenders. Astrophys. J. **597**: 111.
43. LAUER, T., K. GEBHARDT & D. RICHSTONE, et al. 2002. Galaxies with a central minimum in stellar luminosity density. Astron. J. **124**: 1975.
44. FRIDMAN, A.M., O.V. KHORUZHII & E.V. POLYACHENKO. 2002. Observational manifestation of chaos in the gaseous disk of the grand design spiral galaxy NGC 3631. Space Sci. Rev. **102**: 51.
45. KISHEK, R.A., P.G. O'SHEA & M. REISER. 2000. Energy transfer in nonequilibrium space-charge-dominated beams. Phys. Rev. Lett. **85**: 4514.
46. CARLSTEN, B.E. & D.T. PALMER. 1999. Enhanced emittance compensation in a high-frequency rf photoinjector using rf radial focusing. Nucl. Instrum. Methods A **425**: 37.
47. O'SHEA, P.G. 1998. Reversible and irreversible emittance growth. Phys. Rev. E **57**: 1081.
48. BUROV, A., S. NAGAITSEV, A. SHEMYAKIN & YA. DERBENEV. 2000. Optical principles of beam transport for relativistic electron cooling. Phys. Rev. ST Accel. Beams **3**: 094002.
49. BOHN, C.L. 2002. Coherent synchrotron radiation: theory and experiments. AIP Conference Proceedings 647, AIP, New York. 81.
50. VENTURINI, M., R.A. KISHEK & M. REISER. 1999. The problem of dispersion matching in space charge dominated beams. *In* Proceedings of the 1999 Particle Accelerator Conference, IEEE Catalog No. 99CH36366, 3274–3276.
51. KISHEK, R.A., C.L. BOHN, I. HABER, et al. 2001. Computational investigation of dissipation and reversibility of space-charge-driven processes in beams. *In* Proceedings of the 2001 Particle Accelerator Conference, IEEE Catalog No. 01CH37268. 151–154.
52. GROTE, D.P., A. FRIEDMAN, I. HABER & S. YU. 1996. Three-dimensional simulations of high current beams in induction accelerators with WARP3D. Fus. Eng. Design **32–33**: 193.

53. HABER, I., D. KEHNE, M. REISER & H. RUDD. 1991. Experimental, theoretical, and numerical investigation of the homogenization of density nonuniformities in the periodic transport of a space-charge-dominated beam. Phys. Rev. A **44**: 5194.
54. KEHNE, D.M. 1992. Experimental Studies of Multiple Beam Merging, Mismatch, and Emittance Growth in a Periodic Solenoidal Transport Channel. Ph.D. Dissertation, University of Maryland, College Park.
55. REISER, M. 1994. Theory and Design of Charged Particle Beams. Wiley, New York. [see Sec. 6.2.2, cf. Fig. 6.10.]
56. ALLEN, C.K., K.C.D. CHAN, P. COLESTOCK, et al. 2002. Beam-halo measurements in high-current proton beams. Phys. Rev. Lett. **89**: 214802.
57. QIANG, J., P.L. COLESTOCK, D. KILPATRICK, et al. 2002. Macroparticle simulation studies of a proton beam halo experiment. Phys. Rev. ST Accel. Beams **5**: 124201.
58. REISER, M. 1991. Free energy and emittance growth in nonstationary charged particle beams, J. Appl. Phys. **70**: 1919.
59. WANGLER, T.P., K.R. CRANDALL, R. RYNE & T.S. WANG. 1998. Particle-core model for transverse dynamics of beam halo. Phys. Rev. ST Accel. Beams **1**: 084201.
60. BOHN, C.L. & I.V. SIDERIS. 2003. Fluctuations do matter: large noise-enhanced halos in charged-particle beams. Phys. Rev. Lett. **91**: 264801.
61. SIDERIS, I.V. & C.L. BOHN. 2004. Production of enhanced beam halos via collective modes and colored noise. Phys. Rev. ST Accel. Beams **7**: 104202.
62. GERIGK, F. 2004. Beam halo in high-intensity hadron accelerators caused by statistical gradient errors. Phys. Rev. ST Accel. Beams **7**: 064202.
63. SIDERIS, I.V. & H.E. KANDRUP. 2004. Noise-enhanced parametric resonance in perturbed galaxies. Astrophys. J. **602**: 678.
64. VASS, I.M. 2004. Private communication.
65. HEGGIE, D. 1991. Chaos in the N-body problem of stellar dynamics. *In* Predictability, Stability and Chaos in N-Body Dynamical Systems. A. Roy, Ed.: 47–62. Plenum, New York.
66. STRUCKMEIER, J. 1996. Concept of entropy in the realm of charged particle beams, Phys. Rev. E **54**: 830.
67. STRUCKMEIER, J. 2000. Stochastic effects in real and simulated charged particle beams. Phys. Rev. ST Accel. Beams 3, 034202.
68. SIDERIS, I.V. & H.E. KANDRUP. 2002. Chaos and the continuum limit in the gravitational N-body problem. II. Nonintegrable potentials. Phys. Rev. E **65**: 066203.
69. KANDRUP, H.E., I.V. SIDERIS & C.L. BOHN. 2004. Chaos and the continuum limit in charged particle beams. Phys. Rev. ST Accel. Beams **7**: 014202.
70. KANDRUP, H.E. & I.V. SIDERIS. 2003. Smooth potential chaos and N-body simulations. Astrophys. J. **585**: 244.
71. BROWN, N. & M. REISER. 1995. Thermal equilibrium of bunched charged particle beams. Phys. Plasmas **2**: 965.
72. KANDRUP, H.E. & C. SIOPIS. 2003. Chaos and chaotic phase mixing in cuspy triaxial potentials. Mon. Not. R. Astron. Soc. **345**: 727.
73. DEHNEN, W. 1993. A family of potential-density pairs for spherical galaxies and bulges. Mon. Not. R. Astron. Soc. **265**: 250.
74. MERITT, D. & T. FRIDMAN, 1996. Triaxial galaxies with cusps. Astrophys. J. **460**: 136.
75. O'SHEA, P.G. Nonlinear dynamics experiments with electron beams. Ann. N.Y. Acad. Sci. **1045**: this volume.
76. POST, D.E. & L.G. VOTTA. 2005. Computational science demands a new paradigm. Physics Today **58**(1): 35.

Chaotic Dynamics in Charged-Particle Beams

Possible Analogs of Galactic Evolution

COURTLANDT L. BOHN

Northern Illinois University, Department of Physics, DeKalb, Illinois, USA

Fermi National Accelerator Laboratory, Batavia, Illinois, USA

ABSTRACT: During the last couple of years of his life, Henry Kandrup became intensely interested in using charged-particle beams as a tool for exploring the dynamics of evolving galaxies. He and I recognized that both galaxies and charged-particle beams can exhibit collisionless relaxation on surprisingly short time scales, and that this circumstance can be attributed to phase mixing of chaotic orbits. The chaos is often triggered by resonances caused by time dependence in the bulk potential, which acts almost identically for attractive gravitational forces as for repulsive electrostatic forces superposed on external focusing forces. Together we published several papers concerning evolving beams and galaxies, papers that relate to diverse topics such as the physics of chaotic mixing, the applicability of the Vlasov–Poisson formalism, and the production of diffuse halos. We also teamed with people from the University of Maryland to begin designing controlled experiments to be done at the University of Maryland electron ring. This paper highlights our collaborative findings as well as plans for future investigations that the findings have motivated.

KEYWORDS: chaos; N-body problem; nonlinear dynamics; collisionless; halo

INTRODUCTION AND CONTEXT

Consider an example from the charged-particle-accelerator community for which collisionless evolutionary beam dynamics is a central concern. The example is a free-electron laser (FEL) for generating coherent X-rays.[1] This machine generates the coherent radiation by passing an ultrarelativistic electron beam through a long, periodic array of magnets. The magnets deflect the beam, causing the electrons to "wiggle" and thereby radiate. Accordingly, the magnet array is called a "wiggler." Through a process of self-amplified spontaneous emission, the radiation gradually builds its intensity and becomes coherent as the electron beam traverses the wiggler.

The production of coherent X-rays hinges first on the production of a suitable electron beam. This beam consists of electron packets, called *bunches*. Generating high-brightness X-rays requires the bunches to have high charge, at the nanocoulomb (nC) level. In addition, essentially all of the electrons must participate in the lasing process. This means the electrons in each bunch must fit within the optical mode, and in view of the small X-ray wavelength, the optical mode is correspondingly small.

Address for correspondence: Courtlandt L. Bohn, Northern Illinois University, Department of Physics, DeKalb, IL 60115, USA. Voice: 815-753-6473; fax: 815-753-8565
clbohn@niu.edu

This sets a stringent limit on the size of the phase space that the electron bunch can span.

To produce the electron bunches, one uses an external laser to irradiate a photocathode. The profile of the individual bunches and the time structure of the train of bunches hinge on the profile and time structure of the laser radiation. As the bunches traverse the accelerator, various specially designed hardware components further shape their profiles. If all goes according to plan, the electron beam that enters the wiggler is of very high quality, meaning that the electrons in each bunch together span a very small volume of phase space. From a practical perspective, "all goes according to plan" means collisionless processes associated with Coulomb self-forces within the bunch, a phenomenon called *space charge*, have not seriously degraded the beam.

Space charge works against achieving the required quality of the electron beam. Nonlinear collective forces act rapidly to redistribute the electrons. Because the beam from the source is generally far from equilibrium, it carries free energy that, as it redistributes, irreversibly expands the occupied phase space. Space charge is, thus, a key concern with reference to electron sources. In the laboratory frame, the space-charge force decreases inversely with the square of the beam energy. For the transverse component, this arises from the partial cancellation between the self-magnetic and self-electrostatic forces, whereas for the longitudinal component, it is due to Lorentz contraction.[2] Nonetheless, for nC bunch charges, space charge remains *the* important dynamic at beam energies up to about 100 MeV; that is, through a substantial portion of the accelerator.

In addition to degrading the bulk properties of the phase space of the beam, space charge also generates *beam halo*, a diffuse population of electrons that lie far from the bunch centroid. This is of special concern in accelerators that produce a beam with high average current.

EXAMPLES OF RAPIDLY EVOLVING BEAMS

Rapid beam evolution due to space charge is an observed fact. Consider the example depicted in FIGURE 1. This figure concerns beam measurements at the Fermilab photoinjector, a machine that accelerates nC-level bunches to about 15 MeV over a distance of a couple of meters. As FIGURE 1 depicts, the quality of the output beam directly depends on its initial density profile. The beam tries to screen externally applied fields, and because its space charge is strong, the screening length is small compared to the beam size. This means the beam *wants* to be near-uniform, and one thus wants to extract a near-uniform beam from the photocathode to minimize violent evolution. However, one is limited by the quality of the laser beam at the cathode, and this beam can be strongly nonuniform. The consequence of a strongly nonuniform initial density is much more serious beam degradation, which is the message of FIGURE 1. This degradation takes place over just a few meters. What is the physical process that leads to such rapid degradation? Due to the large number of electrons in each bunch (some 10^{10}), evolutionary effects of two-body collisions would require kilometers of propagation distance to manifest themselves. The fact

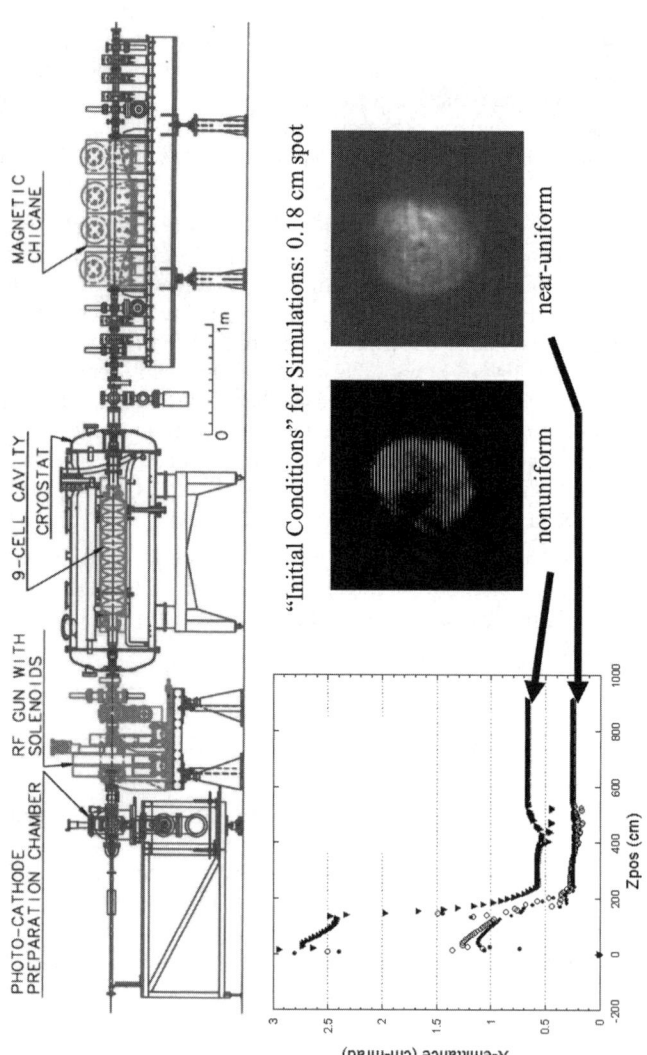

FIGURE 1. (Top) Schematic of Fermilab photoinjector. **(Bottom right)** Nonuniform and nearuniform laser spots on cathode. **(Bottom left)** Computational results for transverse x-emittance (a measure of phase-space area spanned by the beam for the x-axis degree of freedom). The nonuniform beam is seen to degrade twice as badly as the near-uniform beam. Measurements of the x-emittance of the output beams agree with the computed values. (Figure reproduced courtesy of Daniel Mihalcea.)

that degradation appears over just meters means the physics must necessarily be collisionless; it must be a process of phase mixing.

It was considerations like these that led to my collaboration with Henry Kandrup. Back in 2000, I applied the geometrodynamic theory of Pettini and collaborators[3] to estimate evolutionary time scales in both beams and galaxies associated with chaotic mixing. I submitted a paper[4] to *Physical Review Letters* that ultimately was rejected because one of the referees refused to believe that a beam can be chaotic. As a result, I contacted Henry, and thus began a sustained collaboration by which we, together with our colleagues, investigated many facets of the evolution of nonequilibrium N-body Coulomb systems (those in which the interparticle force varies inversely with the square of their separation). Henry came to take a strong interest in the physics of beams over the last two years of his life, and he developed a special interest in doing laboratory experiments with beams that would clarify rapid dynamical processes in large stellar systems. He wanted to author papers on beams; unfortunately this desire came to fruition only posthumously.

A second example of a rapidly evolving beam, one that is perhaps more intuitive and is thoroughly documented, is an experiment conducted some 15 years ago at the University of Maryland that concerned the merging of five nonrelativistic electron beamlets initially constituting a quincunx pattern.[5] These beamlets reappeared only once, after a single dynamical time (orbital period), after which they phase-mixed and completely vanished. This phase mixing, again, took place over just a few meters, that is, in just a few dynamical times. For this beam, the collisional relaxation time corresponds to some 100 km. Recent simulations of this experiment[6] have revealed that the underlying dynamics may be understood in terms of phase mixing of chaotic orbits, that is, *chaotic mixing.*[7]

CHAOTIC MIXING: SOME SALIENT FEATURES

Our work has focused on chaotic orbital dynamics, both in charged-particle beams wherein space charge is important, and in self-gravitating stellar systems (i.e., galaxies). One topic of intensive study was the convergence of an N-body system to the continuum limit in complicated time-independent potentials. Another was the orbital dynamics and associated phase mixing in generic time-dependent potentials. Included in the latter is the formation of diffuse large-amplitude halos in the presence of parametric resonance and colored noise. Progress on each topic is summarized in what follows.

Time-Independent Potentials: the Continuum Limit

A standard, often tacit, assumption in theoretical investigations of charged particle beams (and galaxies as well) is that particle correlations are unimportant. With this assumption, one applies the Vlasov–Poisson equations to calculate the distribution function of the particles in the six-dimensional phase space of a single particle.[5,8] If the system is in static equilibrium, then the distribution function can be expressed as a function of isolating integrals of particle motion in the mean potential. For example, any function of the Hamiltonian derived from the mean space-charge potential is a solution of the Vlasov–Poisson equations, although not all such functions correspond to stable equilibria.[9] A major justification typically provided for using Vlasov–Poisson is that the collisional relaxation time is long compared to, say, the transit time of a beam through any viable linear accelerator (or, for galaxies, long compared to a Hubble time).

As is evident in its derivation from the BBGKY hierarchy,[10] a Vlasov–Poisson system having total charge Q represents a *continuum limit* in which the number of particles $N \to \infty$ and their individual charges $q \to 0$, such that $qN = Q$. By contrast, real systems contain a finite number of particles N, and real charges have nonzero magnitudes. Vlasov–Poisson is, therefore, unrealistic, yet for the reasons stated, it is used to describe systems in the real world. An obvious question, therefore, is (1) to what extent do predictions derived from Vlasov–Poisson adequately describe real finite-N systems? A related question is (2) to what extent do predictions derived from a simulation involving $N < N$ macroparticles adequately describe the real N-body system?

One of our recent investigations[11] concerned how discreteness effects, that is, granularity, influence the answers to both questions. The investigation invoked three qualitatively different time-independent space charge potentials. The first was a uniform density ellipsoid such that, in the corresponding smooth potential, the particles execute simple-harmonic-oscillator orbits; the smooth potential is, thus, completely regular. The second was the same as the first, except a spike of charge was added at the center making the smooth potential entirely chaotic. The third was a configuration of thermal equilibrium[12] for which the smooth potential supports populations of both regular and chaotic orbits. One finding is that, in all three potentials viewed macroscopically, there is a precise sense in which, as the number of macroparticles N increases, trajectories in frozen-N systems converge toward their counterparts in the corresponding smooth potential. For very small N, less than about 10^4, the notion of an average bulk potential fails and orbits in frozen-N systems are very different from smooth-potential characteristics. In particular, the usual distinctions between regularity and chaos that exist in a smooth potential seem completely lost. However, for larger N the distinctions become clearer, and they are manifest in the evolution of *regular* versus *chaotic* clumps of initially localized particles. Just as for clumps evolved in a smooth potential, the *emittance* of a regular clump evolved in a frozen-N potential, that is, the volume of coarse-grained phase space it occupies, was seen to grow as a power law in time, whereas for a chaotic clump it grows roughly exponentially. However, in both cases the growth is more rapid than in the smooth potential. Discreteness effects accelerate emittance growth for *both* regular and chaotic clumps.

In terms of both the statistics of collections of orbits and the complexities of individual orbits, Gaussian white noise was found to mimic well discreteness effects in the context of a Fokker–Planck–Langevin description. This appears true even when considering the short-time behavior of individual orbits. These findings suggest strongly that Langevin simulations are useful for assessing the importance of discreteness effects in real beams for which the constituent number of particles $N \to N$ is too large to allow honest direct-summation integrations. Accordingly, Langevin simulations were applied to the three potentials considered, and the results indicated that discreteness effects can remain important even if N is very large. Such is the case for the physical beam corresponding to the thermal-equilibrium configuration that was considered, for which $N > 10^9$. This is especially true when chaotic orbits are present.

Discreteness effects were also seen to trigger transitions between regular and chaotic behavior. The larger the value of N, the longer it takes for these transitions to become important, and they become impossible in the continuum limit. However, for any finite N there appears to be a finite time beyond which it is unsafe to ignore these discreteness-induced transitions. Even if discreteness effects were too weak to facilitate frequent transitions between regularity and chaos, they may nonetheless play an important role in accelerating diffusion through a complex chaotic phase space. Generic smooth potentials admitting both regular and chaotic orbits have phase spaces in which chaotic regions are partitioned by complex structures associated with cantori in two dimensions and the Arnol'd web in three dimensions. Although they are not absolute obstructions, they serve as *entropy* barriers that impede phase-space transport.[13] However, even very-low amplitude Gaussian white noise has been shown dramatically to accelerate diffusion through such barriers.[14] In view of the finding that discreteness effects can be modeled as Gaussian white noise, they should likewise present a significant source of accelerated phase-space transport.

The meaning of *chaos* in reference to beams is somewhat subtle, just as it is with respect to galaxies. Two distinct sources of chaos were found to exist, associated with physics on different scales. Chaos associated with close encounters between individual charges is always present. They cause nearby orbits to diverge exponentially until a distance comparable to the interparticle spacing separates them. In that context beams are *always locally chaotic*. However, if in the continuum limit the bulk potential admits global stochasticity, then the orbits will continue to separate and exponentially fill global regions of phase space. These two distinct epochs of phase mixing are characterized separately by different sets of Lyapunov exponents. Close encounters trigger an exponential separation of nearby trajectories at a rate χ_N, but the separation saturates at only microscopic scales. The bulk potential triggers an exponential separation at a rate χ_S typically much smaller than χ_N (although much larger than the rate of collisional relaxation), but it saturates on macroscopic scales. Hence, when global stochasticity is present in beams, it leads to macroscopic, operationally irreversible evolution.

All of these considerations have practical implications for beams. In particular, discreteness effects can be important in real beams and over real acceleration time scales, thereby vitiating at least to some extent the Vlasov–Poisson methodology. In turn, simulations that correctly account for the full scale of evolutionary mechanisms may require a huge number of macroparticles, possibly comparable to the number of particles in the real beam bunch itself. Inasmuch as the same phenomenology is reflected in stellar systems,[15] the same conclusion would seem to pertain in that context as well, although application of Langevin techniques to infer discreteness effects in, for example, giant elliptical galaxies with very large N has yet to be done.

Time-Dependent Potentials: Large, Diffuse Halos

Beam loss from impingement of halo particles on accelerator hardware is a major concern in, for example, high-current light-ion accelerators. Just a tiny impingement, about 1 W/m, could generate radioactivation that precludes routine, hands-on maintenance.[16] Given a 1-mA, 1-GeV light-ion beam, for example, for baseline beam parameters of the spallation neutron source (SNS) presently under construction at Oak Ridge National Laboratory, this criterion translates to just one in 10^6 particles lost per meter, a quantity that scales linearly with average beam current. Accordingly, a comprehensive understanding of beam-halo formation is imperative.

Early efforts to identify fundamental mechanisms of beam-halo formation centered on using a *particle–core* model.[17] The basic recognition was that if a uniform-density core is made to pulsate, particles that initially lay outside the core and that resonate with its pulsations could reach large amplitudes and form a *halo*. This led to the identification of parametric resonance as the essential mechanism of halo formation. A key feature of parametric resonance in the context of the particle-core model is a hard upper bound to the amplitude that a halo particle can reach.[18] Because the orbital frequency of the particle is a function of its amplitude, at sufficiently large amplitude the particle falls out of resonance with the core and thereby its amplitude ceases from growing further. The prospect that the beam halo is *self-collimating* had led to hope that aperture requirements for beamline components might be modest.

However, one feature that is unavoidable in real accelerators, but is commonly overlooked in simulations, is the presence of noise. The noise manifests itself by way of imperfections in the electromagnetic fields external to the beam, which then selfconsistently influence the evolving space-charge potential of the beam. A charged particle experiences all of the noise inherent to the total potential. Moreover, the noise generally comprises a superposition of *colored* noise, that is, noise with nonzero autocorrelation time. For example, the autocorrelation time of noise in the collective space-charge potential could be short, say of the order of a plasma period, whereas for hardware irregularities/misalignments it could be long, say several betatron (orbital) periods. By generalizing simple particle–core models to include this noise, we showed that the presence of colored noise can boost statistically rare particles to ever-growing amplitudes by keeping them more in phase with the core oscillation. This leads to rapid formation (within about three to four global pulsations of the beam) of a large halo and removes the fundamental limit on the halo amplitude predicted by the particle–core model.[19,20]

We also pointed out the importance of this mechanism in the context of a self-gravitating stellar system for which environmental noise from surrounding galaxies self-consistently influences the dynamics.[19] By considering a perturbed Plummer sphere, a configuration for which the unperturbed collective potential scales as $(1 + r^2/3)^{-1/2}$, we found that, notwithstanding it is a restoring force, gravity is so weak that, when combined with the noise, only a relatively tiny oscillatory perturbation suffices to pump stars to very large amplitudes. This suggests that the phenomenology applies generically to nonequilibrium Coulomb systems: colored noise combined with parametric resonance will drive a statistically small number of particles to much larger amplitudes than parametric resonance can do on its own, leaving an extended halo as a byproduct.

The results for the Plummer sphere were generated only as a quick *existence proof* of the phenomenology in the context of galactic dynamics. We alerted Henry Kandrup of our findings, at which time Henry teamed with Sideris for a more in-depth, systematic study of noise-induced formation of galactic halos.[21] That study was based on spherically symmetric, time-dependent Dehnen potentials[22] subject to low-amplitude, strictly periodic perturbations. Pseudorandom variations in the pulsation frequency were added, these being modeled as colored noise.

As is true for beams, periodic driving generically tends to pump energy into the stars in a galaxy, thus displacing them toward higher energies and larger radii, an effect resulting from a resonant coupling between the driving frequency and the frequencies of the orbits.[23] Even relatively weak driving can have significant effects within just a few tens of dynamical times (orbital periods), and larger amplitudes can account for violent relaxation (at least in principle). Variations in the driving frequency were found not to vitiate the effects of such resonant couplings. In fact, modest variations in frequency tended on average to increase the maximum radii to which orbits are displaced. However, this does not imply that, on average, such variations result in more energy being pumped into the orbits. To the extent that the orbits have all become *wildly* chaotic, that is, they have one or more positive Lyapunov exponents and they are not locked to the (near-constant) driving frequency, allowing for random frequency variations leads to a Gaussian distribution of maximum energies centered about the maximum energy attained by an orbit subjected to strictly periodic driving. The systematic increase in the average orbital amplitude arises because realistic near equilibrium systems have phase-space distributions that are monotonically decreasing functions of energy, which means a symmetric spread in energies occasions an increase in the number of larger-amplitude orbits at the expense of smaller-amplitude orbits. More importantly, a random frequency can have a very large impact on at least a small number of stars. In particular, it is statistically probable that a few orbits will experience a noisy variable frequency that continually keeps them more in phase with the global oscillation, thereby displacing them to very large radii.

IMPLICATIONS FOR FUTURE RESEARCH

Accounting for intricacies of space charge requires, in principle, an N-body simulation code. For computations that accurately reproduce details of the phase space of the beam, N needs to be large and the computing time correspondingly long. To explore the parameter space of, for example, photoinjectors, fast codes are needed that also encompass sufficiently accurate models of the beam physics to enable them to be used with confidence. For detailed production runs, inasmuch as a nC bunch charge corresponds to 6.25×10^9 electrons, an "exact" simulation is impractical, and accurate alternatives must be found. Analogous statements apply in the context of galactic dynamics; large galaxies comprise some 10^{11}–10^{12} stars.

Preserving a hierarchy of scales in the time-dependent space-charge potential is dynamically important. To reiterate, recent research has revealed that nonlinear, time dependent forces commonly establish large populations of globally chaotic orbits in beams that are out of equilibrium, and such orbits can even be present in thermal equilibrium beams.[12,24] When present, these chaotic orbits mix exponentially throughout their accessible phase space with a time scale of only a few orbital periods, that is, very much faster than collisional relaxation. The presence of colored noise due to space charge fluctuations and/or machine imperfections can, when combined with parametric resonance associated with low-order oscillatory modes, generate much larger halos than would be inferred from parametric resonance alone. Thus, all scales are potentially important to the dynamics.

Wavelets constitute a mathematical tool that is inherently designed to represent a hierarchy of scales. Accordingly, we have recently embarked on the development of a space-charge algorithm based on innovative use of wavelets, as is reported in the accompanying paper by B. Terzić.[25] To start, we are developing a wavelet-based solution of the Poisson equation on a grid. One advantage is simultaneous denoising of macroparticle N-body simulations: because artificial noise arising from the use of macroparticles is present on all scales, using wavelets removes most of the noise without altering the inherent structure of the beam.

In addition to wavelet denoising, we also plan to develop an algorithm for solving the Vlasov–Poisson system using wavelet decomposition. The idea involves the use of a continuous wavelet transform to decompose the six-dimensional phase space of a single electron in terms of wavelets multiplied by time-dependent coefficients. Doing so then reduces the Vlasov–Poisson equations to a set of coupled equations for the time-dependent coefficients. Solving these equations numerically thereby yields the solution for the distribution function, from which any desired beam property can then be calculated. This method may prove to be much faster than large N-body simulations.

A key advantage of wavelet-based solutions is the representation of the potential at each time step in terms of a modest number (about 100) of wavelet coefficients. This *greatly* eases the computational storage of the history of the system: one now does *not* need to store the coordinates and velocities of the (very many) macroparticles at each time step. The net result from a simulation, therefore, becomes a realistic time-dependent potential stored compactly. One can then use this potential to integrate efficiently orbits of very many test particles. Doing so enables in-depth studies of mixing in realistic time dependent systems (an area that has heretofore largely been inaccessible) to include, as just one example, realistic studies of halo dynamics. This is true not only for beams with space charge, but of course also for galaxies.

We have also made significant progress in formulating a rapidly computed measure of orbital chaos, as is reported in the accompanying paper by Sideris.[26] It is based on the existence of morphology patterns in the time series associated with the Poincaré section of the orbit. This gives the technique the important advantage of being applicable, without extra computational effort, to any number of degrees of freedom. For every orbit, and over relatively localized time intervals (a few dynamical times), the code correctly identifies the existing patterns and decides if the orbit is regular or chaotic. If it is chaotic, the code distinguishes between orbit segments that are sticky and segments that are wildly chaotic.

The first major application will be to identify transitions of orbits between regular and chaotic behavior in time-dependent potentials. This way we will be able to build a picture of the evolving phase space and thereby increase dramatically our understanding of the underlying dynamics. The ultimate application is to the self-consistent evolution of N-body systems. Then, not only will we have a new, efficient method for quantifying the dynamics, but also we will be able to compare the performance of various codes based on how their respective outputs, that is, phase spaces, evolve.

EPILOGUE

October 18, 2003, the date of Henry Kandrup's death, was a shocking day for all of us who knew and appreciated him. I personally regard his work and innovative insights to be foundational to the understanding of nonequilibrium N-body systems. I have tried to document here how these ideas have enabled a much-improved understanding of space charge, with considerable interplay back into the field of galactic dynamics. In learning of his death, I resolved to keep Henry's line of investigation alive and growing. This endeavor would seem to be thus far successful. His former student, Ioannis Sideris, has been my postdoctoral fellow since Fall 2002, and his former postdoctoral fellow, Balša Terzić, joined us in April 2004. Henry had a graduate student when he died, Ileana Vass, who has relocated to my group to pursue her Ph.D. on the topic of galactic halos. We are collaborating with Lawrence Berkeley Laboratory in the pursuit of wavelet techniques; Ilya Pogorelov, another former Kandrup doctoral student, being our principal collaborator. The need to preserve hierarchies of scale, something we learned as we worked with Kandrup, is what motivates our pursuit of wavelet techniques for improved simulation codes. Developing a rapidly computed measure of chaos will permit us to quantify the efficacy of the wavelet code *vis-à-vis* "conventional" codes, like those based on Green functions and fast Fourier transforms, for modeling time-dependent N-body systems. Although we launched these new pursuits subsequent to Henry's death, we are confident he would have been intensely interested in them.

There is an additional collaboration formed some two years ago, while Henry was alive, and it is a collaboration he valued highly. He and we teamed with the University of Maryland to begin devising controlled experiments with intense nonrelativistic electron beams as laboratory analogs of galactic systems. Plans remain in place to do these experiments using the University of Maryland electron ring, a facility described in the accompanying paper by the UMER group.[27] We have documented this collaboration in a paper published in these *Annals*,[6] a paper that was started with Henry prior to his death, and that we subsequently completed, retaining him as coauthor. We fondly remember Henry and his influence as we pursue the new experimental and theoretical investigations mentioned herein; we sorely miss him.

ACKNOWLEDGMENTS

This work was supported by the U.S. Departments of Education under Grant P116Z010035, Energy under Grant DE-FG02-04ER41323, and Defense under Air Force Contract FA9451-04-C-0199.

The author declares that he has no competing financial interest.

REFERENCES

1. O'SHEA, P.G. & H.P. FREUND. 2001. Free-electron lasers: status and applications. Science **292:** 1853.
2. CHAO, A.W. 1993. Physics of Collective Beam Instabilities in High Energy Accelerators. Wiley, New York. 26.

3. CASETTI, L., C. CLEMENTI & M. PETTINI. 1996. Riemannian theory of Hamiltonian chaos and Lyapunov exponents. Phys. Rev. E **54:** 5969.
4. BOHN, C.L. 2000. Rapid irreversible mixing in charged-particle beams. FERMILAB-Pub-00/052-T.
5. REISER, M. 1994. Theory and Design of Charged Particle Beams, Chapter 6. Wiley, New York.
6. KANDRUP, H.E., C.L. BOHN, R.A. KISHEK, et al. 2005. Chaotic collisionless evolution in galaxies and charged-particle beams. Ann. N.Y. Acad. Sci. **1045:** 12–33.
7. MERRITT, D. & M. VALLURI. 1996. Chaos and mixing in triaxial stellar systems. Astrophys. J. **471:** 82.
8. DAVIDSON, R.C. 1990. Physics of Nonneutral Plasmas. Addison-Wesley, Redwood City.
9. DAVIDSON, R.C. 1998. Nonlinear stability theorem for high-intensity charged particle beams. Phys. Rev. Lett. **81:** 991.
10. BALESCU, R. 1997. Statistical Dynamics: Matter Out of Equilibrium. Imperial College Press, London.
11. KANDRUP, H.E., I.V. SIDERIS & C.L. BOHN, 2004. Chaos and the continuum limit in charged particle beams. Phys. Rev. ST Accel. Beams **7:** 014202.
12. BOHN, C.L. & I.V. SIDERIS. 2003. Chaotic orbits in thermal-equilibrium beams: existence and dynamical implications. Phys. Rev. ST Accel. Beams **6:** 034203.
13. LICHTENBERG, A.J. & M.A. LIEBERMAN. 1992. Regular and Chaotic Dynamics. Springer, Berlin.
14. POGORELOV, I.V. & H.E. KANDRUP. 1999. Noise-induced phase space transport in two-dimensional Hamiltonian systems. Phys. Rev. E **60:** 1567.
15. KANDRUP, H.E. 2002. Should elliptical galaxies be idealised as collisionless equilibria? Space Sci. Rev. **102:** 101.
16. JAMESON, R. 1996. Beam losses and beam halos in accelerators for new energy sources. Fusion Eng. Des. **32–33:** 149.
17. GLUCKSTERN, R.L. 1994. Analytic model for halo formation in high current ion linacs. Phys. Rev. Lett. **73:** 1247.
18. WANGLER, T.P., K.R. CRANDALL, R. RYNE & T.S. WANG. 1998. Particle-core model for transverse dynamics of beam halo. Phys. Rev. ST Accel. Beams **1:** 084201.
19. BOHN, C.L. & I.V. SIDERIS. 2003. Fluctuations do matter: large noise-enhanced halos in charged-particle beams. Phys. Rev. Lett. **91:** 264801.
20. SIDERIS, I.V. & C.L. BOHN. 2004. Production of enhanced beam halos via collective modes and colored noise. Phys. Rev. ST Accel. Beams **7:** 104202.
21. SIDERIS, I.V. & H.E. KANDRUP. 2004. Noise-enhanced parametric resonance in perturbed galaxies. Astrophys. J. **602:** 678.
22. TERZIĆ, B. & H.E. KANDRUP. 2004. Orbital structure in oscillating galactic potentials. Mon. Not. R. Astron. Soc. **347:** 957.
23. KANDRUP, H.E., I.V. SIDERIS, B. TERZIĆ & C.L. BOHN, 2003. Supermassive black hole binaries as galactic blenders. Astrophys. J. **597:** 111.
24. KISHEK, R.A., S. BERNAL, C.L. BOHN, et al. 2003. Simulations and experiments with space-charge-dominated beams. Phys. Plasmas **10:** 2016.
25. TERZIĆ, B. & I.V. POGORELOV. 2005. Wavelet-based Poisson solver for use in particle-in-cell simulations. Ann. N.Y. Acad. Sci. **1045:** 55–67.
26. SIDERIS, I.V. 2005. Characterization of chaos: a new fast and effective method. Ann. N.Y. Acad. Sci. **1045:** 79–92.
27. KANDRUP, H.E., C.L. BOHN, R.A. KISHEK, et al. 2005. Chaotic collisionless evolution in galaxies and charged-particle beams. Ann. N.Y. Acad. Sci. **1045:** 12–33.

The University of Maryland Electron Ring
A Platform For Study of Galactic Dynamics on a Laboratory Scale

R.A. KISHEK, P.G. O'SHEA, S. BERNAL, I. HABER,
J.R. HARRIS, Y. HUO, H. LI, AND M. REISER

*Institute for Research in Electronics and Applied Physics,
University of Maryland, College Park, Maryland, USA*

ABSTRACT: The University of Maryland electron ring (UMER) is a novel experimental storage ring designed to investigate the dynamics of large systems of collisionless particles that nevertheless interact via collective mechanisms. Heavily diagnosed and designed to circulate a 100 mA electron beam at 10 keV for several turns, UMER allows us to follow the evolution of the beam over a large number of dynamical periods. Given the similarity of dynamics between the Coulomb forces and gravitational forces, it is possible to design beam experiments that will simulate such astrophysical events as galactic merger, for instance. The cross-comparison of beam experiments with accelerator simulation codes provides invaluable benchmarks of transient processes akin to, for example, violent relaxation in stellar systems that are otherwise impossible to obtain.

KEYWORDS: chaos; N-body problem; nonlinear dynamics; collisionless

INTRODUCTION

A long-standing mystery in the evolution of galaxies is the fast mixing time scale relative to predictions from collisions or regular phase mixing.[1,2] This has suggested a mechanism whereby chaotic orbits lead to accelerated mixing in galaxies. To test this idea, however, the only available resource to astrophysicists to date has been computer simulations. Unlike other sciences, astrophysics is somewhat hampered by the lack of possibility to perform controlled experiments. Recently, we have begun to take advantage of the close connection between galactic dynamics and charged-particle-beam dynamics[3] to explore the correspondences and the possibility of setting up beam experiments to investigate questions in galactic dynamics.[4,5] This paper describes the University of Maryland electron ring (UMER), the facility on which we intend to perform such experiments.

As discussed elsewhere,[3] large collections of stars resemble large collections of charged particles in that they both interact via inverse-square-law forces, where typically an external focusing force is superimposed on the charged particle collection to overcome the mutual repulsion. Most importantly, the interaction between

---

Address for correspondence: P.G. O'Shea, Institute for Research in Electronics and Applied Physics, University of Maryland, College Park, MD, USA.
poshea@umd.edu

individual stars or particles is very small compared to the interaction between those particles/stars and the collective fields, the one difference being that charged-particle beams adjust themselves to screen the external focusing force. Because of this Debye shielding, or screening, the bulk potential of a charged particle beam (including external focusing) cannot generally be molded to match precisely that of an evolving stellar system. For example, structures mimicking the presence of central density cusps or black holes in galaxies cannot be preserved in a beam because space charge is repulsive. However, phenomenology inherent to time-dependent collective dynamics in galaxies *can* be mimicked with beams. Gravitational examples (and their beam analogs), include colliding and merging galaxies (merging beamlets), collapsing galaxies (rms-mismatched beams), or perturbed but comparatively quiescent galaxies (beams with evolving density inhomogeneities). Thus, the key transient dynamics underlying galactic systems, that which arises as a consequence of the collective influence of the long-range (inverse-square) interparticle force, can be studied in the laboratory.

Accordingly, we are planning a series of experiments to study phase mixing and attendant collisionless relaxation using UMER, a facility that is just now coming on line.[6-9] UMER is a 10-keV electron ring designed to transport the beam through many turns spanning over 1 km, corresponding to some 500–1,000 plasma periods, and the relative strength of the collective space-charge force is adjustable over a wide range. The following sections provide an overview of this effort. After introducing concepts in beam physics, we introduce the UMER design and relevant parameters. We then review beam merging experiments and the applicability of that technique to investigating galactic mergers, and finally, we discuss a novel perturbation generation technique that can be used to model density inhomogeneities.

INTRODUCTION TO BEAM PHYSICS

A charged particle beam is a collection of particles that have a component of momentum in one direction that is much greater than momentum components in orthogonal directions. Since particles of like charge repel each other, external focusing forces need to be applied to keep the beam confined. This inward external force needs to balance, on average, the outward forces from space charge repulsion and from thermal pressure that results from the random nature of particle velocities. This condition is what is generally known in the beam community as *matching*. Consequently, a *mismatched* beam experiences a force imbalance that leads to violent oscillations in beam size, orientation, density profile, and so forth. Such oscillations can, furthermore, resonate with individual particle trajectories, leading to a deterioration of beam quality. Thus, to characterize a beam, we typically specify one parameter that quantifies the intensity of the space charge forces, and another that quantifies beam quality. These two parameters are defined below. More information on beam dynamics can be found elsewhere.[10] Note that over the length of most accelerators, charged-particle beams experience relatively few close-range Coulomb collisions. Instead, the forces affecting a particle trajectory derive mostly from the long-range collective potential of the entire beam. This circumstance is analogous to the collisionless nature of galaxies over a Hubble time. For example, the characteristic scale for interparticle collisions in UMER is at least 1km, corresponding to 1,000 plasma periods—the equivalent of 100 turns around the ring and the ultimate goal for the operation of UMER. The transverse motion of a typical particle exhibits oscillations, called *betatron oscillations*. Space charge depresses the frequency of these oscillations. For UMER, the space-charge-depressed betatron oscillation, which is now a measure of the *dynamical time* of the beam, is about 10 plasma periods, corresponding roughly to one transit time around the ring. Thus, the collisional relaxation time of a UMER electron bunch is at least 100 dynamical times.

The intensity of the beam can be quantified in a number of ways, for example using the dimensionless intensity parameter, χ (the ratio of the space-charge force to the external focusing force evaluated at the beam edge[6,7]). Space charge depresses the betatron oscillation frequency by a ratio

$$\frac{\omega_\beta}{\omega_{\beta_0}} = \frac{k}{k_0} = \sqrt{1-\chi},$$

whereas the plasma frequency ω_p is enhanced,

$$\frac{\omega_\beta}{\omega_{\beta_0}} = \sqrt{2\chi}.$$

Another parameter of importance, and one that is related directly to the value of χ, is the ratio of beam radius, r_b, to the Debye length $\lambda_D = v_{th}/\omega_p$, where the beam kinetic energy is $1/2 m v_{th}^2$ and the plasma frequency in gaussian units is $\omega_p^2 = 4\pi n e^2/m$, with n denoting the particle density, e the particle charge, and m the particle rest mass. For intensities exceeding $\chi = 0.5$, the Debye length (over which the external focusing force is screened from the beam interior) becomes significantly smaller than the beam radius, and the beam is able to support collective density oscillations, since those oscillations usually do not occur at wavelengths less than λ_D.

The most popular measure of beam quality is *emittance* (in this paper, the normalized effective emittance to be precise), which essentially measures the total thermal spread in particle momenta,

$$\varepsilon_{xn}^2 = \frac{16}{m^2 c^2}(\langle x^2 \rangle \langle p_x^2 \rangle - \langle x p_x \rangle^2), \tag{1}$$

where $\langle \cdot \rangle$ indicates a moment over the four-dimensional transverse beam distribution $f(x, p_x, y, p_y)$ orthogonal to the direction of motion of the beam. Here x is the particle position in the horizontal direction (referenced to the beam centroid), p_x and p_y are the particle momenta with respect to the reference trajectory, and c is the speed of light. A similar expression holds for the y emittance, ε_{yn}. The unit of emittance is that of length. Physically, emittance corresponds to the total thermal energy in a beam. More precisely, the geometric emittance $\varepsilon = (\varepsilon_{xn}\varepsilon_{yn}\varepsilon_{zn})^{1/3}$ is a measure of the volume of the six-dimensional phase space subtended by the beam bunch.

FIGURE 1 compares the plasma frequency, related to collective oscillations, and the space-charge-depressed betatron frequency, related to single particle oscillations, as a function of the intensity parameter χ. The range of UMER intensities is compared on the same scale to that of existing rings, indicating the importance of collective effects on the beam dynamics of UMER. The experimental and theoretical study of beam dynamics in UMER will have important applications to existing and future high-current circular accelerators, such as high-energy physics booster

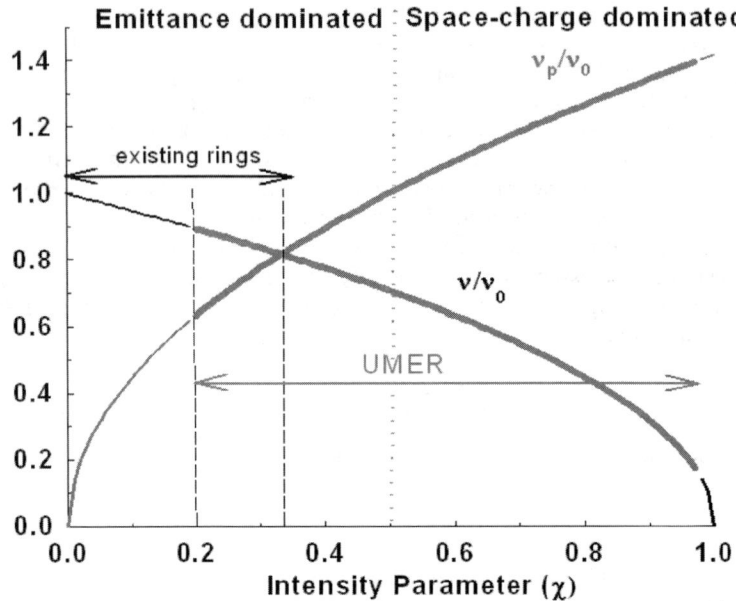

FIGURE 1. Emittance-dominated and space-charge-dominated regimes, showing the space-charge-depressed betatron frequency (*black*) and plasma frequency (*grey*), both normalized to the betatron frequency corresponding to the bare external focusing force, with increasing intensity parameter χ. The range of χ values accessible to UMER is indicated by the *bold* line.

synchrotrons, muon colliders, heavy-ion fusion recirculators, and spallation neutron sources.

OVERVIEW OF UMER

UMER builds on a tradition of experiments at the University of Maryland using low energy, nonrelativistic electron beams to model inexpensively higher energy hadron and ion beams. With the design beam current of 100 mA, an initial emittance of $\varepsilon_{xn} = \varepsilon_{yn} = 12$–$15\,\mu$m, and an average beam radius $r_b = 1$ cm, UMER can achieve an intensity of $\chi = 0.97$, which is near the extreme space-charge limit ($\chi = 1$). Furthermore, this intensity can be varied over a wide range (down to $\chi \sim 0.25$) by changing to different aperture sizes in the beam collimator at the exit of the gun, by changing the bias voltage on the cathode grid, or by changing the anode–cathode spacing.

FIGURE 2 shows a recent photograph of the closed ring and TABLE 1 lists some of the relevant parameters. More detail can be found elsewhere[7] and on the website <http://www.ireap.umd.edu/umer>. Current experiments employ a drifting 10-keV (nonrelativistic) electron beam. Three induction gaps distributed equidistantly around the ring provide longitudinal focusing and will eventually be used to accelerate the beam to 50 keV in a future stage. Transverse focusing is provided by 36

FIGURE 2. Photograph of UMER, as of October 2004.

TABLE 1. UMER Design specifications at 10 keV

Beam Energy	10 keV
β (= v/c)	0.2
Beam Current	$\leq$ 100 mA
Generalized Perveance	$\leq$ 0.0015
Emittance, 4× rms, normalized	10 μm
Pulse Length	50–100 nsec
Ring Circumference	11.52 m
Lap Time	197 nsec
Pulse Repetition Rate	60 Hz
Mean Beam Radius	$\leq$ 1 cm
FODO Period	0.32 m
Zero-Current Phase Advance, σ_0	76°
Zero-Current Betatron Tune, v_0	7.6
Tune Depression	0.2

alternating-gradient quadrupole (focusing–drift–defocusing–drift, called FODO) cells around the 11.52-meter circumference of the ring. Each pair of FODO cells, along with two 10° bends, is installed on one out of 18 mechanical sections. Longitudinally, the UMER beam consists of a single pulse whose length can be varied from 50 to 100 nsec. Since the circulation time for a 10 keV electron around the 11.52-m UMER circumference is 200 nsec, only a single pulse is injected at any one time, at a repetition rate of 10–60 Hz.

UMER was constructed during the past three years. During that time, we performed a wide range of experiments over its first turn.[8,9] As of the time of this writing, we have propagated the beam all the way around the machine and into the second turn, registering a second beam pulse on the beam position monitor in the first ring chamber. Thus, UMER has entered the multiturn commissioning phase, opening the way for investigating phenomena over longer time scales (path lengths).

FIVE-BEAMLET EXPERIMENT

An example of experiments that can be performed on UMER related to galactic dynamics is a multibeamlet experiment to emulate a galactic merger. FIGURE 3 shows phosphor screen images of the beam measured 99 cm downstream from the anode plane when a five-beamlet mask is applied, with holes arranged in a quincunx pattern (number '5' on a die). This photograph is typical of the whole series of pictures taken along the injector using the moveable P-screen. Also shown in the same figure are simulations performed using the WARP PIC code in single-slice mode, that is, wherein the same transverse cross-section of the beam is monitored as it transits through the ring.[11] The emittances labeled on the simulation curves shown in FIGURE 2 are calculated by multiplying the thermal velocity assumed for the beamlets by the total area the beam would have at the anode plane in the absence of the mask. It is clear that the shape of the pattern downstream depends on the beam emittance. From a comparison between the simulated and measured pattern one infers that the best match occurs when the initial normalized emittance of the total beam is

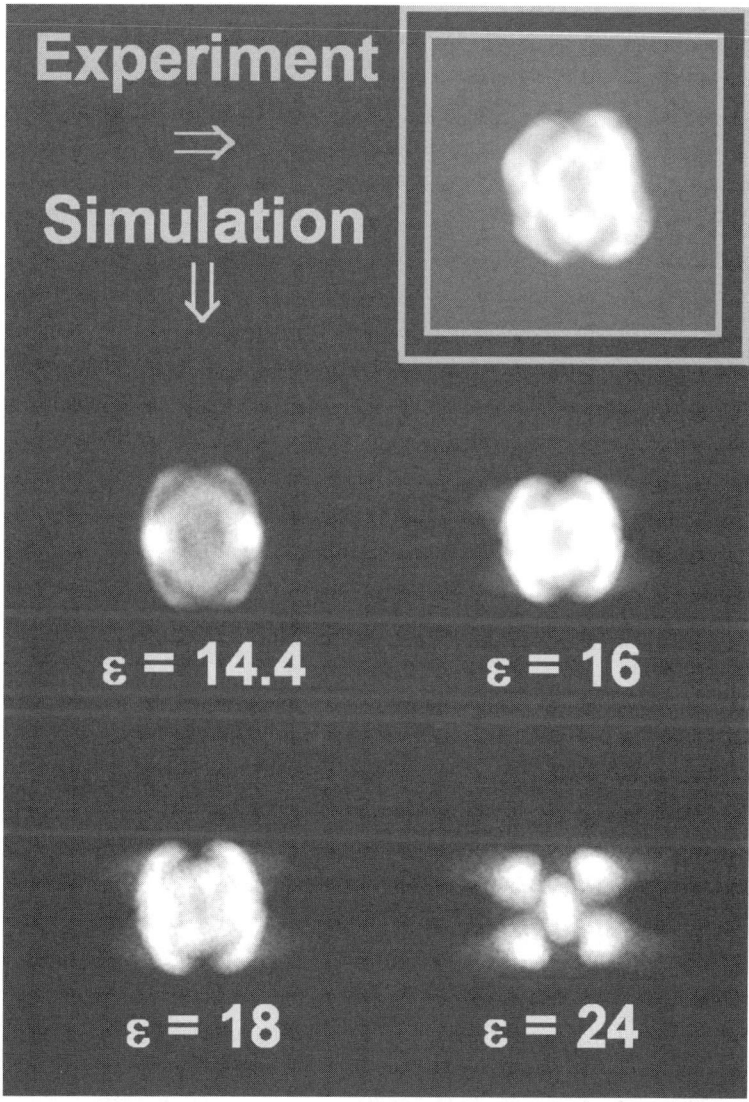

FIGURE 3. Photograph of beam incident on a phosphor screen 99 cm downstream from the anode while the five-beamlet mask is applied near the anode. The experiment is compared with the results of four simulations starting with a semi-Gaussian distribution at the anode, but with different initial emittances as indicated in the figure.

between 16mm and 18mm. In simulations and experiments conducted over a longer distance,[3,12,13] the individual beamlets are observed to merge, leaving no trace of the previous structure. In cases where we inject a mismatched beam, a halo is seen to develop during the merger.

Such laboratory experiments can be very useful in benchmarking computer codes that can afterwards be used to cross-compare with stellar dynamical codes or be modified (by eliminating the external focusing force and reversing the sign of the interparticle force) and confidently used to simulate galactic mergers directly. In addition to the macroscopic features evident in FIGURE 3, the computer codes can also provide valuable trajectory information that can provide a tremendous amount of information on the presence of transient chaos in the system. An example of simulations of this five-beamlet configuration using the WARP code is discussed in detail elsewhere.[3] Such a study of the merger of several beamlets is quite similar, in essence, to studies of galactic mergers.[14]

PERTURBATIONS

In addition to exploring effects of enormous nonlinearities as can be generated through a five-beamlet system, we have a wide suite of techniques for generating minor nonlinearities and perturbations to the ideal beam potential. This we can accomplish by (1) using the induction gaps, (2) using the cathode grid, or (3) using photoemission from a five-nanosecond laser pulse directed at the cathode. The laser method is the most flexible of the three, as it allows us to arbitrarily vary the relative

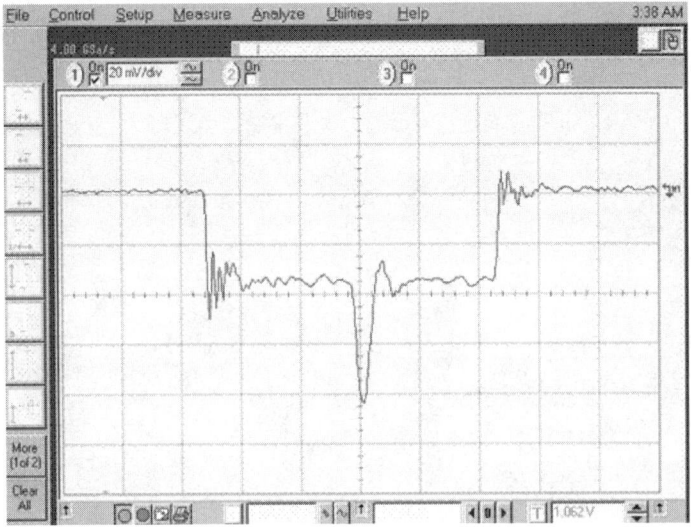

FIGURE 4. Oscilloscope trace of beam current versus time from UMER showing a typical five-nanosecond laser-induced perturbation (in this case highly nonlinear) superimposed over a 100nsec rectangular thermionic main beam.

intensities of the main beam and the perturbation, and also to combine longitudinal and transverse perturbations (by including masks in front of the laser).[15] FIGURE 4 shows a typical laser-induced perturbation (in this case a highly nonlinear one) superimposed over a rectangular main beam pulse. Observing its evolution over the first turn of UMER, we see the current pulse split into a forward-traveling "fast" wave, and a backward-traveling "slow" wave. By measuring the rate of separation of those pulses, we get an accurate measurement of the wave speed, and it agrees nicely with the theoretical predictions[10] for sufficiently small initial perturbations. This experiment has been simulated using the WARP code and the actual experimental measurement of the initial pulse. The simulation agrees well with the experiment. Recently, by applying laser interferometry to the perturbing laser pulse, we have been able to extend this technique to generating a train of perturbations on a long-pulse beam.[16]

The evolution and mixing of initial perturbations can be tracked using the comprehensive suite of diagnostics incorporated into UMER. As a whole, the diagnostic suite permits direct measurement of mixing time scales in units of the characteristic dynamical time, and the degree of mixing in both configuration space and in energy, by enabling the evolution of macroscopic features to be observed and quantified. It should thereby be possible to distinguish observationally between chaotic (i.e., exponential or global) phase mixing versus regular (i.e., secular or more local) phase mixing. We also plan, of course, to confirm our interpretations using simulation codes. We project an added benefit, as well: establishing the phenomenology of phase mixing in time-dependent beam potentials both experimentally and numerically should likewise provide an unambiguous mechanism for validating codes and simulation techniques in both beam physics and galactic dynamics.

We conclude by reviewing the four categories of questions relevant to evolving galaxies that UMER will likely be able to answer, as posited in Reference 3. (1) Do initially localized *clumps* mix as simulations predict? Specifically, do they clearly mix exponentially in regimes where one would predict a preponderance of chaotic orbits and secularly otherwise? Do the mixing time scales agree with code predictions? (2) How important are collisions during the early evolution of an inverse-square-law system? (3) Do nonequilibrium inverse-square-law systems evolve so as to reduce the population of chaotic orbits? (4) For systems initially far from equilibrium, what is the relative importance of *violent* short-time evolution versus *calm* long-time evolution? Can the latter significantly alter the macroscopic properties of the system that one measures in the laboratory in the case of beams or observes on the sky in the case of galaxies? In other words, is what one observes really the product of violent relaxation exclusively?[3]

Questions like these are precisely what the UMER experiments are designed to answer. The UMER beam diagnostics suite is well configured to provide measurements at the level of detail necessary to gain the respective insight. Consequently, the prognosis for success is good. However, with every answer come more questions, that will translate into stricter specifications for the diagnostics, so the process will no doubt be an ongoing one.

ACKNOWLEDGMENTS

This work was supported by the U.S. Department of Energy Grant DEFG02-94ER40855 and Grant DEFG02-92ER54178, and by the Office of Naval Research Grant N000140210919. We thank our collaborator Court Bohn for his careful reading and comments during the preparation of this manuscript.

REFERENCES

1. MERITT, D. & M. VALLURI. 1996. Chaos and mixing in triaxial stellar systems. Astrophys. J. **471**: 82.
2. KANDRUP, H.E., I.M. VASS & I.V. SIDERIS. 2003. Transient chaos and resonant phase mixing in violent relaxation. Mon. Not. R. Astron. Soc. **341**: 927.
3. KANDRUP, H.E., BOHN, C.L., KISHEK, R.A., et al. 2005. Chaotic collisionless evolution in galaxies and charged-particle beams. Ann. N.Y. Acad. Sci. **1045**: 12–33.
4. KISHEK, R.A., C.L. BOHN, I. HABER, et al. 2001. Computational investigation of dissipation and reversibility of space-charge-driven processes in beams. Proceedings of the 2001 Particle Accelerator Conference, IEEE Catalog No. 01CH37268, 151–154.
5. BOHN, C.L., I.V. SIDERIS, H.E. KANDRUP & R.A. KISHEK. 2003. Mixing of regular and chaotic orbits in beams. Proceedings of the LINAC 2002, Gyeongju, August 2002, CD-ROM ISBN: 89-954175-0-1 98420, 391.
6. REISER, M., P.G. O'SHEA, R.A. KISHEK, et al. 1999. The Maryland electron ring for investigating space-charge dominated beams in a circular FODO system. Proceedings of the 1999 IEEE Particle Accelerator Conference, New York City, NY, 234.
7. O'SHEA, P.G., M. REISER, R.A. KISHEK, et al. 2001. The University of Maryland electron ring (UMER). Nucl. Instr. Meth. **A464**: 646–652.
8. KISHEK, R.A., S. BERNAL, Y. CUI, et al. 2005. HIF Research on the University of Maryland electron ring (UMER). Nucl. Instr. Meth. A. In press.
9. BERNAL, S., H. LI, T. GODLOVE, et al. 2004. Beam experiments in the extreme space-charge limit on the University of Maryland electron ring (UMER). Phys. Plasmas **11**(5): 2907.
10. REISER, M. 1994. Theory and Design of Charged Particle Beams. Wiley, New York.
11. GROTE, D.P., A. FRIEDMAN, I. HABER & S. YU. 1996. Three-dimensional simulations of high current beams in induction accelerators with WARP3D. Fus. Eng. Design **32–33**: 193.
12. HABER, I., D. KEHNE, M. REISER & H. RUDD. 1991. Experimental, theoretical, and numerical investigation of the homogenization of density nonuniformities in the periodic transport of a space-charge-dominated beam. Phys. Rev. A **44**: 5194.
13. KEHNE, D., M. REISER & H. RUDD. 1992. Experimental studies of emittance growth in a nonuniform, mismatched, and misaligned space-charge-dominated beam in a solenoid channel. In High-Brightness Beams for Advanced Accelerator Applications. W.W. Destler and S.K. Guharay, Eds.: 47. AIP Conference Proceedings No. 253.
14. NIPOTI, C., P. LONDRILLO & L. CIOTTI. 2003. Galaxy merging, the fundamental plane of elliptical galaxies, and the M_{BH}–σ relation. Mon. Not. R. Astron. Soc. **344**: 748–760.
15. HUO, Y. 2004. Private communication.
16. NEUMANN, J.G., J.R. HARRIS, B. QUINN & P.G. O'SHEA. 2005. Production of photoemission-modulated beams in a thermionic electron gun. Rev. Sci. Instr. **76**: 033303.

Wavelet-Based Poisson Solver for Use in Particle-in-Cell Simulations

BALSA TERZIĆ[a] AND ILYA V. POGORELOV[b]

[a]*Northern Illinois University, Department of Physics, DeKalb, Illinois, USA*
[b]*Accelerator and Fusion Research Division,
Lawrence Berkeley National Laboratory, Berkely, California, USA*

ABSTRACT: We report on a successful implementation of a wavelet-based Poisson solver for use in three-dimensional particle-in-cell simulations. Our method harnesses advantages afforded by the wavelet formulation, such as sparsity of operators and data sets, existence of effective preconditioners, and the ability simultaneously to remove numerical noise and additional compression of relevant data sets. We present and discuss preliminary results relating to the application of the new solver to test problems in accelerator physics and astrophysics.

KEYWORDS: multiscale dynamics; wavelets; N-body simulations; particle-in-cell method; Poisson equation

INTRODUCTION

The possibility of gaining insight into the dynamics of multiparticle systems, such as charged particle beams or self-gravitating systems, relies heavily on N-body simulations. The many-body codes can be grouped into three main categories: (1) direct summation, (2) tree, and (3) particle-in-cell (PIC). Direct summation codes are prohibitively expensive for large systems, since their computation cost scales as N^2. Tree codes use direct summation for nearby particles and evoke statistical arguments for contribution from particles further away. The PIC codes incorporate a computational grid into which particles are binned, thus resulting in a coarse-grained, discretized particle distribution. The potential associated with the discretized particle distribution is computed by solving the Poisson equation on a grid. Finally, the forces needed to advance each individual particle are computed by interpolation from the discretized potential on the grid. A detailed treatment of computational methods to simulate multiparticle systems is given by Hockney and Eastwood.[1] In this paper we outline a wavelet-based algorithm for solving the Poisson equation for use in PIC codes.

It is important that the algorithms used in solving the Poisson equation:

1. account for *multiscale dynamics*, because even fluctuations on the smallest scales can lead to global instabilities and fine-scale structure formation, as exemplified by halo formation and microbunching instability observed in beam dynamics experiments;[2–5]

Address for correspondence: Balsa Terzić, Northern Illinois University, Department of Physics, DeKalb, IL 60115, USA. Voice: 815-753-1821; fax: 815-753-8565.
bterzic@nicadd.niu.edu

2. *minimize the numerical noise*, due to the fact that the number of particles used to sample the phase-space distribution function in the *N*-body simulation is several orders of magnitude smaller than the number of particles in the physical system that is being modeled; and

3. be as *efficient* as possible in terms of computational speed and storage requirements, without compromising accuracy.

Furthermore, for some important applications, such as coherent synchrotron radiation (CSR) in beam dynamics, it is necessary to have a compact representation of the history of the system.[5,6]

The wavelet-based solver we present here has the potential to satisfy all of the requirements listed above. In the next section we briefly outline the concept of wavelets and wavelet transforms. We then report on our formulation of the Poisson equation on the grid and its solution using the wavelet-based approach. Subsequently, we apply the wavelet-based solver to model two density distributions of interest in beam dynamics and astrophysics. We then replace (for testing purposes) the FFT/Green's function-based Poisson solver in the ImpactT3D beam dynamics code[7,8] with the wavelet based solver, and compare results produced by the two versions of ImpactT3D evolving the same initial distributions. We conclude with a summary of the work presented here and a description of work in progress.

WAVELETS AND WAVELET TRANSFORMS

Wavelets and wavelet transforms are a relatively new concept, introduced in the 1980s.[9–12] The discrete wavelet transform can be viewed in two different ways:

1. As the discrete analog of the continuous wavelet transform of a function $f(t)$ given by

$$g(s, x) = \int_{-\infty}^{\infty} f(t) P_{s,x}(t) dt, \quad (1)$$

$$P_{s,x}(t) = \frac{1}{\sqrt{s}} P\left(\frac{t-x}{s}\right), \quad (2)$$

where $P(t)$ is a mother wavelet, from which the wavelet basis functions are computed by scaling and/or translation.[12]

2. As a family of perfect reconstruction high- and low-pass filters that can extract information from the signal at varying scales. Then the forward and inverse discrete wavelet transforms can be represented by filtering and down- or up-sampling, respectively.[10,11]

There are three main reasons that a wavelet-based Poisson solver is of interest:

1. Solving the problem in wavelet space enables information to be retained about the dynamics on different scales spanned by the wavelet expansion.

2. Wavelet formulation allows for numerical noise to be removed naturally (denoising) by thresholding the wavelet coefficients. This also reduces the

effective dimensionality of the problem, thereby reducing the computational load.

3. Formulating and solving the Poisson boundary-value problem in wavelet space provides considerable computational speedup; both the Laplacian and the inverse Laplacian operators are *sparse*. Furthermore, the iterative methods can be accelerated using preconditioners that are effectively diagonal in wavelet bases.[13–15]

WAVELET-BASED POISSON SOLVER

The Poisson equation solved by the PIC codes is defined on a computational grid that contains all the particles used in the simulation. Discretization takes the Poisson equation with Dirichlet boundary conditions (BCs), that is, for which the value of the function is specified on the boundary, from its continuous form

$$\nabla^2 u = f, \qquad u_{\text{bnd}} = h, \tag{3}$$

where ∇^2 is the continuous Laplacian derivative, to

$$LU = F, \qquad U_{\text{bnd}} = H, \tag{4}$$

where the Laplacian operator L, potential U, and density F are all defined on the computational grid, and H is specified on the surface of the grid. Equation (4) represents a well-known problem in numerical analysis. It can be solved using several iterative methods, such as multigrid, successive over-relaxation, steepest descent, or conjugate gradient. For the work presented here, we generalized to three dimensions the preconditioned conjugate gradient (PCG) method.[16,17] The PCG method iteratively updates the initial solution along the conjugate directions until the exit requirement

$$|LU - F|_2 \le \varepsilon^2 |F|_2 \tag{5}$$

is satisfied in some norm $|\cdot|_2$. The convergence rate of the method depends on the condition number (k) of the operator L,

$$|U - U^i|_2 \le \left(\frac{\sqrt{k}-1}{\sqrt{k}+1}\right)^i |U|_2, \tag{6}$$

where U^i is the approximation to the exact solution U after the ith iteration. The smaller the condition number, the faster the approximation U^i approaches to the exact solution U. The condition number k of the Laplacian operator L on a grid is proportional to the square of the grid resolution, $k(L) \sim O(N^2)$, where N is the number of cells in each coordinate. This large condition number results in a slowly-converging scheme. However, in wavelet space, there is an effective diagonal preconditioner P for the wavelet-transformed Laplacian operator L_w that reduces the condition number of the preconditioned operator to $k(PL_wP) \sim O(N)$.[13–15] This provides a significant computational speedup.

Whereas the rate of convergence is set by the relation given in Equation (6), the number of iterations needed to attain a certain predefined accuracy also depends on how close the initial guess is to the solution. Since one does not expect significant changes in the potential from one instant in time $t = t_0$ to the next $t = t_0 + \delta t$, the

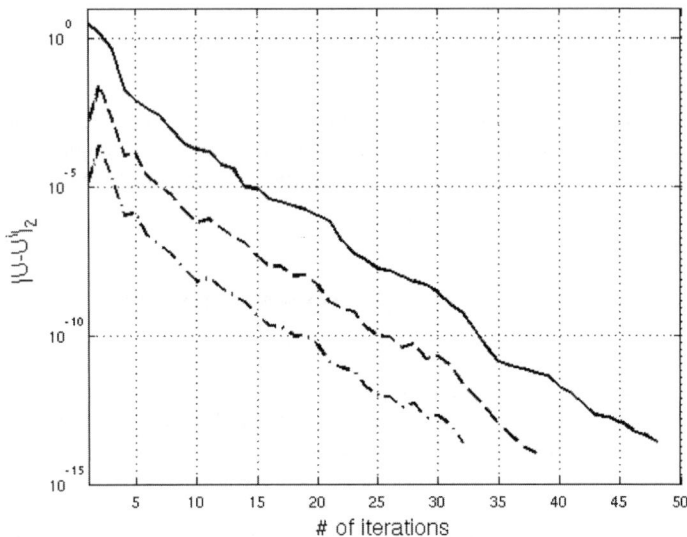

FIGURE 1. Effect of the initial guess on the preconditioned conjugate gradient method. The rate of convergence remains the same, given by Equation **(6)**, but it is obvious that a better initial guess leads to convergence to the desired accuracy within significantly fewer steps: ———, $u = 0$; – – –·, good and –·–·, better initial guesses.

potential at $t = t_0$ serves as a good initial guess for the conjugate gradient iteration at $t = t_0 + \delta t$. The importance of having a "smart" initial guess is illustrated in FIGURE 1.

Thresholding the wavelet coefficients (setting them equal to zero if their magnitudes are below a certain predefined threshold) can be used to remove the smallest scale fluctuations usually associated with numerical noise. It is worth recalling, however, that essential physics that must be captured in a typical PIC simulation includes various instabilities and fine structure/substructure formation. These processes owe their existence to the coupling between multiple spatial scales on which the dynamics of the system unfolds. Uncontrolled denoising carries with it the obvious danger of smoothing out the fine-scale details that serve as seeds for the onset of these processes. Nevertheless, there are numerous indications that properly implemented adaptive denoising can enable significant reduction in the size of the relevant data sets without compromising the ability of the solver to resolve the physically important multiscale aspects of the system dynamics. Denoising-by-thresholding effectively reduces (constrains) the search space for the iterative PCG method. Our implementation of the constrained PCG (CPCG) will be reported elsewhere.

A schematic representation of the solver is given in FIGURE 2.

Implementing Boundary Conditions

In the current implementation, we take the beam to pass through a grounded rectangular pipe. Over the four walls of the pipe, $U = 0$, and the two open ends through which the beam passes have open boundary conditions. We choose the computational grid to have dimensions 3–10 times smaller than those of the pipe, and we compute the potential on the six surfaces of this grid using the Green's function corresponding to this particular choice of BCs, while satisfying the constraints on U that the pipe imposes. Accordingly, the computation of BCs reduces to solving the following system of equations:[8]

$$\rho^{lm}(z) = \frac{4}{ab}\int_0^a\int_0^b \rho(x,y,z)\sin(\alpha_l x)\sin(\beta_m y)dx dy \quad (7)$$

$$\frac{\partial^2 \phi^{lm}(z)}{\partial z^2} - \gamma_{lm}^2 \phi^{lm}(z) = -\frac{\rho^{lm}(z)}{\varepsilon_0} \quad (8)$$

$$\phi(x,y,z) = \sum_{l=1}^{N_x}\sum_{m=1}^{N_y} \phi^{lm}(z)\sin(\alpha_l x)\sin(\beta_m y), \quad (9)$$

where ρ is the charge distribution, ϕ is the potential, $\alpha_l = l\pi/a$, $\beta_m = m\pi/b$, $\gamma_{lm}^2 = \alpha_l^2 + \beta_m^2$, and ε_0 is the permittivity of vacuum and the geometry of the pipe is given by $0 \le x \le a$ and $0 \le y \le b$.[8] Equation (9) is evaluated only on the surface of the computational grid, and for the predefined number of expansion coefficients N_x and N_y, thus yielding U_{bnd} from (4). This is only one of the ways to compute the potential on the surface of the grid. Other, more efficient and computationally cheaper, methods will be implemented in the future editions of the code.

The PCG solves Equation (5) by assuming that $U = 0$ outside the computational grid. The inhomogeneous Dirichlet boundary-value problem in (4) has been made

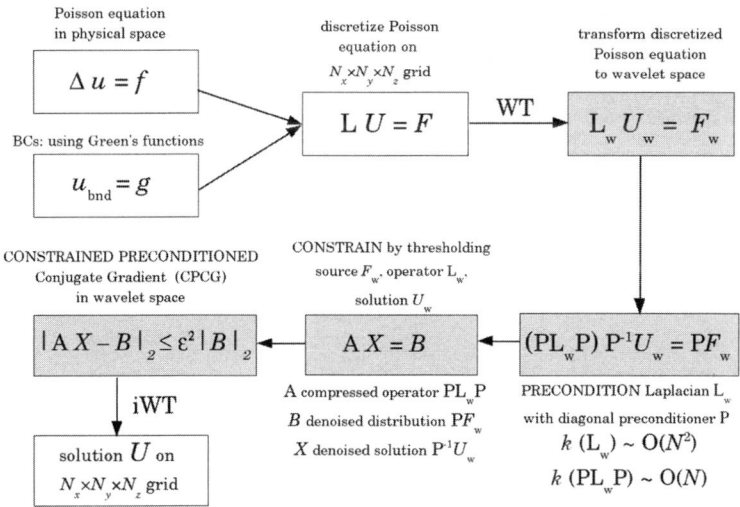

FIGURE 2. Flow-chart outline of the wavelet-based Poisson solver using the (constrained) PCG method. The *gray boxes* represent the wavelet-space; the physical space is in *white*. Constraining of the PCG method is shown in the *bottom middle box*. The current version of the code does not yet have this step implemented.

equivalent to the homogeneous problem by transferring the inhomogeneous boundary value terms to the source. In the simplest case, where the spacing of the computational grid Δ is the same in x-, y-, and z-directions, the altered source becomes $F \rightarrow F - H/\Delta^2$.[18]

APPLICATIONS

Our goal has been to develop a wavelet-based Poisson solver that can be easily merged into existing PIC codes designed for multiparticle dynamics simulations. As the first step toward that goal, we tested the solver on two idealized particle distributions, one from astrophysics and the other from beam dynamics. We used the PCG solver to compute the potential associated with the Plummer spherical stellar distribution (see FIGURE 3). Both the potential and density are analytically known and are given by

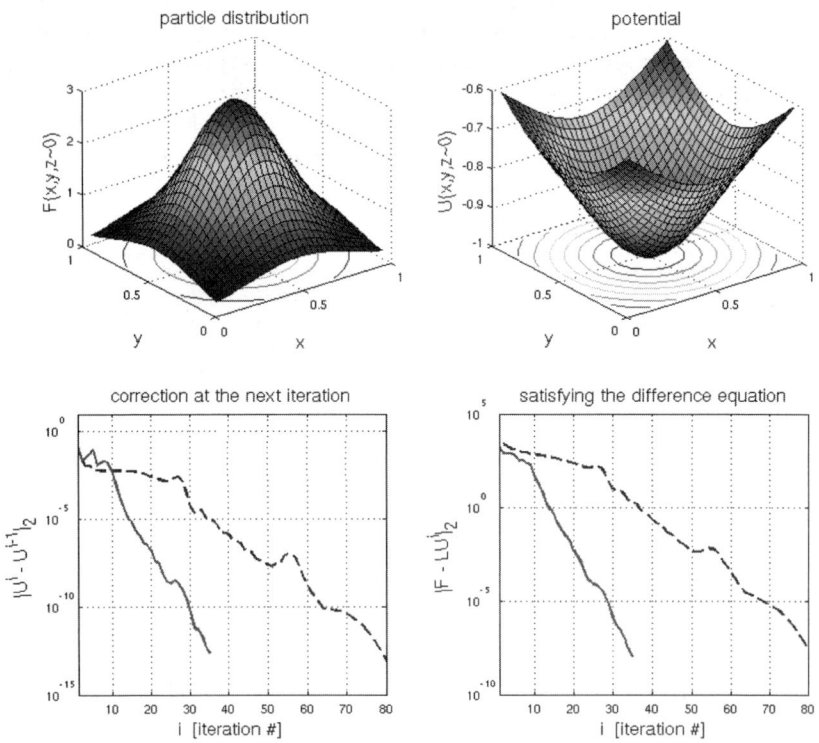

FIGURE 3. Plummer sphere particle distribution (**top left**) and corresponding potential (**top right**) obtained using the PCG. The **lower panels** show two convergence criteria—correction at the next iteration (**bottom left**) and how well the difference equation is satisfied (**bottom right**), with (*solid line*) and without (*dashed line*) the preconditioner.

$$F(r) = \frac{3}{(1+r^2)^{5/2}}, \quad U(r) = -\frac{1}{\sqrt{1+r^2}}, \tag{11}$$

where $r = \sqrt{x^2 + y^2 + z^2}$. The potential on the surface of the computational grid is specified analytically. The bottom panels of FIGURE 3 demonstrate the substantial computational speedup gained by preconditioning. Here we applied open BCs, $U(r \to \infty) \to 0$, which is a natural choice for self-gravitating systems.

We then applied the algorithm to a more realistic setting in which only the particle distribution is analytically known, and where the potential on the surface of the computational grid is computed using the analytically known Green's function (see FIGURE 4). It is an axially symmetric "fuzzy cigar-shaped" configuration of charged particles (a beam bunch) given by

$$F(x, y, z) = d_1(R)d_2(z) \tag{12}$$

$$d_1(R) = \begin{cases} 1 & 0 \le R \le R_1 \\ \frac{1}{4(R_1 - R_2)^4}(R - R_2)^2[R - (3R_1 - 2R_2)]^2 & R_1 \le R \le R_2 \\ 0 & \text{otherwise} \end{cases} \tag{13}$$

$$d_2(z) = \begin{cases} 1 & z_{1,2} \le z \le z_{2,1} \\ \frac{1}{4(z_{1,2} - z_1)^4}(z - z_1)^2[z - (3z_{1,2} - 2z_1)]^2 & z_1 \le z \le z_{1,2} \\ \frac{1}{4(z_{2,1} - z_2)^4}(z - z_2)^2[z - (3z_{2,1} - 2z_2)]^2 & z_{2,1} \le z \le z_2 \\ 0 & \text{otherwise} \end{cases} \tag{14}$$

where $R = \sqrt{x^2 + y^2}$ and the beam parameters $R_1, R_2, z_1, z_2, z_{1,2}, z_{2,1}$ are chosen so that $0 \le R_1 \le R_2$ and $z_1 \le z_{1,2} \le z_{2,1} \le z_2$. We applied BCs of a grounded rectangular pipe in the transverse direction (i.e., $U = 0$ on the pipe walls), and open in the longitudinal (z) direction. Similarly to the case of the Plummer sphere, a high accuracy solution is obtained in about 30 iterations of the algorithm with a preconditioner, or about 60 iterations without a preconditioner.

Upon successfully testing the PCG as a stand-alone Poisson solver, we replaced the standard Green's function-based Poisson solver in the PIC code ImpactT3D[7,8] with the PCG. For algorithm testing purposes, our solver uses Green's functions to evaluate the potential only on the surface of the computational grid, and then proceeds with the PCG algorithm to compute the potential on the interior. This introduces a certain computational inefficiency that will be eliminated at the stage of optimizing the solver for performance by using a different approach for computing BCs. The details of this optimization will be reported elsewhere. The parameters N_x and N_y specify the number of Green's function expansion coefficients in x- and y-directions. The BCs, again, correspond to a grounded rectangular pipe in the transverse directions and open in the longitudinal direction. In a typical simulation, the

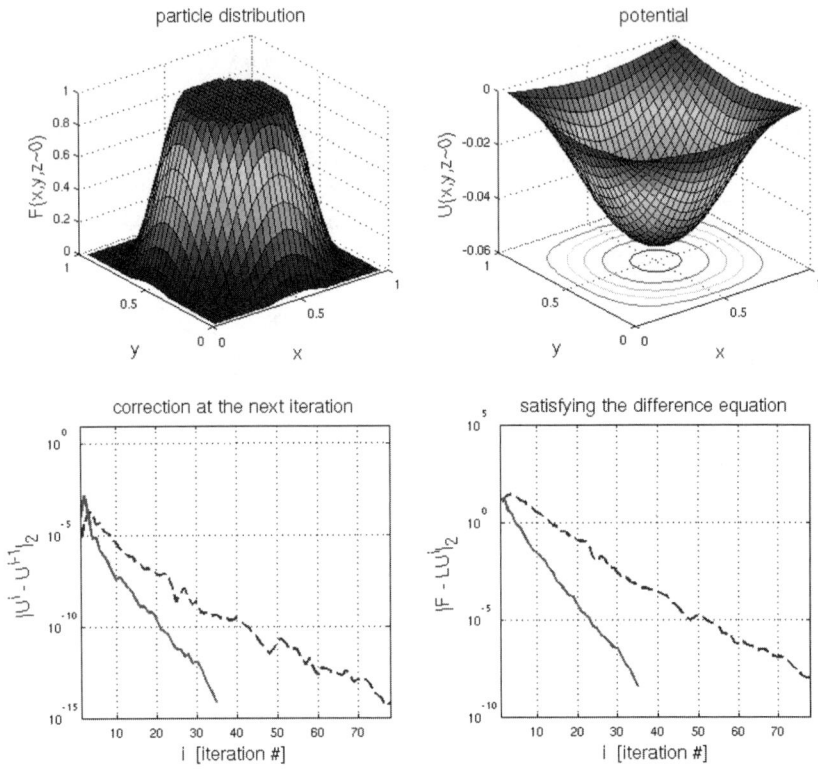

FIGURE 4. Particle distribution (**top left**) and corresponding potential (**top right**) for the "fuzzy cigar" obtained using the PCG. The **lower panels** show two convergence criteria—correction at the next iteration (**bottom left**) and how well the difference equation is satisfied (**bottom right**), with (*solid line*) and without (*dashed line*) the preconditioner.

cross-section of the pipe is larger than the cross-section of the computational grid by a factor of 3–10.

In FIGURE 5, we compare the Green's function-based Poisson solver used by ImpactT3D with the PCG by plotting the potential each algorithm computes from the same initial particle distribution. For the simulations done with the wavelet-based Poisson solver, no wavelet coefficient thresholding was done, that is, the full wavelet expansion was retained. The thresholding and the resulting denoising will be developed and implemented in the final version of the code.

We tested our wavelet-based code in the context of the Fermilab/NICADD photoinjector using 200,000 simulation particles and a nonuniform initial particle distribution at the cathode. It appears that not specifying the potential on the surface of the grid accurately enough causes considerable smoothing of the smaller scale features as computed by the wavelet algorithm. This, however, does not significantly affect the root mean square (rms) properties of the beam (see FIGURES 6 and 7), as confirmed by the excellent agreement between simulation runs done with standard

ImpactT3D (solid lines) and ImpactT3D with a wavelet-based Poisson solver with $N_x = N_y = 30$, that is, 900 expansion coefficients (dashed line). The difference between the simulations with the standard ImpactT3D (solid line) and ImpactT3D with a wavelet-based Poisson solver with $N_x = N_y = 100$ is almost imperceptible (FIG. 6).

These results clearly demonstrate that the simulations using wavelet-based Poisson solver and the standard ImpactT3D are in excellent agreement with respect to the computation of beam moments. This establishes the present study as an important proof-of-concept.

DISCUSSION AND CONCLUSION

We formulated and implemented a prototype, three-dimensional, wavelet-based Poisson solver that uses the preconditioned conjugate gradient method. The idea of combining the wavelet formulation and PCG to solve the Poisson equation is not

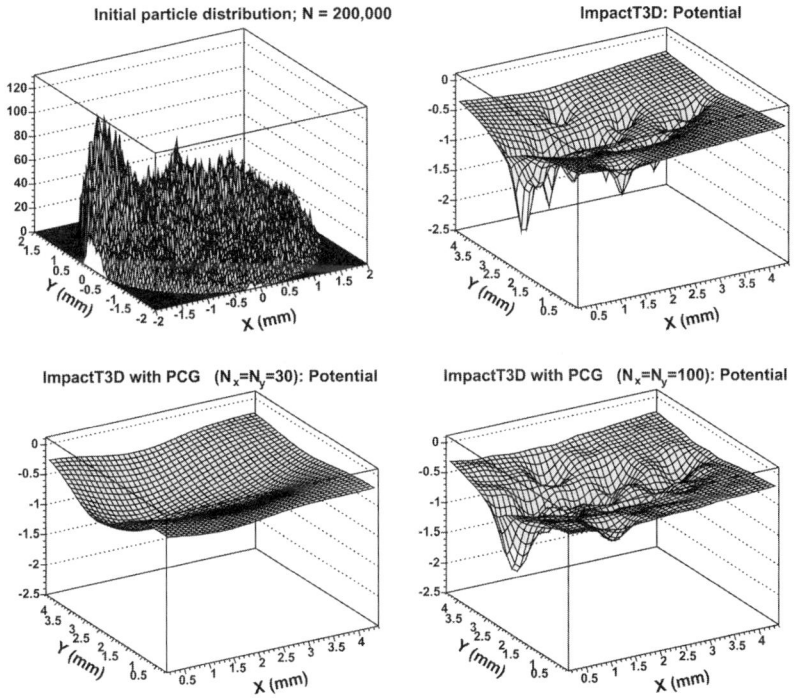

FIGURE 5. The z-integrated non-axisymmetric particle distribution (**top left**) for a beam. The other panels show the corresponding potential computed using the Green's function method used in the standard ImpactT3D (**top right**), ImpactT3D with PCG and $N_x = N_y = 30$, that is, 900 coefficients in the Green's function expansion of the potential on the surface of the computational grid (**bottom left**), and ImpactT3D with PCG and $N_x = N_y = 100$ (**bottom right**).

new: there are pioneering implementations of wavelet-based solvers for the Poisson equation with homogeneous ($U = 0$) Dirichlet BCs in one dimension[13] and periodic BCs in one, two, and three dimensions.[14,15] We built on this earlier work to design and implement a solver for the one-dimensional Poisson equation with *general* (inhomogeneous) Dirichlet BCs. This constitutes an original contribution on our part, since the formulation of the discretized problem, which includes the treatment of the BCs and the Laplacian operator, differs significantly from the periodized problem.

Having first tested our method as a stand-alone solver on two model problems, we then merged it into ImpactT3D to obtain a fully functional serial PIC code. We found that simulations performed using ImpactT3D with the native Poisson solver (based on Green functions and fast Fourier transforms) and ImpactT3D with the PCG solver described in this paper produce essentially equivalent outcomes (in terms of a standard set of rms diagnostics). This result enables us to move from the proof-of-concept stage to an advanced optimization and application-specific

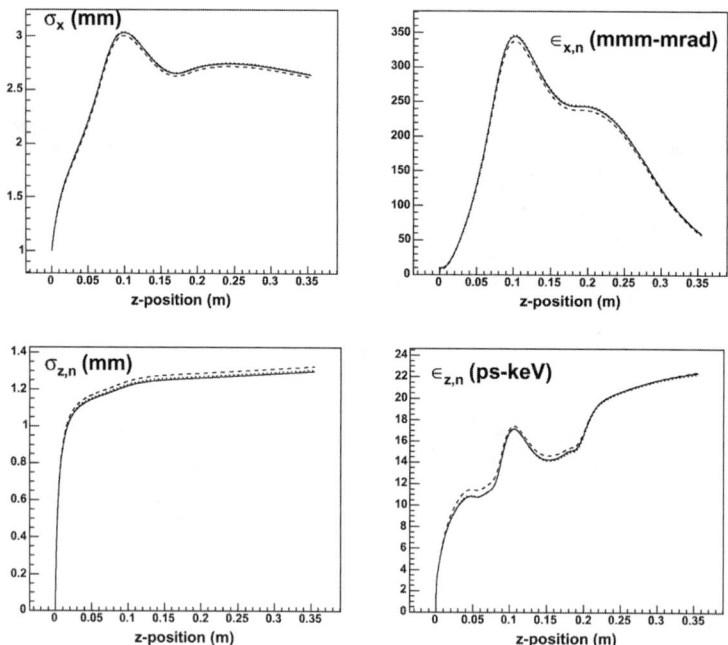

FIGURE 6. Simulation results for the radio frequency gun of the Fermilab/NICADD photoinjector done with the standard version of ImpactT3D (*solid lines*), ImpactT3D with PCG with $N_x = N_y = 30$ (*dashed line*) and ImpactT3D with PCG with $N_x = N_y = 100$ (*dotted line*): rms beam radius (**top left**), rms normalized transverse emittance (**top right**), rms bunch length (**bottom left**), rms normalized longitudinal emittance (**bottom right**). The close agreement indicates that both codes launch the beam in essentially identical fashion; this is critically important because the output of the full photoinjector depends sensitively on the initial conditions.

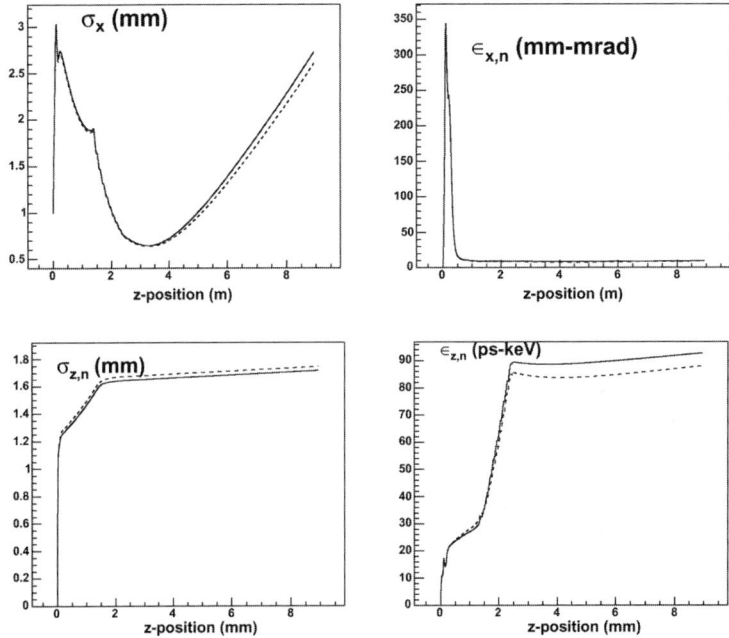

FIGURE 7. Simulation results for the full Fermilab/NICADD photoinjector done with the standard version of ImpactT3D (*solid lines*) and ImpactT3D with PCG with $N_x = N_y = 30$ (*dashed line*): rms beam radius (**top left**), rms normalized transverse emittance (**top right**), rms bunch length (**bottom left**), rms normalized longitudinal emittance (**bottom right**).

algorithm design. To our knowledge, the work reported here constitutes the first application in beam dynamics of the wavelet-based multiscale methodology to three-dimensional computer simulations.

Our current efforts are focused on several areas that encompass both algorithm optimization and applications work. On the optimization side, the top priority is to enable efficient computation of the potential over the boundary of the computational grid (as distinct from the physical boundaries of the system). Another priority is incorporation into the solver of state-of-the-art routines for efficient storage and multiplication of multidimensional sparse arrays. Next, adaptive denoising (and simultaneous compression) in the context of PIC modeling presents us with a unique set of technical challenges and a wealth of complex and engaging multiscale physics. Finally, we have not yet addressed the complex issues of solver parallelization for use with the parallel version of Impact3D on multiprocessor machines.

On the side of applications, we are working on leveraging the advantages afforded by multiscale wavelet formulation to tackle the previously all but intractable—in the sense of being prohibitively expensive to compute—problem of high-precision three-dimensional modeling of CSR and its effects on the dynamics of beams in a variety of accelerator systems. The details of our approach will be reported, together with the first results, in the near future.

ACKNOWLEGMENTS

Both authors are thankful to Henry Emil Kandrup for the role he played in their lives. His thoughtful guidance and patient mentorship greatly influenced our scientific careers. It is through him that this collaboration was established in the first place.

The authors are grateful to Daniel Mihalcea for running the numerical simulations and for generating FIGURES 5–7. Ji Qiang provided valuable help in integrating the solver into the ImpactT3D suite. We are thankful to Courtlandt Bohn for useful and stimulating discussions and comments. Balša Terzić was supported by the Air Force contract FA9471-040C-0199. The work of Ilya V. Pogorelov was supported by the Director, Office of Science, of the U.S. Department of Energy under Contract DE-AC03-76SF00098.

The authors declare that they have no competing financial interests.

REFERENCES

1. HOCKNEY, R.W & J.W. EASTWOOD. 1988. Computer Simulations Using Particles. Institute of Physics Publishing, London.
2. ALLEN, C.K., et al. 2002. Beam-halo measurements in high-current proton beams. Phys. Rev. Lett. **89:** 214802.
3. BOHN, C.L. & I.V. SIDERIS. 2003. Fluctuations do matter: large noise-enhanced halos in charged-particle beams. Phys. Rev. Lett. **91:** 264801.
4. QIANG, J., R.D. RYNE & I. HOFMANN. 2004. Space-charge driven emittance growth in a 3D mismatched anisotropic beam. Phys. Rev. Lett. **92:** 174801.
5. HUANG, Z., M. BORLAND, P. EMMA, et al. 2004. Suppression of microbunching instability in the linac coherent light source. Phys. Rev. ST Accel. Beams **7:** 074401.
6. BOHN, C. 2002. Coherent synchrotron radiation: theory and experiments. AIP Conf. Proc. **647:** 81.
7. QIANG, J., R.D. RYNE, S. HABIB & V. DECYK. 2000. An object-oriented parallel particle-incell code for beam dynamics simulation in linear accelerators. J. Comp. Phys. **163:** 434–451.
8. QIANG, J. & R.D. RYNE. 2001. Parallel 3D Poisson solver for a charged beam in a conducting pipe. Comp. Phys. Comm. **138:** 18.
9. GOEDECKER, S. 1998. Wavelets and Their Applications. Presses Polytechniques et Universitaires Romandes, Lausanne.
10. MISITI, M., Y. MISITI, G. OPPENHEIM & J. POGGI. 1997. Wavelet Toolbox For Use with MATLAB. The MathWorks, Inc., Natick, Massachusetts.
11. WICKERHAUSER, M.V. 1994. Adaptive Wavelet Analysis From Theory To Software. A.K. Peters, Wellesley, Massachusetts.
12. DAUBECHIES, I. 1992. Ten Lectures on Wavelets. SIAM, Philadelphia.
13. BEYLKIN, G. 1993. On wavelet-based algorithms for solving differential equations. *In* Wavelets: Mathematics and Applications. J. Benedetto & M. Frazier, Eds. CRC Press LLC, Boca Raton.
14. AVERBUCH, A., G. BEYLKIN, R. COIFMAN, et al. 2003. Adaptive solution of multidimensional PDEs via tensor product wavelet decomposition. Unpublished notes. <http://www.cs.tau.ac.il/~amir1/PS/poisson.pdf>.
15. AVERBUCH, A., G. BEYLKIN, R. COIFMAN & M. ISRAELI. 1998. A wavelet-based constrained preconditioned conjugate gradient for elliptic problems. *In* Signal and Image Representation in Combined Spaces. Y. Zeevi & R. Coifman, Eds. Academic Press, San Diego.
16. GOLUB, G.H. & C.F. VAN LOAN. 1996. Matrix Computations. Johns Hopkins University Press, Baltimore and London.

17. HESTENES, M.R. & E. STIEFEL. 1952. Methods of conjugate gradients for solving linear systems. J. Res. Nat. Bur. Stand. **49:** 409.
18. PRESS, W.H., S.A. TEUKOLSKY, W.T. VETTERLING & B.P. FLANNERY. 1992. Numerical Recipes in FORTRAN: The Art of Scientific Computing, 2nd edit.: 851. Press Syndicate of the University of Cambridge, Cambridge.

Energy Trapping in Loaded String Models with Long- and Short-Range Couplings

ILYA V. POGORELOV[a] AND HENRY E. KANDRUP[b,c,d]

[a]*Accelerator and Fusion Research Division, Lawrence Berkeley National Laboratory, Berkeley, California, USA*

[b]*Department of Astronomy, University of Florida, Gainesville, Florida, USA*

[c]*Department of Physics, University of Florida, Gainesville, Florida, USA*

[d]*Institute for Fundamental Theory, University of Florida, Gainesville, Florida, USA*

ABSTRACT: This paper illustrates the possibility, in simple loaded string models, of trapping most of the system energy in a single degree of freedom for very long times, demonstrating in particular that the robustness of the trapping is enhanced by increasing the *connectance* of the system, that is, the extent to which many degrees of freedom are coupled directly by the interaction Hamiltonian and/or the strength of the couplings.

KEYWORDS: Hamiltonian dynamics; long-range interactions; energy localization

INTRODUCTION

The work described here was motivated by a desire to understand the qualitative difference in dynamical behavior in systems interacting via short- and long-range couplings. In particular, how do bulk properties involving mixing and relaxation vary as the strength or the range of the interaction is increased? Considerable work has been done on systems, like the Fermi–Pasta–Ulam (FPU) model, that involve only nearest neighbor couplings.[1] However, comparatively little is known about the phenomenology of systems that manifest longer range couplings. How does the dynamics change for high-connectance systems,[2] which couple directly all, or almost all, the degrees of freedom?

When considering systems with longer range couplings, even the natural language in which to describe the dynamics changes. Most work on FPU-type models has involved an analysis formulated in terms of the modes of the system. However, for systems with longer range interactions it often becomes less obvious how to identify a natural set of modes. More natural, perhaps, is to consider individual degrees of freedom as fundamental and to interpret the observed behavior in terms of those degrees of freedom.

The work described here can be viewed as a prolegomenon toward more systematic work underway that aims to study mixing properties in self-consistent systems

Address for correspondence: Ilya V. Pogorelov, M/S 71J 100A, Lawrence Berkeley National Laboratory, 1 Cyclotron Rd., Berkeley, CA 94720, USA. Voice: 510-495-2408; fax: 510-495-2323.
  ivpogorelov@lbl.gov

described in the continuum limit by partial differential equations, such as the Vlasov–Poisson system of galactic astronomy, plasma physics, or charged-particle beams.

MODELS AND NUMERICAL SIMULATIONS

As a simple example, consider a loaded string-type model consisting of $N = 16$ identical nonlinear oscillators arranged in a one-dimensional closed chain with dynamics generated by the Hamiltonian

$$H = \sum_{i=0}^{N-1} \left(\frac{1}{2}p_i^2 + \frac{1}{2}q_i^2 + \frac{1}{4}q_i^4\right) + \frac{1}{4}\sum_{i \neq j = 0}^{N-1} c_{ij}(q_i - q_j)^4 \equiv \sum_{i=0}^{N-1} H_i + \sum_{i \neq j = 0}^{N-1} H_{ij},$$

with $q_N \equiv q_0$, and view the system as a sum of N individual one-degree-of-freedom Hamiltonians H_i plus an interaction Hamiltonian $H_{\text{int}} = \Sigma H_{ij}$. To explore how the dynamics depends on both the range of the interaction and the number of degrees of freedom that are coupled directly, one can then allow for three different types of couplings:

1. a model with nearest neighbor couplings, with c_{ij} nonzero only for $|i - j| = 1$, and with all nonzero c_{ij} assuming the same value;

2. a maximally connected model, with all pairs coupled identically; and

3. another long-range model in which the coupling constant decreases linearly from a fixed value c for nearest neighbors to zero strength for oscillators separated by $N/2$ "spaces".

Unless otherwise specified, the computations described here assumed a coupling strength $c_{i,i+1} = 1/3$.

As a highly special, albeit interesting, initial condition, suppose that, at $t = 0$, all the system energy is deposited into a single degree of freedom, setting the kinetic energy for one of the oscillators, say oscillator 0, equal to the total system energy E. The obvious question then is how long it takes before a sizeable fraction of the energy is transferred to the other oscillators and/or the interaction Hamiltonian.

Consider first models with nearest neighbor couplings. In this case, one discovers that a significant fraction of the total energy can remain trapped in a single degree of freedom for times as long as 10^2–10^3 periods of uncoupled oscillations. This is, for example, illustrated in FIGURE 1 and FIGURE 2, that exhibit the distribution of kinetic energies among the various oscillators as a function of time for four different energies. In these figures, as in FIGURES 3–5, the kinetic energies have been smoothed from raw data recorded at intervals $\delta t = 0.1$ by a boxcar averaging over 400 adjacent points. The initial localization involves, of course, the oscillator in which the energy was originally deposited. Eventually, however, various degrees of freedom emerge as the location of this energy localization, the transitions occurring quite abruptly and with no obvious correlation between the locations of the localization prior to and following the transitions. The sojourn times also appear to be random, except that for higher energies and/or weaker coupling, the intervals between the successive transitions become shorter overall. For example, in the long-range model with linear decrease in coupling strength, reducing the value of c from 1/3 to 1/9 can decrease the typical duration of these intervals by as much as a factor of five. For very high energies, the localized configuration breaks down rapidly, never to appear again (see, e.g., FIG. 2 bottom panel).

The obvious question is how this conclusion is altered for the case of longer range interactions. In particular, does increasing the range of the coupling make the energy "disperse" more quickly? Here the answer is a resounding: *no*! Far from providing new channels for energy transfer between the degrees of freedom, *long-range couplings make this energy localization, or trapping, even more robust.* For every value of energy that was explored—$0.02 \leq E \leq 327.68$—localization was more robust for the models that coupled together all the degrees of freedom than for the nearest neighbor model, with the maximally connected model allowing trapping for the longest time. The degree to which a long range coupling enhances trapping can be inferred from FIGURE 3, which exhibits the distribution of kinetic energies for the various oscillators in the three models, in each case allowing for an energy

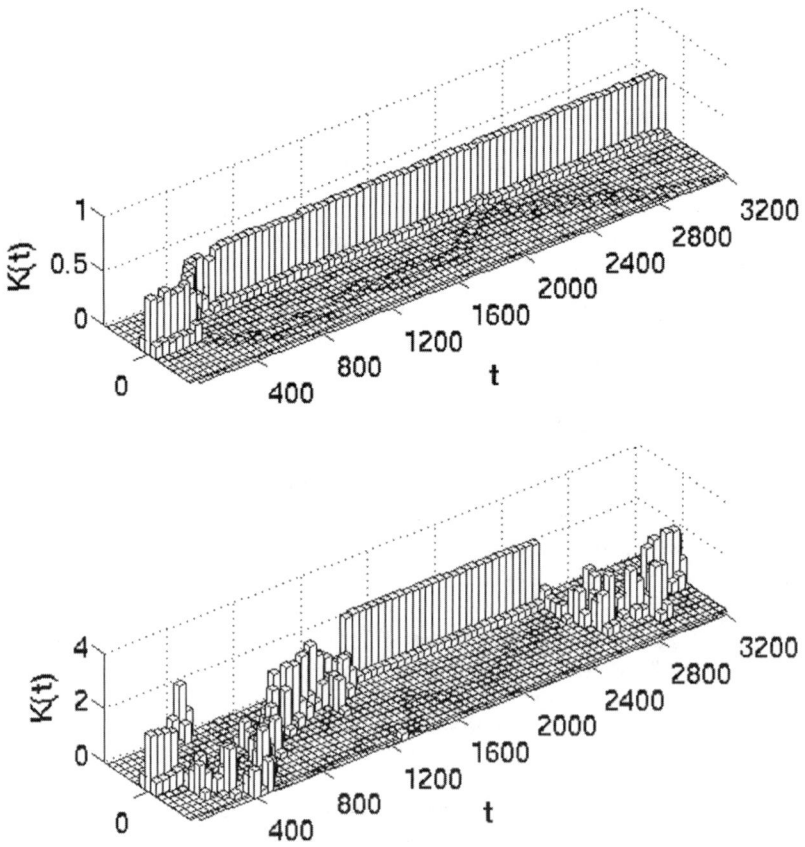

FIGURE 1. Energy trapping in the nearest neighbor model: distribution of kinetic energies as a function of time for an initial condition with all the energy $E = 1.28$ (**top**) and $E = 5.12$ (**bottom**) originally in oscillator 0.

$E = 327.68$. Also evident from FIGURES 1–3 is the fact that, for all three models, the "trapping time" decreases with increasing energy, a relatively sharp decrease being observed for energies greater than $E \sim 5$–20, the value depending on the model.

However, as illustrated in FIGURE 4, localization does not emerge for "more generic" initial conditions with energy distributed randomly amongst all the oscillators. Thus, one might perhaps conjecture that this localization is a fluke reflecting a very nongeneric initial condition. It is, therefore, important to determine the stability of these localized configurations. In particular, one needs to determine whether, for initial conditions "less singular" than $E_0 = E$, energy trapping persists and, if so, for how long.

These issues were addressed by considering alternative sets of initial conditions, where the initial kinetic energy of the zeroth oscillator was assigned smaller fractions of the energy, $E_0 = 0.85E$ and $E_0 = 0.7E$, and the remaining energy was

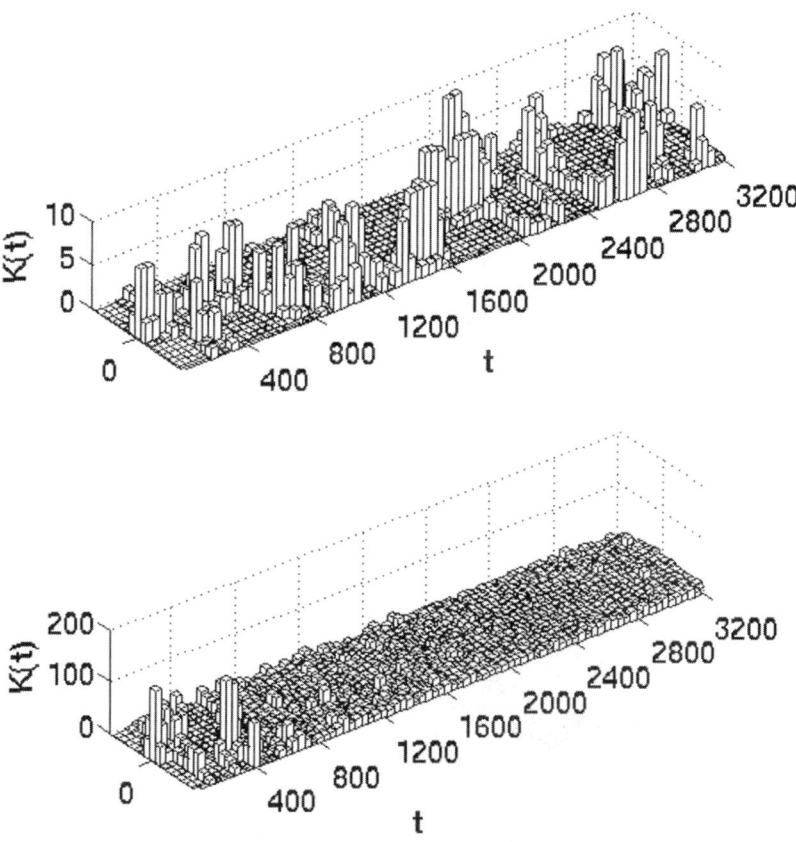

FIGURE 2. Energy trapping in the nearest neighbor model: distribution of kinetic energies as a function of time for an initial condition with all the energy $E = 20.48$ (**top**) and $E = 327.68$ (**bottom**) originally in oscillator 0.

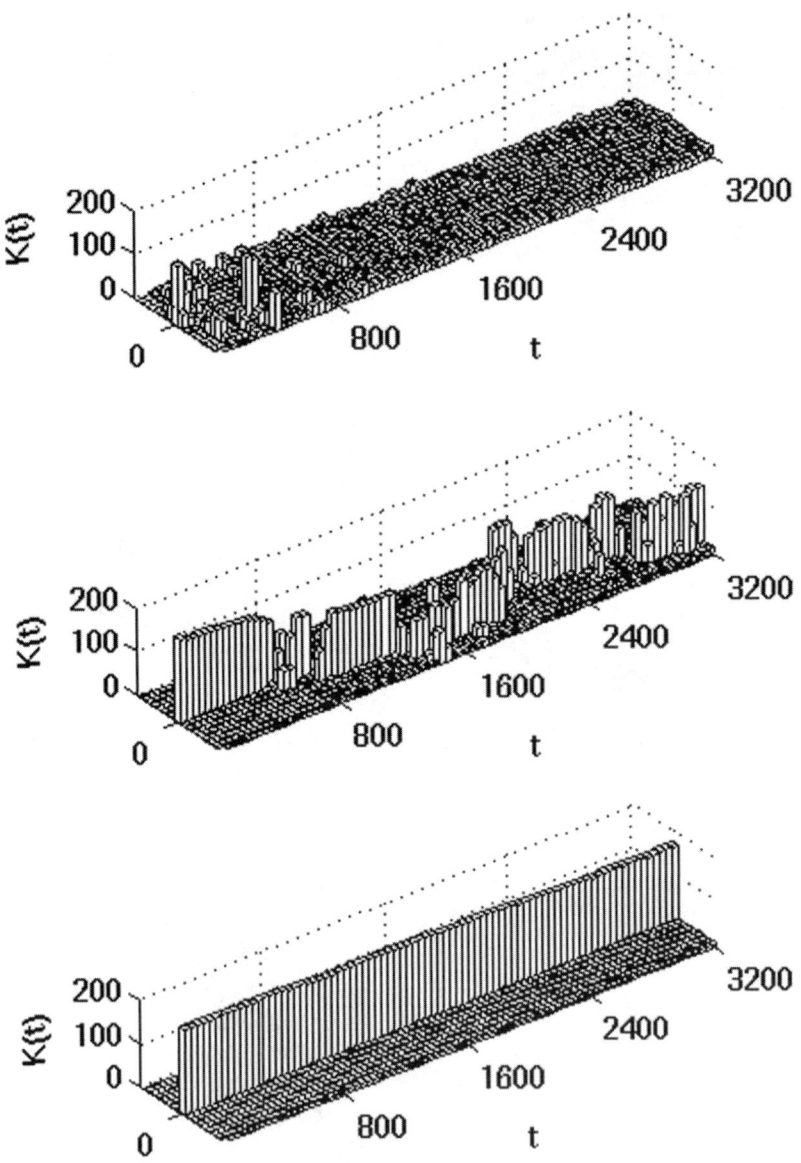

FIGURE 3. Distributions of kinetic energies as a function of time for an initial condition with the energy $E = 327.68$ all originally in oscillator 0: (**top**) the nearest neighbor model; (**middle**) the long range model with coupling decreasing linearly; (**bottom**) the maximally connected model.

apportioned randomly among the other degrees of freedom. Putting energy into the other degrees of freedom does indeed tend to make localization less robust. However, even for $E_0 = 0.7E$ one can see distinct localization patterns that persist over hundreds of natural oscillation periods. Moreover, as before, stronger and/or longer range couplings mean that the localization persists for longer times. Decreasing the fraction of the total energy placed into the "trap" results in shorter trapping times. This behavior is illustrated in FIGURE 5, where various coupling types and values of E are chosen so as to demonstrate the competing effects of longer range coupling and higher energy. In all three cases, the initial condition is one with 70% of the total energy deposited in oscillator 0.

The middle panel of FIGURE 5 also illustrates another interesting point. It is possible for most of the energy localized initially in a single oscillator to be deposited in two other oscillators, where it remains localized for a comparatively long time. Alternatively, as illustrated in the bottom panel of FIGURE 1, an initial state that would appear to have become largely delocalized can "relocalize" with most of the energy concentrated in a different oscillator.

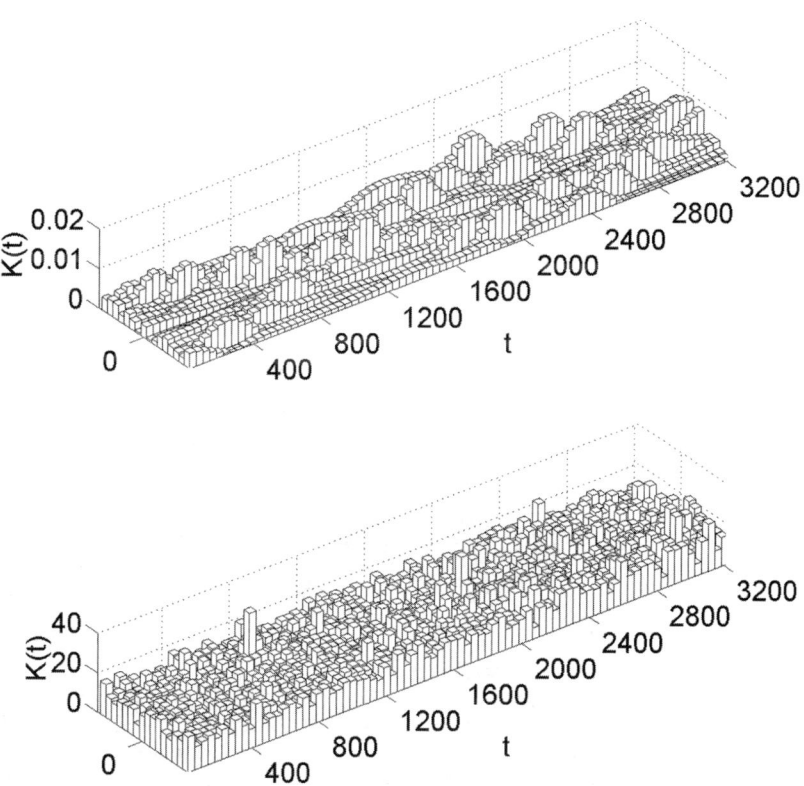

FIGURE 4. Distribution of kinetic energies as a function of time for a random initial condition with $E = 0.08$ (**top**) and $E = 327.68$ (**bottom**) evolved in the nearest neighbor model.

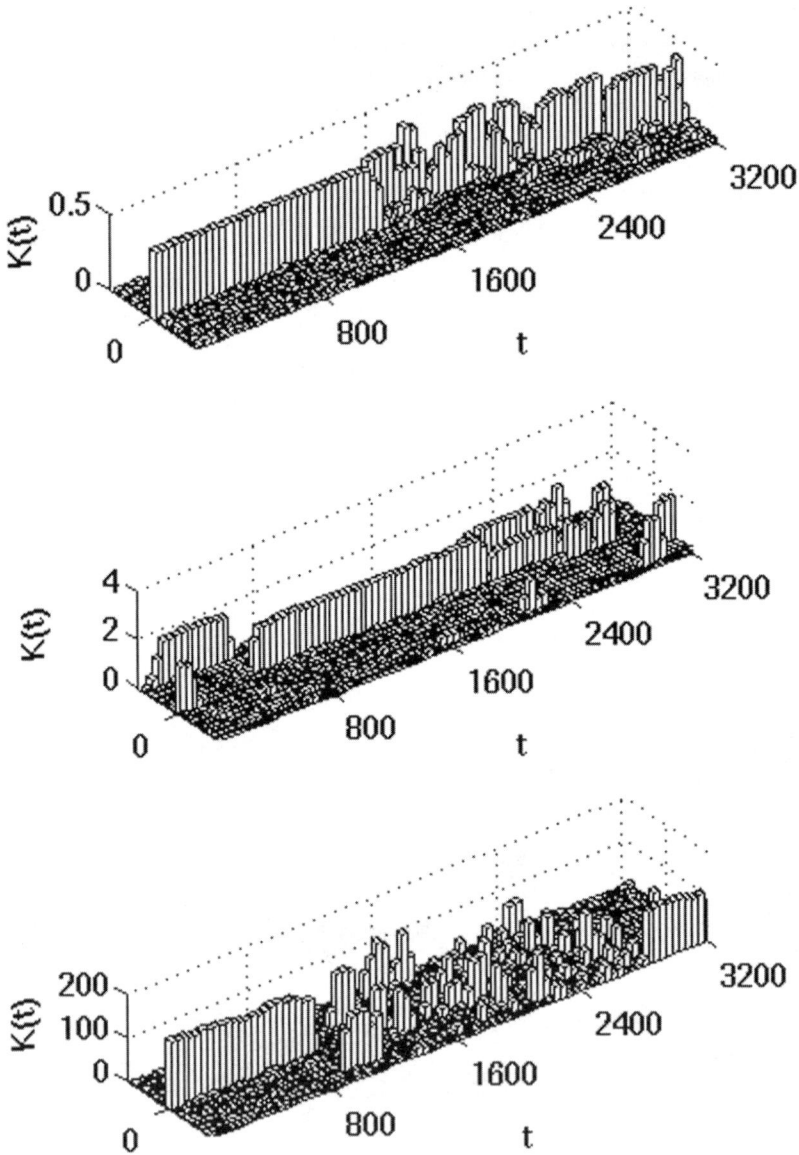

FIGURE 5. Top. Distribution of kinetic energies as a function of time for an initial condition with $E_0 = 0.70E$ evolved in the linearly decreasing long-range coupling model with total energy $E = 1.28$. The remaining energy is initially distributed randomly amongst the other degrees of freedom. **Middle.** The same for an initial condition with $E = 5.12$ evolved in the maximally connected model. **Bottom.** An initial condition with $E = 327.68$ evolved in the maximally connected model.

Since the efficacy of mixing is well known to differ for regular and chaotic dynamics, it is also natural to ask whether the comparatively abrupt transition from robust to less stable trapping with increasing energy E correlates with a transition from (near-) regular to (more strongly) chaotic dynamics. One useful diagnostic in addressing such a transition is provided by examining the "complexity" of the time series associated with some phase space variable, such as the position, momentum, or energy of one of the oscillators. One can, for example, determine the number of frequencies required, on average, to capture some significant fraction, (say) 95%, of the total power in the Fourier spectrum associated with a time series.

Such an analysis reveals that, for lower energies, the time series are very close to periodic whereas, for higher energies, any quasiperiodic approximation requires an enormous number of frequencies. There is, moreover, a strong correlation between the magnitude of the time series complexity and the rate of delocalization. The transition from very slow to considerably more rapid delocalization and the transition from very long-lived to considerably less long-lived trapping are comparably abrupt. Furthermore, the complexity for the shorter range models, where localization is less robust, tends to be somewhat larger than for the time series of the longer range models. Finally, it is evident that, at least while energy remains trapped in one oscillator, the time series for the coordinate or momentum corresponding to that oscillator is typically much less complex than the time series for the other degrees of freedom. Alternatively, at least for the case of the maximally connected model, the oscillator in the chain directly opposite from the oscillator in which the energy is localized tends to have an especially complex time series.

Examples of this behavior are exhibited in FIGURE 6, which derives from orbital data recorded at intervals $\delta t = 0.1$ for a total time $t = 3200$. In each case, $n(0.95)$ represents the fraction of frequencies required to capture 95% of the power in a time series for phase space coordinates q_i and p_i. The bottom curve exhibits data for the oscillator into which all the energy was originally deposited, the diamonds representing mean values obtained by computing complexities individually for q_0 and p_0 and then averaging. The top curve exhibits data for the oscillator separated by $N/2$ spaces. The middle curve was derived by computing complexities analogously for all 16 oscillators and then constructing an average over the oscillators.

Presuming that the flow is ergodic, one would expect an eventual evolution toward a well mixed state; and, for that reason, it is natural to determine the extent to which there is an asymptotic approach toward equipartitioning of energy at late times. For the case of random initial conditions, there are in fact clear indications of such an approach, at least in a time-averaged sense, although the time scale for this approach can be very long, exceeding 10^3 orbital periods. Obviously, though, for localized initial states there can be no such approach as long as trapping persists. However, even here there are indications for an eventual approach toward equipartitioning for higher energies, weaker couplings, and/or shorter range interactions. Examples of an approach toward equipartitioning, or lack thereof, are provided in FIGURE 7, which tracks the time-averaged kinetic energies[3]

$$\langle K(t) \rangle = \frac{1}{n} \sum_{i=1}^{n} K(t_i),$$

with $t_1 = 0$ and $t_n = t$, of several oscillators in three representative integrations.

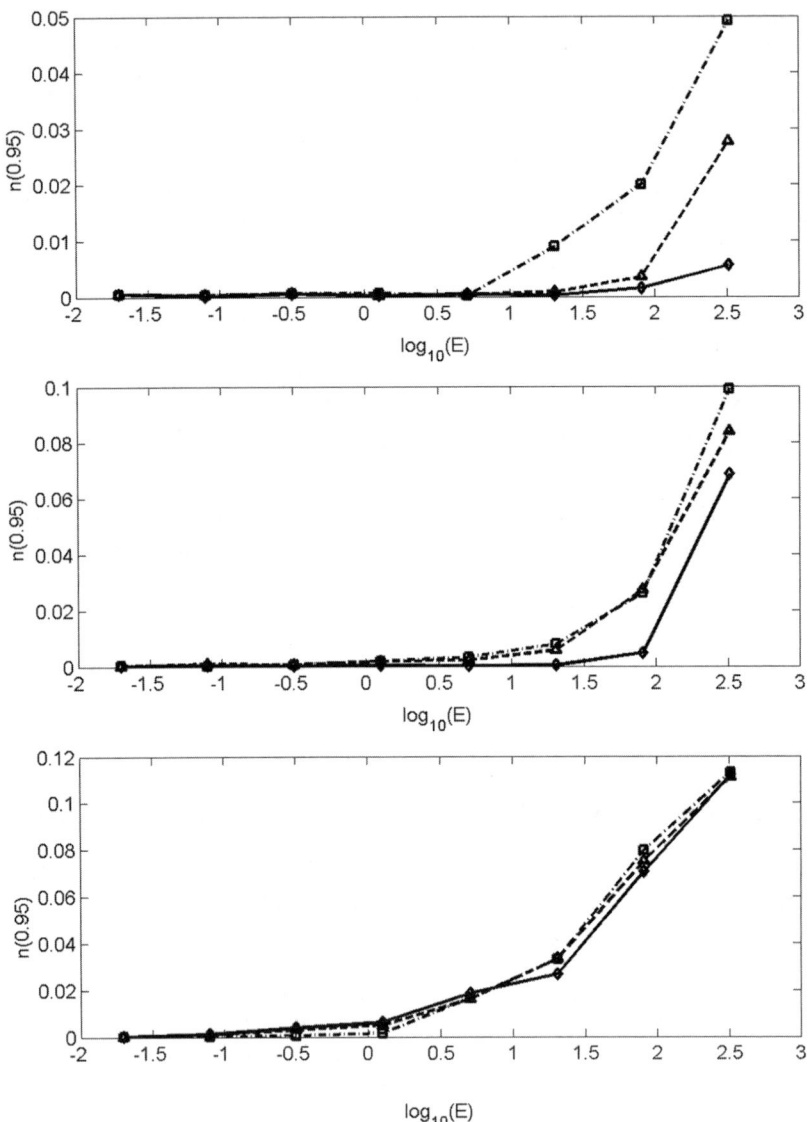

FIGURE 6. Top. Orbital complexities as a function of energy E, computed for various degrees of freedom in the maximally connected model from initial conditions in which all the energy is initially localized in a single degree of freedom. The *bottom curve* (◆) is for the oscillator that receives the initial energy, the *top curve* (□) is for the oscillator separated by $N/2$ "spaces", and the *middle curve* (▲) represents an average over all 16 oscillators. **Middle.** The same for the linear long range model. **Bottom.** The same for the nearest neighbor model.

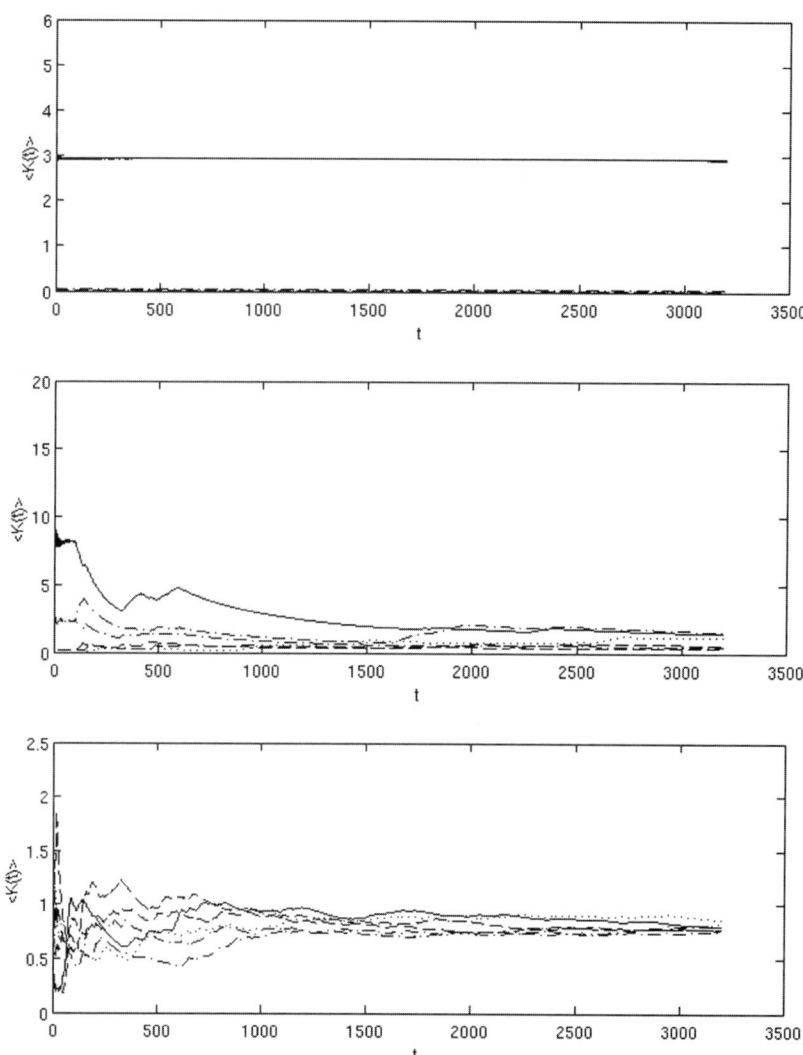

FIGURE 7. Top. Time-averaged kinetic energies $\langle K(t) \rangle$ for oscillators in the model with long-range couplings that decrease linearly, evolved from an initial state in which all the energy $E = 5.12$ is deposited in one oscillator. **Middle.** The same for the model with nearest neighbor couplings, now constructed for an initial state with $E = 20.48$. **Bottom.** The same for a random initial condition with $E = 20.48$ evolved in the model with maximal coupling. ———, reference oscillator; — · —, nearest neighbors; — — —, second nearest neighbors; and ········, most distant.

CONCLUSIONS

A number of interesting questions remain to be addressed. For each of the different types of coupling, is there, for example, a threshold value of energy or coupling strength signaling a transition from regularity to chaos and a comparatively rapid breakdown of localization? How does the behavior observed for $N = 16$ vary as the number of degrees of freedom increases? Perhaps most importantly, is this localization unique to one-dimensional chains? Will a similar localization persist on a two-dimensional lattice configured as a torus? Work on these questions is currently underway.

In any event, the numerical experiments performed hitherto yield three unambiguous conclusions: (1) FPU-type systems with both nearest-neighbor and longer range couplings admit states in which most of the energy remains trapped in a single degree of freedom for relatively long times. (2) The stronger the coupling in terms of range, connectance, or size of the coupling constant c, the more robust is this energy trapping. (3) The trapping is more robust at lower energies, where the dynamics seems more nearly regular.

ACKNOWLEDGMENTS

HEK was supported in part by NSF-AST-0070809. IVP is grateful to Henry Kandrup for many years of interesting, stimulating discussions and joint collaborations. The author declares that he has no competing financial interest.

REFERENCES

1. FORD, J. 1992. Phys. Rep. **213**: 271.
2. FROESCHLÉ, C. 1978. Phys. Rev. A **18**: 277.
3. BATT, J. 1987. Transport Theory Stat. Phys. **16**: 763.

Characterization of Chaos

A New, Fast, and Effective Measure

IOANNIS V. SIDERIS

*Department of Physics, Northern Illinois University,
DeKalb, Illinois, USA*

ABSTRACT: A new technique for characterization of the regular or chaotic nature of dynamical orbits has been discovered. It takes advantage of morphological and dynamical properties of orbits, and is very effective, at least for time-independent systems with two degrees of freedom. The new technique was initially designed with time-dependent and N-body systems in mind. For this reason one of its main goals is to provide straightforward information about the transient chaos associated with such regimes. Equally important is the distinction it can provide between sticky and wildly chaotic epochs during the evolution of chaotic orbits. The most important advantage over the existing methods is that it can characterize an orbit using a very small number of orbital periods. For these reasons the new method is extremely promising to be useful and effective in a broad spectrum of disciplines.

KEYWORDS: chaotic dynamics; chaotic measures; nonlinear dynamics; stickiness; transient chaos

INTRODUCTION

Characterization of the periodic or chaotic nature of dynamical orbits has been an important issue in many different disciplines. Correct characterization can contribute to critical understanding and subsequent intuition about the dynamics of a system. A number of techniques have been contrived to pursue the problem.[1–4] Most of these methods are based either on the convergence of a measure, the Lyapunov exponents[5] providing a typical example, or on a frequency analysis scheme, such as Laskar's method.[6]

There is no doubt that these methods have proved valuable in the past and that they can provide critical information. Nevertheless, they have limitations. The typical complaint is that they require long evolution times to provide reliable results. The best methods usually need a minimum of tens, or even hundreds, of orbital periods to provide confident characterizations. This means that in contexts like galactic dynamics, or even worse, charged-particle beams, where the evolution time is relatively short, these methods may be problematic. Many researchers have achieved serious breakthroughs in order to address this problem,[1–6] but generally the problem persists.

Address for correspondence: Ioannis V. Sideris, Department of Physics, Northern Illinois University, DeKalb, IL 60115, USA. Voice: 815-753-0787; fax: 815-753-8565.
  sideris@nicadd.niu.edu

A second problem is that the current methods have not been created by design to distinguish between the sticky[7–8] and the wildly chaotic epochs of a chaotic orbit. It is well known that a chaotic orbit may stay trapped close to a regular island for a long time and then escape into a broader chaotic sea, where it can move without constraints. Such escapes are associated with motion through cantori (in two dimensions), or in the Arnold web (in three or more dimensions). The need for distinction between these two regimes and the computation of diffusion times has appeared in several contexts,[9] including the structure and evolution of galaxies.[10]

A third, more complex, problem is related to the phenomenon of transient chaos, that appears in time-dependent systems. In these regimes the time dependence may cause an orbit to move from regular into chaotic (and *vice versa*) several times during its evolution.[11] The established methods can give information in an average manner, but they have not been designed to easily provide additional information about when these transitions happen. Extraction of this information using the existing measures, although not impossible, is usually tedious to achieve.

A new method has been devised that attempts to address all of these problems. It is based on a new approach only loosely connected to previous techniques. In the next section, the new technique is presented for a typical example. Subsequently, advantages and disadvantages of the new method, as well as plans for future work, are discussed.

DESCRIPTION OF THE METHOD

A random set of 500 initial conditions, with energy 0.125, were integrated in the Hénon–Heiles potential[12]

$$V(x, y) = \frac{1}{2}\left(x^2 + y^2 + 2x^2y - \frac{2}{3}y^3\right).$$

Their orbital data y, v_x, and v_y were recorded, in the typical Poincaré section fashion, when the orbits were crossing the plane y ($x = 0$). For convenience, hereafter, these orbital data series y, v_x, and v_y (recorded when $x = 0$) are called *signals* or *time-series*. The largest Lyapunov exponent of each orbit was also computed for comparison.

When one plots the Poincaré section of an orbit, morphological patterns emerge. One can visually identify regions of regularity and chaos. The periodic patterns of regular orbits on a Poincaré section are consequences of the existence of global and local integrals of motion, associated with underlying dynamical symmetries. When the number of underlying global and local symmetries is smaller than the number of degrees of freedom of the system, chaotic motion emerges and the periodic patterns are absent.

Similarly one can identify an obvious difference between regular and chaotic orbits when the points that comprise the recorded time-series are plotted without the lines that connect them (see FIGURES 1 and 2). For regular orbits there are similar morphological patterns that succeed each other (FIG. 1). On the other hand morphological patterns are absent in chaotic orbits in the strict, obvious, way they exist in regular orbits (FIG. 2). However, one can still recognize some loose repetition in parts of a chaotic time-series. It is this loose repetition that often corresponds to a sticky epoch of a chaotic orbit. A striking observation is that the signal of a chaotic

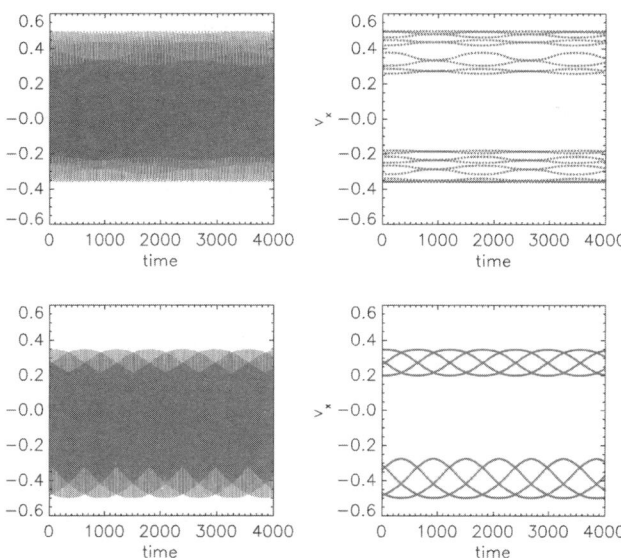

FIGURE 1. Examples of the velocity v_x signals for two typical regular orbits evolved in the Hénon–Heiles potential. The **left panels** show the recorded points with the lines that connect them. The **right panels** show the same signals but without the connecting lines between the points. It is obvious that repetitive patterns emerge.

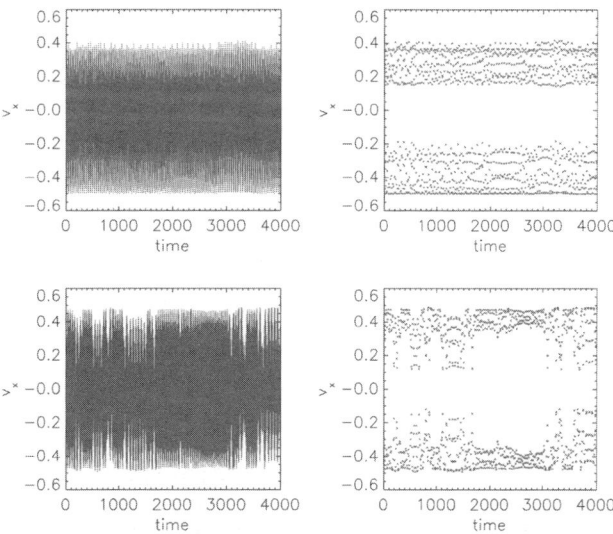

FIGURE 2. Same as in FIGURE 1 but for chaotic orbits. There are no patterns in the strict obvious sense that appear for regular orbits.

orbit provides hints about its evolutionary history, revealing visits to a number of stochastic regions that are characterized by different properties.

One may notice that a signal reveals not just spatial patterns (as a Poincaré section does), but *spatiotemporal* patterns instead. The difference is important. Not only should one search for order in a restrained region of the phase space (where only positions and velocities are involved), but one also needs to investigate if the recorded points repeat themselves in an ordered way in time. For example, if one shuffles the recorded times of the data points of a regular orbit, its Poincaré section will look exactly the same, because the Poincaré section is a plot of the position and velocity variables only and does not involve time. On the other hand, when the recorded times of the data are shuffled, the plot of the time-series points show no patterns, exactly because time is involved in this plot. Therefore, the patterns one recognizes in a signal are not just spatial/morphological but are spatiotemporal in nature. The spatial data do relate to and reveal part of the dynamics, but since signals involve both space and time one can hope that they may associate with dynamics even more precisely.

Several questions immediately arise: What is an effective algorithm to identify spatiotemporal, sequential patterns? How does analysis based on sequential patterns compare to the characterization of orbits by established measures, such as Lyapunov exponents? How can one use the loose patterns in the signals of the chaotic orbits to identify their epochs of stickiness? Are sequential patterns always adequate for characterization of the nature of an orbit?

To zeroth order the sequential patterns of signals can be defined in a rather straightforward manner. Generally, a sequential pattern exists in a signal when, starting at a point y_n of a signal, the conditions $|y_k - y_n| < \varepsilon$, $|y_{k+1} - y_{n+1}| < \varepsilon$, ..., and $|y_{2k-n-1} - y_{k-1}| < \varepsilon$ (where ε represents a small number), are all true. What this means, in practical terms, is that k successive points in the signal almost repeat themselves sequentially. Below, we address what "almost" means and how ε can be computed effectively.

A generalization of the aforementioned inequality to condition all three available signals y, v_x, and v_y, was applied to search for sequential patterns in the evolution of orbits in the Hénon–Heiles potential. Specifically, a sequential pattern of order k, exists when

$|y_k - y_{n1}| < \varepsilon$, or $|v_{xk} - v_{xn}| < \varepsilon$, or $|v_{yk} - v_{yn}| < \varepsilon$, and

$|y_{k+1} - y_{n+1}| < \varepsilon$, or $|v_{xk+1} - v_{xn+1}| < \varepsilon$, or $|v_{yk+1} - v_{yn+1}| < \varepsilon$, and

...

$|y_{2k-n-1} - y_{n+1}| < \varepsilon$, or $|v_{x2k-n-1} - v_{xn+1}| < \varepsilon$, or $|v_{y2k-n-1} - v_{yk-1}| < \varepsilon$.

(We mention that, since the time-series can be considered a map, the discussion directly applies to maps also.)

Every pattern corresponds to a time segment of the orbital data. After all the existing patterns are identified, the union of their corresponding time segments is computed. If this union spreads over the entire evolution time of the orbit, the orbit is characterized as regular, otherwise it is characterized as chaotic (see FIGURE 3).

The same scheme is capable of recognizing loose sequential patterns in chaotic orbits. These patterns are associated with the time a chaotic orbit spends in a sticky regime. Intuitively one expects there should be underlying near-symmetries that

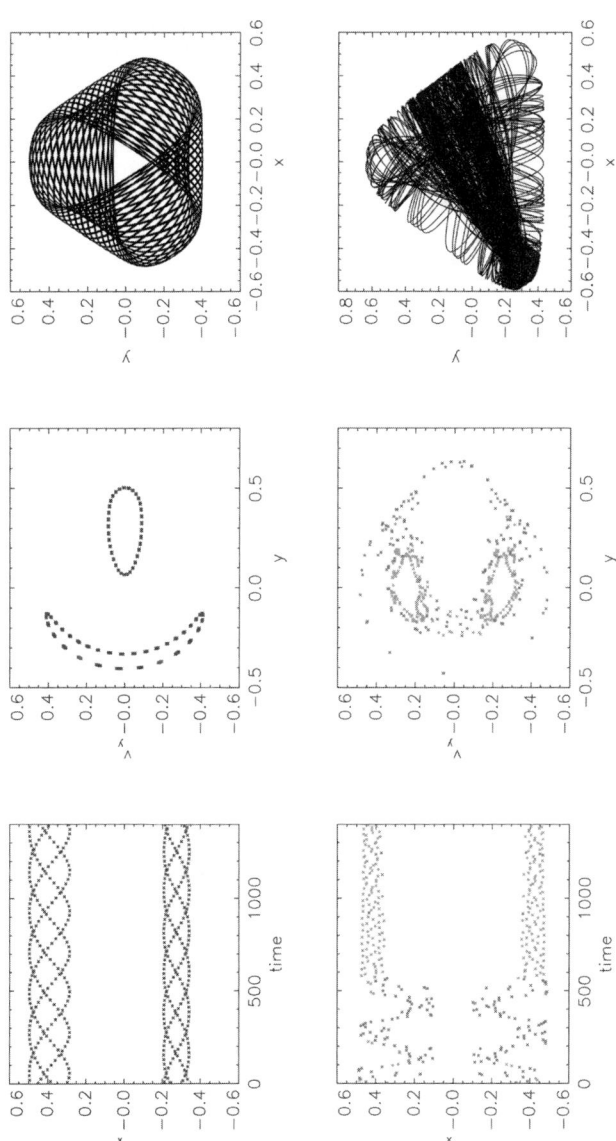

FIGURE 3. Top left panel: the v_x signal of a regular orbit evolved in the Hénon–Heiles potential. The algorithm identifies the patterns in these signals and the time segment that corresponds to each pattern. The union of these time segments spreads over the entire evolution time; therefore, this orbit is characterized as regular. **Top middle panel**: the Poincaré section of this orbit. **Top right panel**: a plot of the orbit. **Bottom right panel**: the v_x signal of a chaotic orbit evolved in the same potential. There are parts with no repetition (*dark dots*), and parts with loose repetition (*light dots*) that correspond to sticky epochs of the evolution. **Bottom middle panel**: the Poincaré section of the orbit. The *light dots* correspond to the light (sticky) dots of the signal. It is obvious that the *light dots* are localized in a sticky region. **Bottom left panel**: plot of the orbit.

characterize the orbit when it is sticky, and they are reflected in the existence of loose patterns. When only a part of a signal has loose patterns, it is identified as sticky chaotic; the rest is identified as wildly chaotic (FIG. 3).

Although an intelligent choice for the value of ε can be made without employing a rigid scheme, that could lead to confusion. A more rigid technique can be formulated. Assume that one wants to identify the sequential patterns of an orbit and determine their union for various values of ε, starting at zero. For an individual orbit the first pattern appears at some $\varepsilon = a_0$. As the value of ε increases the union becomes larger and eventually, at some $\varepsilon = a_1$ it spreads over the entire evolution time. If one plots the union of the time segments associated with patterns versus ε, a striking observation can be made. The slope of the curve for a typical regular orbit is very steep, whereas for a chaotic orbit it is usually much shallower (see FIGURE 4). If one computes the maximum values of a_1 achieved by the orbits with very steep slope, one can determine a value for ε to serve as a criterion in the algorithm for the distinction between sticky and wildly chaotic epochs of chaotic orbits.

It needs to be stressed here that an exact distinction between the sticky and the wildly chaotic regions of the phase space may not be possible. The sticky regime moves continuously into the wildly chaotic regime and gradually disappears where the stochastic sea dominates. Therefore, the aforementioned computation of the value for ε to can serve as a criterion for distinction between the two regimes, is only a suggestion, and different researchers may treat the problem differently. Nevertheless, the suggested scheme seems to give sensible results.

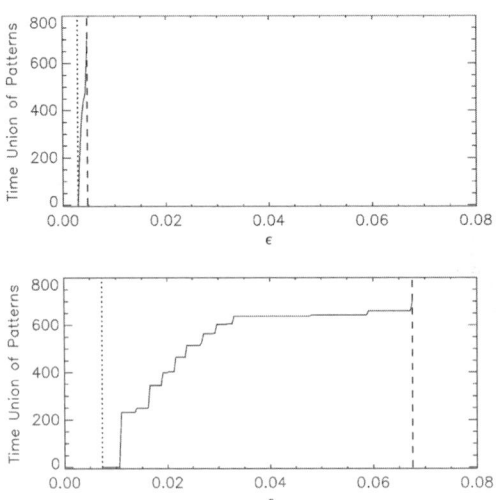

FIGURE 4. The time-union of patterns versus ε for a regular orbit (**top panel**) and a chaotic orbit (**bottom panel**) evolved in the Hénon–Heiles potential. The *dotted line* signifies the smallest $\varepsilon = a_0$ for which a pattern appears and the *dashed line* signifies $\varepsilon = a_1$ for which the time-union of patterns is equal to the entire evolution time, which in this example was 700 DE units (about 100 orbital periods).

Having established the largest value of a_1 for a sample of regular orbits, one may apply this criterion to the chaotic orbits as is illustrated in FIGURE 5. The parts of a chaotic orbit that have patterns (sticky) are on the left of this criterion line and those that do not exhibit patterns are on the right (wildly chaotic). It is to be noted that it may be possible to take advantage of the information included in FIGURE 5 and connect it to diffusion times for sticky chaotic epochs, a topic that will be part of future investigation and design.

The new method was able to easily identify the regular orbits and, for the chaotic orbits, the sticky and wildly chaotic parts (see FIGURE 6). It is obvious from the bottom left panel of FIGURE 6 that the sticky regime does not separate abruptly from the wildly chaotic regime. Instead, there is an overlapping region between the two regimes. At larger distances from a regular island, the sticky region gradually becomes less dense and wildly chaotic behavior starts to emerge. Eventually the wildly chaotic regime dominates. This is rather natural; it reveals the existence of the continuum that describes the passage from stickiness to wildly chaotic behavior.

It is also interesting that two kinds of sticky epochs appear. The first is characterized by few long patterns and they can be termed *strong* sticky chaotic regions. However, it was possible to identify a second type of sticky epochs (usually not identified by other measures) that are characterized by many and short patterns. They are

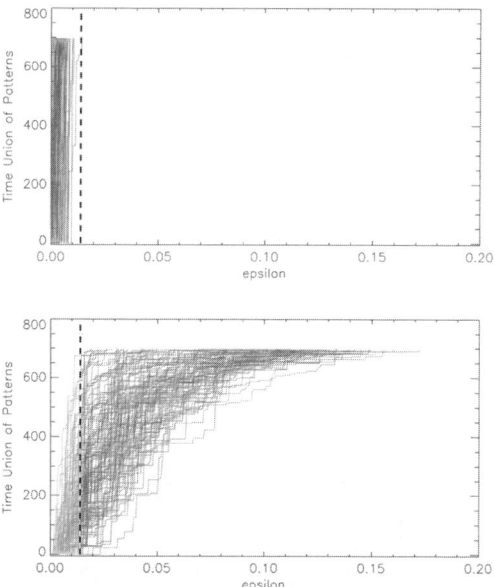

FIGURE 5. Union of the time-segments associated with patterns for several orbits evolved in the Hénon–Heiles potential. **Top panel**, regular orbits. **Bottom panel**, chaotic orbits. The *dashed line* signifies the largest value for ε achieved by any orbit with slope of time-union greater than 89 degrees. Extending this line into the chaotic orbit plot ca serve as a criterion for distinguishing between sticky (*light*) and wildly chaotic (*dark*) epochs of chaotic orbits.

located in the region of the main broken separatrices of the system and the main three hyperbolic points of the Hénon–Heiles potential. This type of sticky epochs are called *weak*. Although an arbitrary criterion that distinguishes between weak and strong sticky chaotic epochs is necessary, an intelligent choice is usually very straightforward.

How well do the characterizations from the new technique correlate with the largest Lyapunov exponents of the orbits? The answer is: amazingly well. The same initial conditions were integrated for various evolution times, from 25 differential equations (DE) units (corresponding to about four orbital periods) to 1,200DE units (corresponding to about 190 orbital periods). The number of the orbits for which the characterization agreed with their largest Lyapunov exponent was computed (see FIGURE 7). For evolution time greater than 90 orbital periods the results agree remarkably; 99% of the orbits are characterized correctly. However, the crucial point here is that between 10 and 90 orbital periods, the success is in the 90% regime. Even for only four orbital periods (at least for the Hénon–Heiles potential) this method is able to identify correctly about 75% of the orbits.

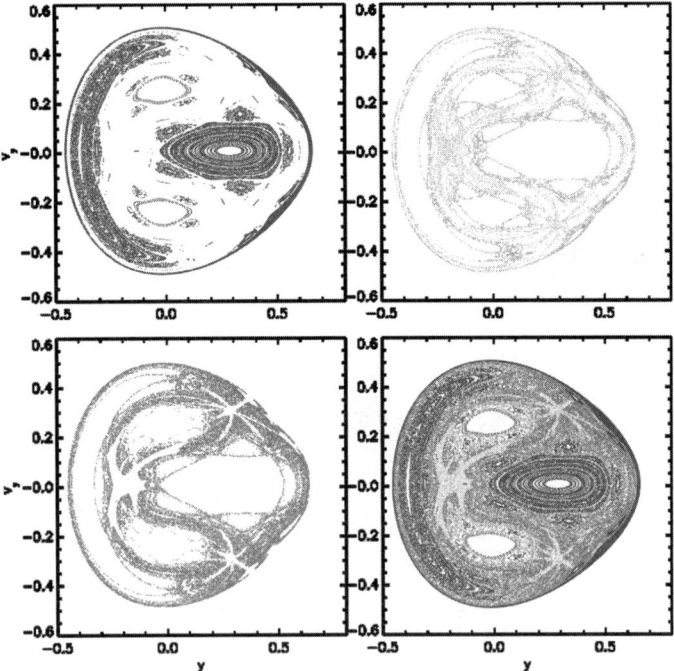

FIGURE 6. Phase space of the Hénon–Heiles potential for a sample of 500 orbits of energy 0.125, integrated for about 100 orbital periods. **Top left panel**, regular orbits. **Top right panel**, strong sticky epochs of chaotic orbits are black and weak sticky epochs are light grey. **Bottom left panel**, wildly chaotic epochs of chaotic orbits. **Bottom right panel**, all orbits.

This is a result of major importance. Although for short time evolution not all the orbits are characterized correctly (compared to their Lyapunov exponents), many of them are, and this is sufficient to provide adequate information for a crude view of the structure of the phase space (see FIGURE 8). This is beyond the capabilities of the preexisting methods. When the evolution time is longer, not only does this method agrees almost completely with the largest Lyapunov exponent, but it also provides additional information about the sticky epoch of the chaotic orbits.

A comment is in order concerning the orbits that are not characterized correctly (about 10% of the orbits) when one uses data from only 10 orbital periods. These orbits are usually the very sticky ones that, during 10 orbital periods, behave in a manner very similar to that of a regular orbit (see FIGURE 9). If one plots these orbits, their physiognomy is regular. However, dynamically they are chaotic and they escape to the stochastic sea later in their evolution. Nevertheless, when the evolution time of interest of a real system is as short as 10 orbital periods, one may be less interested in the dynamical nature of the orbits than in their physiognomy. This is something to be decided by the individual researcher in keeping with the nature of the problem.

How effective this method is can be seen if one plots the slope of the time-union curve (FIGS. 4 and 5) versus the number of orbital periods. Steep slopes mean that the orbit is regular, whereas shallow slopes mean that the orbit is chaotic. In FIGURE 10 examples for three typical regular and three typical chaotic orbits appear. It is clear that the slopes converge to a value close to 90 degrees in about 10 orbital periods. On the other hand, the chaotic orbits wander away from a steep angle and eventually converge to some value very different than 90 degrees.

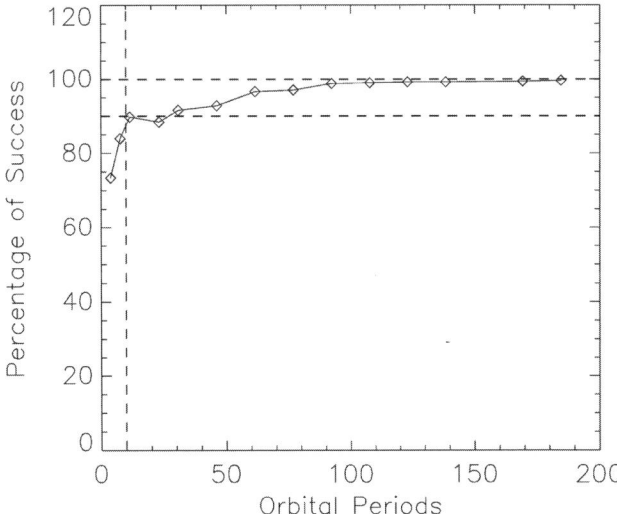

FIGURE 7. The success percentage for 500 initial conditions evolved in the Hénon–Heiles potential for a number of different evolution times, was computed by comparing how many of the orbits agree with the characterization provided by their largest Lyapunov exponents.

Motion in a time-dependent potential may be characterized by transient chaos. Because the energy is not constant it is possible for an orbit to move from regular to chaotic and *vice versa*. In a time-dependent regime the described method can recognize when the orbit has repetitious epochs (regular) and when it does not. When repetitious patterns do not emerge the epoch is usually chaotic, although it may also be regular but so short-lived that patterns did not have sufficient time to form. To this extent this method is able to distinguish automatically between the two regimes without any extra effort or analysis. In this sense, it can be valuable to investigators interested in time-dependent systems.

The toy model used as a first application to a time-dependent regime was the Plummer potential plus a time-dependent sinusoidal perturbation of strength m_0.[11]

$$V(x, y, z, t) = -\frac{1}{(1 + x^2 + y^2 + z^2)^{1/2}} - \frac{m_0 \sin(\omega t)}{(1 + x^2 + \alpha^2 y^2 + z^2)^{1/2}},$$

with $\alpha^2 = 1.1$ and $\omega = 0.5$.

The same 100 initial conditions were evolved in this potential for four different values of m_0. Then the orbital data were searched for repetitious patterns. It was

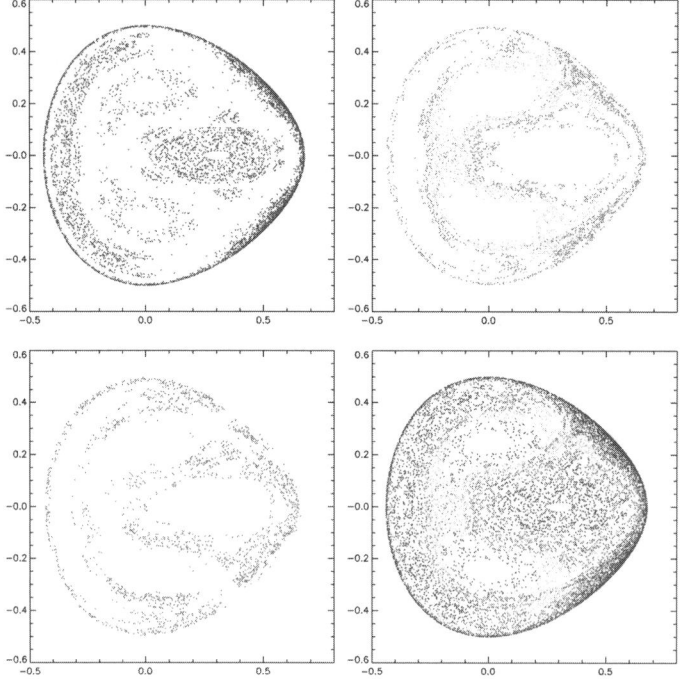

FIGURE 8. Phase space of the Hénon–Heiles potential for a sample of 500 orbits of energy 0.125, integrated for only 10 orbital periods. **Top left panel**, regular orbits. **Top right panel**, strong sticky epochs of chaotic orbits are *black* and weak sticky epochs are *light grey*. **Bottom left panel**, wildly chaotic epochs of chaotic orbits. **Bottom right panel**, all orbits.

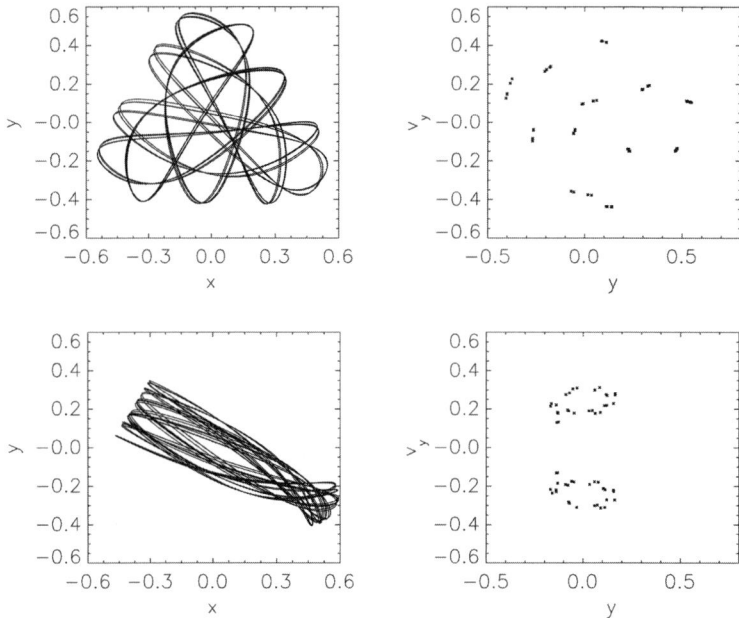

FIGURE 9. Two examples of sticky orbits and their Poincaré sections (evolved in the Hénon–Heiles potential) that are characterized as regular when the data from only 10 orbital periods are used. Later these orbits escape to the stochastic sea.

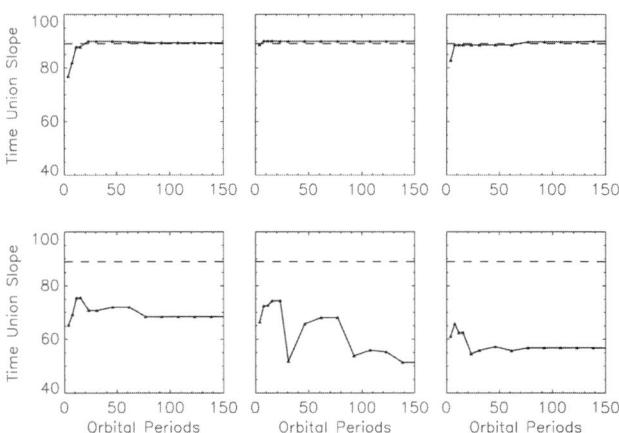

FIGURE 10. Slope of time-union of patterns versus number of orbital periods. The **top panels** correspond to typical regular orbits evolved in the Hénon–Heiles potential. All three converge to a value close to 90 degrees (the *dashed line* is located at 89 degrees). The **bottom panels** correspond to typical chaotic orbits evolved in the same potential. Their curves wander well away from 90 degrees. The distinction is clear.

found, as expected, that as the perturbation increases the number of orbits that stay regular throughout the entire evolution decreases (see FIGURE 11). Most orbits become chaotic and a small halo appears. What is striking, however, is that there is a repetitive region in the phase space that is located in the middle plane (axis *y*). When a particle finds itself in that region it will follow an ordered motion. This kind of information could not be extracted easily with previously available measures.

DISCUSSION AND CONCLUSIONS

The new method discussed in this paper has been carefully designed and engineered to provide new pieces of information that, until now, were either not accessible at all, or at best were difficult to compute. The advantages of this method are: (1) it applies to time-dependent as well as time-independent systems, (2) it applies

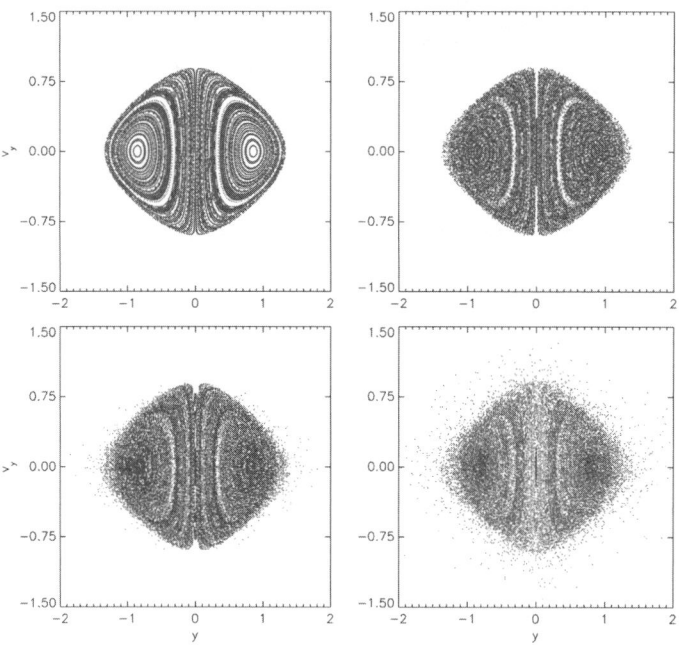

FIGURE 11. Phase space of the Plummer plus time-dependent perturbation potential for a sample of 100 orbits, integrated for about 100 orbital periods (in average). **Top left panel**, perturbation $m_0 = 0$, all orbits are regular. **Top right panel**, perturbation $m_0 = 0.10$. There is still a large number of orbits that stay regular throughout their entire evolution, but some wildly chaotic epochs start to appear (shown in *grey*). The algorithm distinguishes between the chaotic and the regular epochs of orbits that move from regular to chaotic and *vice versa*. **Bottom left panel**, perturbation $m_0 = 0.17$. More chaotic epochs appear. **Bottom right panel**, perturbation $m_0 = 0.25$. Even more chaotic epochs appear. There is a *light grey area* that surrounds the *y* axis, which shows a regular epoch of transient orbits localized in a region of the phase space.

to either maps or flows, (3) it is algorithmically straightforward and easily programmable, (4) it is neither computationally expensive nor memory intensive; it required less than two minutes to characterize 500 orbits evolved for 100 orbital periods (this is equivalent to about 200 points in the time-series) in a 3.2GHz AMD processor, (5) the characterization is computed on output data and not in real time during the evolution of an orbit, which can save considerable computer time, and (6) most *importantly it converges very rapidly* (for most of the orbits this was within about 10–15 orbital periods).

The main drawback of this method is that, since it is based on the existence of patterns in the signals associated with Poincaré sections, it is not possible to provide a valid characterization before some patterns have already been formed. There may be ways to work around this problem in future by taking advantage of other kinds of symmetries associated with the signal. It is true, however, that this problem did not appear (or at least did not seem to be very important) for the case of the Hénon–Heiles potential.

Furthermore, the sequential patterns method becomes computationally expensive for orbital data with many recorded points, such as those in celestial mechanics, or in storage rings in accelerators. However, in these contexts one may be in the position to trust the results from the analysis of the first few hundreds of points. Nonetheless, there could be cases for which careful analysis of all the points is necessary. In these cases parallelization of the algorithm is required and this appears to be feasible.

Our immediate future plans are to (1) solve the aforementioned problems, and (2) generalize this method to systems with three degrees-of-freedom as well as N-body systems. There the time-series may be more complicated because a larger number of frequencies may characterize the motion. In that case one may need to define and include a small additional number of different patterns (other than sequential). It will be of major physical interest to analyze more effectively, and thereby understand the time-dependent dynamics involved in various systems; for example, beam dynamics in accelerators. Careful investigation may reveal underlying physics that was inaccessible or tediously accessible by means of previous techniques.

ACKNOWLEDGMENTS

The author expresses his deepest gratitude to his mentor Henry E. Kandrup. Henry's ethos, values, sense of humor, originality, knowledge, and intuition gave motivation and shape, to those who had the greatest luck to have met and/or worked with him. People like Henry are the "salt and pepper" of science and are literally irreplaceable. He is missed not only as a teacher but also, even more importantly, as a close and trusted friend.

This work was supported by the U.S. Departments of Education under Grant P116Z010035, Energy under Grant DE-FG02-04ER41323, and Defense under Air Force Contract FA9451-04-C-0199.

The author declares that he has no competing financial interest.

REFERENCES

1. CONTOPOULOS G. & N. VOGLIS. 1996. Spectra of stretching numbers and helicity angles in dynamical systems. Cel. Mech. Dyn. Astron. **64:** 1.
2. SKOKOS, C. 2001. Alignment indices: a new, simple method for determining the ordered or chaotic nature of orbits. J. Phys. A **34:** 10029.
3. FROESCHLE, C., E. LEGA & R. GONCZI. 1997. Fast Lyapunov indicators. Application to asteroidal motion. Cel. Mech. Dyn. Astron. **67:** 41.
4. CONTOPOULOS G. 2002. Order and Chaos in Dynamical Astronomy. Springer-Verlag, Berlin.
5. BENETTIN, G., L. GALGANI, A. GIORGILLI & J.-M. STRELCYN. 1980. Lyapunov characteristic exponents for smooth dynamical systems and for Hamiltonian systems—a method for computing all of them. I—Theory. II—Numerical application. Meccanica **15:** 9, 21.
6. LASKAR, J. 1990. The chaotic motion of the solar system—a numerical estimate of the size of the chaotic zones. Icarus **88:** 266.
7. CONTOPOULOS, G. 1971. Orbits in highly perturbed dynamical systems. III. Nonperiodic orbits. Astron. J. **76:** 147.
8. SHIRTS R.B. & W.P. REINHARDT, 1982, Approximate constants of motion for classically chaotic vibrational dynamics: vague tori, semiclassical quantization and classical intramolecular energy flow. J. Chem. Phys. **77:** 5204.
9. KANDRUP, H.E., I.V. POGORELOV & I.V. SIDERIS. 2000. Chaotic mixing in noisy Hamiltonian systems. Mon. Not. Roy. Astr. Soc. **311:** 719.
10. ATHANASSOULA, E., O. BIENYAM, L. MARTINET & D. PFENNIGER. 1983. Orbits as building blocks of a barred galaxy model. Astron. Astrophys. **127:** 349.
11. KANDRUP, H.E., I.M. VASS & I.V. SIDERIS. 2003. Transient chaos and resonant phase mixing in violent relaxation. Mon. Not. Roy. Astr. Soc. **341:** 927.
12. HÉNON, M. & C. HEILES. 1964. The applicability of the third integral of motion: some numerical experiments. Astron. J. **69:** 73.

Hard Sphere Dynamics for Normal and Granular Fluids

JAMES W. DUFTY AND APARNA BASKARAN

Department of Physics, University of Florida, Gainesville, Florida, USA

ABSTRACT: A fluid of N smooth, hard spheres is considered as a model for normal (elastic collision) and granular (inelastic collision) fluids. The potential energy is discontinuous for hard spheres so that the pairwise forces are singular and the usual forms of Newtonian and Hamiltonian mechanics do not apply. Nevertheless, particle trajectories in the N particle phase space are well defined and the generators for these trajectories can be identified. The first part of this presentation is a review of the generators for the dynamics of observables and probability densities. The new results presented in the second part refer to applications of these generators to the Liouville dynamics for granular fluids. A set of eigenvalues and eigenfunctions of the generator for this Liouville dynamics system is identified in a special *stationary representation*. This provides a class of exact solutions to the Liouville equation that are closely related to hydrodynamics for granular fluids.

KEYWORDS: hard sphere dynamics; granular fluid; Liouville equation

INTRODUCTION

The properties of simple atomic fluids are typically well-described by classical mechanics of point particles with phenomenological pairwise additive potentials. These potentials are chosen to capture the quantitative features of the actual quantum mechanical description of electronic interactions between pairs of nuclei: strongly repulsive effects at short distances and weak attraction at large distances. Except at low temperatures the short range repulsion dominates and the classical potentials behave qualitatively as $(\sigma/r)^{-n}$ where r is the distance between a pair, σ is a characteristic force range, and n is a large integer (e.g., $n = 12$ for the Lennard–Jones potential). The primary effect of the phenomenological potentials is, therefore, one of excluded volume, dominating all thermodynamic, structural, and transport properties. This suggests a final idealization to consider, the hard sphere limit $n \to \infty$, where the potential is unbounded for $r < \sigma$ and zero for $r > \sigma$. The single parameter of the hard sphere fluid, σ, can then be chosen to give a reasonable quantitative description of real fluids.[1]

The equilibrium statistical mechanics of hard spheres reduces to a determination of configurations for non-overlapping spheres, a geometry problem. All thermodynamic and structural properties of the hard sphere fluid can be understood simply as a limiting form for strongly repulsive potentials. In contrast, the dynamical properties

---

Address for correspondence: James W. Dufty, Department of Physics, University of Florida, Gainesville, FL 32611, USA. Voice: 352-392-6693; fax:352-846-0295.

dufty@phys.ufl.edu

of hard spheres provide new problems due to the singular forces and vanishing time scale for pair collisions. Consequently, the description of trajectories in the N particle phase space cannot be given directly by the Newton or Hamilton equations.[2–7] Nevertheless, it is clear that such trajectories exist and are uniquely defined for the usual initial data. All particles move freely until a given pair is at contact, then the momenta of that pair is changed instantaneously according to some chosen collision rule (different for elastic and inelastic collisions, as described in the next section). Subsequently, the particles continue to move freely until the next pair at contact occurs, the collision rule is applied instantaneously to that pair, and free streaming is again resumed. The resulting trajectories are unique, deterministic, and reversible up to the initial data. Some of the complexities of the crossover from strongly repulsive continuous potentials to the hard sphere limit have recently been described.[8,9]

In the next section, a representation of the generators for hard sphere dynamics is recalled.[2–7] Such hard sphere generators have been applied with great success during the past thirty years to explore many of the subtleties of transport phenomena in normal fluids (e.g., nonanalytic density dependencies and algebraic time decay).[5,7] More recently, interest in hard sphere fluids has been revived as models for granular flows.[10,11] Real granular media consist of mesoscopic sized grains (e.g., sand, seeds, and pharmaceutical pills) which, when activated, behave like complex fluids, that is, fluids whose constituent particles have non-trivial internal degrees of freedom. It turns out that many of the qualitative and quantitative differences between normal and granular fluids are captured by a hard sphere fluid model whose constituents undergo *inelastic* or dissipative collisions. For an overview of recent developments, see the edited volumes of References 12 and 13, and the new text by Poschel.[14]

Following the definitions of generators, attention is focused on their application to the statistical mechanics of granular media. The Liouville equation is described and shown to have no stationary solutions (equilibrium) for an isolated system. Nevertheless, a change of variables to accommodate collisional *cooling* due to the inelastic collisions leads to a representation supporting a stationary solution. Subsequently, it is shown that the existence of the stationary solution implies certain solutions to the eigenvalue problem for the Liouville operator in this representation. In this way, a class of special solutions to the Liouville equation is identified. It is noted that the eigenvalues are the same as those for the exact macroscopic balance equations for mass, energy, and momentum in the long wavelength limit. This suggests that the microscopic excitations are the precursors of macroscopic hydrodynamics in a granular fluid.

GENERATORS FOR HARD SPHERE DYNAMICS

The system of interest is a one-component fluid of N identical smooth hard disks or spheres (with mass m and diameter σ). The position and velocity coordinates of the fluid particles is denoted by $\{\mathbf{q}_i, \mathbf{v}_i\}$. The state of the system at time t is completely characterized by the positions and velocities of all particles at that time and is represented by a point $\Gamma_t \equiv \{\mathbf{q}_1(t), ..., \mathbf{q}_N(t), \mathbf{v}_1(t), ..., \mathbf{v}_N(t)\}$ in the associated $2dN$ dimensional phase space, where $d = 2$ for hard disks and $d = 3$ for hard spheres. The dynamics consists of free streaming (straight line motion along the direction of the velocity at time t), until any pair of particles, say i and j, is in contact. At the contact time the relative velocity $\mathbf{g}_{ij} = \mathbf{v}_i - \mathbf{v}_j$ of that pair changes instantaneously according to the collision rule

$$\tilde{\mathbf{g}}_{ij} = \mathbf{g}_{ij} - (1 + \alpha)(\hat{\sigma} \cdot \mathbf{g}_{ij})\hat{\sigma}, \qquad (1)$$

where $\hat{\sigma}$ is a unit vector directed from the center of particle j to the center of particle i through the point of contact. The parameter α (the coefficient of normal restitution) is chosen *a priori* in the range $0 < \alpha \leq 1$ and remains fixed for a given system. The value $\alpha = 1$ corresponds to elastic, energy conserving collisions, whereas $\alpha < 1$ describes an inelastic collision with a corresponding energy loss for the pair. However, the center of mass velocity is unchanged, so that the total mass and momentum of the pairs are conserved for all values of α. Subsequent to the change in relative velocity for the pair i,j the free streaming of all particles continues until another pair is in contact, and the corresponding instantaneous change in their relative velocities is performed. The sequence of free streaming and binary collisions uniquely determines the positions and velocities of the hard particles at time t for given initial conditions.

Statistical Mechanics

The statistical mechanics for a fluid of inelastic hard spheres is described elsewhere.[10,11] It comprises the dynamics just described, a state specified in terms of a probability density $\rho(\Gamma)$, and a set of observables denoted by $A(\Gamma)$. The expectation for an observable at time $t > 0$ for a state $\rho(\Gamma)$ given at $t = 0$ is defined by

$$\langle A(t); 0 \rangle \equiv \int \rho(\Gamma) A(\Gamma_t) d\Gamma, \qquad (2)$$

where $A(\Gamma, t) = A(\Gamma_t)$ and $\Gamma = \Gamma_{t=0}$. The dynamics can be represented in terms of the generator L defined by

$$\langle A(t); 0 \rangle = \int \rho(\Gamma) e^{tL} A(\Gamma) d\Gamma. \qquad (3)$$

For a continuous potential, the generator is easily recognized from the Hamilton equation as a Poisson bracket operation with the corresponding Hamiltonian. However, its identification for the discontinuous hard sphere potential is less direct.

There are two components to the generator, corresponding to the two steps, free streaming and velocity changes at contact. The first part is the same as for continuous potentials; the second part replaces the contribution from the singular force by a *binary collision operator* $T(i,j)$ for each pair i,j

$$L = \sum_{i=1}^{N} \mathbf{v}_i \cdot \nabla_i + \frac{1}{2} \sum_{i=1}^{N} \sum_{j \neq i}^{N} T(i, j). \qquad (4)$$

The binary collision operator $T(i,j)$ for continuous potentials is identified directly from the Poisson bracket of the Hamilton equations

$$T(i, j) \to \theta_{ij} = -m^{-1}((\nabla_{q_i} V(|\mathbf{q}_i - \mathbf{q}_j|)) \cdot (\nabla_{v_i} - \nabla_{v_j})). \qquad (5)$$

For hard spheres, the position variables remain continuous functions of time, but the momenta are piecewise constant (in the absence of external forces) and discontinuous. Thus, a general phase function has the form

$$A(\Gamma_t) = \Theta(t_1 - t)A(\{\mathbf{q}_1(t), ..., \mathbf{q}_N(t), \mathbf{v}_1(0), ..., \mathbf{v}_N(0)\}) \\ + \sum \Theta(t - t_p)\Theta(t_{p+1} - t)A(\{\mathbf{q}_1(t), ..., \mathbf{q}_N(t), \mathbf{v}_1(p), ..., \mathbf{v}_N(p)\}), \quad (6)$$

where $\{t_p\}$ are the times for the colliding pairs and $\{\mathbf{v}_i(p)\}$ are the velocities in the time interval between collisions p and $p + 1$. The Heaviside theta functions identify the time intervals between collisions. Direct differentiation of this form leads to an identification of the binary collision operator for hard spheres

$$T(i, j) = \Theta(-\mathbf{g}_{ij} \cdot \hat{\mathbf{q}}_{ij})|\mathbf{g}_{ij} \cdot \hat{\mathbf{q}}_{ij}|\delta(q_{ij} - \sigma)(b_{ij} - 1), \quad (7)$$

where $\mathbf{q}_{ij}$ is the relative position vector of the two particles, Θ is the Heaviside step function, and b_{ij} is a substitution operator,

$$b_{ij}A(\mathbf{g}_{ij}) = A(b_{ij}\mathbf{g}_{ij}) = A(\tilde{\mathbf{g}}_{ij}), \quad (8)$$

that changes the relative velocity $\mathbf{g}_{ij}$ into its scattered value $\tilde{\mathbf{g}}_{ij}$, given by Equation (1). The theta function and delta function in (7) assure that a collision takes place, that is, the pair is in contact and directed toward each other. Additional details of this approach can be found in Appendix A of Reference 15, by Lutsko. Alternatively, the same result can be obtained through a limiting procedure starting with a continuous potential whose slope increases without bound.

An alternative equivalent representation of the dynamics is to transfer it from the observable $A(\Gamma)$ to the state $\rho(\Gamma)$, by means of the definition

$$\int \rho(\Gamma)e^{tL}A(\Gamma)d\Gamma \equiv \int (e^{-t\bar{L}}\rho(\Gamma))A(\Gamma)d\Gamma. \quad (9)$$

The representation in terms of a dynamical state is referred to as Liouville dynamics. Implicit in the analysis of the previous paragraph for hard spheres is the restriction of the phase space to nonoverlapping configurations. This is assured when the generator L is used in the context of averages, such as (3), since all acceptable probability densities $\rho(\Gamma)$ must include a singular factor excluding the domain of any overlapping pair. However, the right side of (9) no longer has that restriction and, consequently, the generator for Liouville dynamics is not the same as that for observables (as in the case of continuous potentials). Instead, direct analysis of (9) leads to the result

$$\bar{L} = \sum_{i=1}^{N} \mathbf{v}_i \cdot \nabla_i - \frac{1}{2}\sum_{i=1}^{N}\sum_{j \neq i}^{N} \bar{T}(i, j), \quad (10)$$

with the new binary collision operator

$$\bar{T}_-(i, j) = \delta(q_{ij} - \sigma)|\mathbf{g}_{ij} \cdot \hat{\mathbf{q}}_{ij}|(\Theta(\mathbf{g}_{ij} \cdot \hat{\mathbf{q}}_{ij})\alpha^{-2}b_{ij}^{-1} - \Theta(-\mathbf{g}_{ij} \cdot \hat{\mathbf{q}}_{ij})), \quad (11)$$

where b_{ij}^{-1} is the inverse of the operator b_{ij} in (8)

$$b_{ij}^{-1}\mathbf{g}_{ij} = \mathbf{g}_{ij} - \frac{1+\alpha}{\alpha}(\hat{\boldsymbol{\sigma}} \cdot \mathbf{g}_{ij})\hat{\boldsymbol{\sigma}}. \quad (12)$$

In summary, the problems presented by the singular forces for a fluid of hard spheres are resolved if the Hamilton equations for observables are replaced by

$$(\partial_t - L)A(\Gamma, t) = 0, \quad (13)$$

and the Liouville equation for probability densities is replaced by

$$(\partial_t + \bar{L})\rho(\Gamma, t) = 0, \quad (14)$$

with the respective generators given by **(4)** and **(10)**. An important observation emphasized by Lutsko[15] is that the form of the generator L and corresponding binary collision operator $T(i,j)$ does not depend on the details of the collision rule defining the operator b_{ij}. In particular, the result applies to both elastic and inelastic collisions. In contrast, the generator for Liouville dynamics is obtained by a change of variables that introduces the Jacobian of the transformation between the variables $\mathbf{g}_{ij}$ and $b_{ij}\mathbf{g}_{ij}$. Hence, it depends explicitly on the collision rule or, for granular media, on the restitution coefficient α.

LIOUVILLE EQUATION FOR GRANULAR FLUIDS

In the remainder of this presentation attention is focused on the case of granular fluids, $\alpha < 1$, for an isolated system. In contrast to normal fluids, there is no stationary solution to the Liouville equation for an isolated system. This follows by calculating from it the average *thermal* speed of the particles $v(t)$; defined in terms of the average kinetic energy

$$v^2(t) \equiv \frac{4}{3mN}\left\langle \sum_{i=1}^{N} \frac{1}{2}mv_i^2(t); 0 \right\rangle, \qquad (15)$$

where m is the mass and the second term of the equality defines the average. Using either representation **(13)** or **(14)**, the average speed is found to be monotonically decreasing

$$\partial_t \ln v(t) = -\frac{1}{2}\zeta(t), \qquad (16)$$

where $\zeta(t) > 0$ is the *cooling* rate due to inelastic collisions,

$$\zeta(t) = (1-\alpha^2)\frac{N}{6v^2(t)}\int (\mathbf{g}_{12}\cdot\hat{\mathbf{q}}_{12})^3 \Theta(\mathbf{g}_{12}\cdot\hat{\mathbf{q}}_{12})\delta(q_{12}-\sigma)\rho(\Gamma,t)d\Gamma. \qquad (17)$$

This shows that there is no *approach to equilibrium* for a granular fluid, since there is no such stationary equilibrium state. Nevertheless, it is expected (on both theoretical grounds and from computer simulation results) that there is a universal state that is approached for a wide range of initial preparations.

This universal state, the homogeneous cooling solution (HCS), is special in the sense that it is spatially homogeneous (translationally invariant) and all of its time dependence occurs through the average speed $v(t)$. If there are no other externally imposed energy scales, dimensional analysis requires that the dependence on $v(t)$ occurs only through normalization and by scaling the velocities,

$$\rho_{\text{hcs}} = (\ell v_{\text{hcs}}(t))^{-3N}\rho^*_{\text{hcs}}\left(\left\{\frac{\mathbf{q}_{ij}}{\ell},\frac{\mathbf{v}_i-\mathbf{u}}{v_{\text{hcs}}(t)}\right\}\right), \qquad (18)$$

where $v_{\text{hcs}}(t)$ is the thermal speed evaluated for the HCS, $\mathbf{u}$ is the constant average velocity of the system, and ℓ is a constant characteristic length (e.g., the mean free path). In this state, collisional cooling due to inelastic collisions amounts only to a monotonic decrease in all velocities. Use of **(18)** in **(17)** shows that the cooling rate in the HCS has the time dependence

$$\zeta_{hcs}(t) = \frac{v_{hcs}(t)}{\ell}\zeta^*_{hcs}, \tag{19}$$

where ζ^*_{hcs} is a dimensionless constant. Then, (16) can be solved directly to get the explicit time dependence of $v_{hcs}(t)$

$$v_{hcs}(t) = v_{hcs}(0)\left(1 + \frac{v_{hcs}(0)}{2\ell}\zeta^*_{hcs}t\right)^{-1} \to \frac{2\ell}{\zeta^*_{hcs}t}. \tag{20}$$

Also given is the limiting behavior at long times, showing that the scaling becomes independent of the initial conditions.

Stationary Representation

More generally, collisional cooling affects both the form of the distribution function as well as scaling the velocities. It is useful to account for the latter by making a time dependent change of variables to obtain a dimensionless form of the Liouville equation

$$\mathbf{q}^*_i = \frac{\mathbf{q}_i}{\ell}, \qquad \mathbf{V}^*_i = \frac{\mathbf{v}_i - \mathbf{u}}{\omega(t)}, \qquad ds = \frac{\omega(t)}{\ell}dt, \tag{21}$$

$$\rho^*(\{\mathbf{q}^*_i, \mathbf{V}^*_i\}, s) = (\ell\omega(t))^{3N}\rho(\Gamma, t), \tag{22}$$

where $\omega(t)$ is a characteristic velocity whose form remains to be chosen, see below. In terms of these variables the dimensionless Liouville equation becomes

$$\partial_s \rho^* - \ell\frac{\dot\omega(t)}{\omega^2(t)}\sum_{i=1}^N \nabla_{\mathbf{V}^*_i} \cdot (\mathbf{V}^*_i \rho^*) + \bar{L}^* \rho^* = 0, \tag{23}$$

with the time independent dimensionless generator

$$\bar{L}^* = \frac{\ell}{\omega(t)}\bar{L}_- = [\bar{L}_-]_{\{\mathbf{v}_i = \mathbf{V}^*_i\}}. \tag{24}$$

Next, $\omega(t)$ is chosen to make the coefficients of the Liouville equation independent of s

$$-\ell\frac{\dot\omega(t)}{\omega^2(t)} = \omega^* \equiv 1, \tag{25}$$

where ω^* is an arbitrary dimensionless constant, taken here to be unity. The solution to this equation is

$$\omega(t) = \omega(0)\left(1 + \frac{\omega(0)}{\ell}t\right)^{-1}, \tag{26}$$

and, consequently, the new time scale is

$$s = \ln\left(1 + \frac{\omega(0)}{\ell}t\right). \tag{27}$$

The final form for the dimensionless Liouville equation is then

$$\partial_s \rho^* + \bar{L}^* \rho^* = 0, \tag{28}$$

with the modified generator

$$\overline{\mathcal{L}}^* \rho^* = \overline{L}^* \rho^* + \sum_{i=1}^{N} \nabla_{\mathbf{v}_i^*} \cdot (\mathbf{V}_i^* \rho^*). \tag{29}$$

This time dependent change of variables was first suggested by Lutsko,[16] motivated by the objective of molecular dynamics simulation of the HCS. Direct simulation in the original variables is difficult at long times since the velocities monotonically scale toward zero. In contrast, the corresponding solution to (28) is now a stationary solution ρ_0^* given by

$$\overline{\mathcal{L}}^* \rho_0^* = 0. \tag{30}$$

Consequently, the corresponding thermal speed for the new variables approaches a constant

$$\begin{aligned} v^{*2}(t) &= \frac{v^2(t)}{\omega^2(t)} \\ &= \frac{2}{3N} \int \rho^*(\Gamma^*, s) \left(\sum_{i=1}^{N} \mathbf{v}_i^{*2} \right) d\Gamma^* \\ &\to \frac{2}{3N} \int \rho_0^*(\Gamma^*) \left(\sum_{i=1}^{N} \mathbf{v}_i^{*2} \right) d\Gamma^* \equiv v^{*2}(\infty). \end{aligned} \tag{31}$$

Since both $v_{\text{hcs}}(t)$ and $\omega(t)$ decay as t^{-1} at long times (see (20) and (26)), it follows that the dimensionless cooling rate ζ_{hcs}^* is simply related to $v^*(\infty)$

$$\zeta_{\text{hcs}}^* = \frac{2}{v^*(\infty)}. \tag{32}$$

Since (28) supports a stationary state it is referred to here as the stationary representation of the Liouville equation.

Special Solutions to the Liouville Equation

The existence of the stationary state ρ_0^* of the generator $\overline{\mathcal{L}}^*$ given by (30) has some interesting consequences. The stationary solution is assumed known. In the original variables it satisfies the equations

$$\rho_0(\Gamma, t) = (\ell \omega(t))^{-3N} \rho_s^* \left(\left\{ \mathbf{q}_i^*, \frac{\mathbf{v}_i - \mathbf{u}}{\omega(t)} \right\} \right) \tag{33}$$

$$\overline{L} \rho_0 + \frac{\omega(t)}{\ell} \sum_{i=1}^{N} \nabla_{\mathbf{v}_i} \cdot ((\mathbf{v}_i - \mathbf{u}_i) \rho_0) = 0. \tag{34}$$

Differentiating (34) with respect to $\omega(t)$ at constant $\{\mathbf{q}_i, \mathbf{v}_i\}$ gives

$$\overline{L} \partial_\omega \rho_0 + \frac{\omega(t)}{\ell} \sum_{i=1}^{N} \nabla_{\mathbf{v}_i} \cdot ((\mathbf{v}_i - \mathbf{u}_i) \partial_\omega \rho_0) = -\ell^{-1} \sum_{i=1}^{N} \nabla_{\mathbf{v}_i} \cdot ((\mathbf{v}_i - \mathbf{u}_i) \rho_0). \tag{35}$$

However, it also follows from (33) that

$$\partial_\omega \rho_0 = -\omega^{-1} \sum_{i=1}^{N} \nabla_{\mathbf{v}_i} \cdot ((\mathbf{v}_i - \mathbf{u}_i)\rho_0). \tag{36}$$

Substitution of (36) into (35) and transformation back to dimensionless variables gives the result

$$\overline{\mathcal{L}}^* \Psi^{(1)} = \Psi^{(1)}, \qquad \Psi^{(1)} = \sum_{i=1}^{N} \nabla_{\mathbf{v}_i^*} \cdot (\mathbf{V}_i^* \rho_0^*). \tag{37}$$

Thus, the stationary state implies directly one solution to the eigenvalue problem for $\overline{\mathcal{L}}^*$.

Repeating this analysis, but differentiating with respect to the total number of particles N and the flow velocity $\mathbf{u}$ at constant ω gives the additional eigenvalues and eigenvectors

$$\overline{\mathcal{L}}^* \Psi^{(2)} = 0, \qquad \Psi^{(2)} = \partial_N \rho_0^* \tag{38}$$

$$\overline{\mathcal{L}}^* \Psi^{(3)} = -\Psi^{(3)}, \qquad \Psi^{(3)} = \sum_{i=1}^{N} \nabla_{\mathbf{v}_i^*} \rho_0^*. \tag{39}$$

The eigenvalue -1 is threefold degenerate, since the equation holds for each component of the vector $\Psi^{(3)}$.

These eigenvalues and eigenvectors allow construction of a special class of exact solutions to the Liouville equation (28)

$$\begin{aligned}\rho^*(\Gamma^*, s) &= e^{-\overline{\mathcal{L}}^* s}(\rho_0^*(\Gamma^*) + c_1 \Psi^{(1)}(\Gamma^*) + c_2 \Psi^{(2)}(\Gamma^*) + c_3 \Psi^{(3)}(\Gamma^*)) \\ &= \rho_0^*(\Gamma^*) + c_1 \Psi^{(1)}(\Gamma^*) e^{-s} + c_2 \Psi^{(2)}(\Gamma^*) + c_3 \Psi^{(3)}(\Gamma^*) e^s,\end{aligned} \tag{40}$$

where $\{c_i\}$ are arbitrary constants. Note that normalization is preserved since the integral over phase space of all eigenfunctions vanishes. In terms of the original variables, this class of solutions can also be written

$$\rho(\Gamma, s) = \rho_0(\Gamma) + \sum_{n=1}^{5} e^{-\lambda_n s} c_n \partial_{y_n} \rho_0(\Gamma) \tag{41}$$

with

$$\lambda_n \leftrightarrow (1, 0, -1, -1, -1), \qquad y_n \leftrightarrow (\omega, N, \mathbf{u}). \tag{42}$$

Relationship to Hydrodynamics

To provide an interpretation to the dynamical effects represented by these eigenfunctions and eigenvalues it is useful to consider the exact macroscopic balance equations for density n, momentum density $mn\mathbf{u}$, and energy density $(3/2)nk_B T$. These follow directly by using either (13) or (10) to calculate the averages of the corresponding local microscopic densities, with the results

$$D_t n + n\nabla \cdot \mathbf{u} = 0, \tag{43}$$

$$D_t u_i + (mn)^{-1}\partial_i p + (mn)^{-1}\partial_j P_{ij} = 0, \tag{44}$$

$$(D_t + \zeta)T + \frac{2}{3n}p\nabla \cdot \mathbf{u} + \frac{2}{3n}(P_{ij}\partial_j u_i + \nabla \cdot \mathbf{q}) = 0, \tag{45}$$

where $D_t = \partial_t + \mathbf{u} \cdot \nabla$ is the material derivative, $T(\mathbf{r}, t)$ is the temperature, $\mathbf{u}(\mathbf{r}, t)$ is the flow velocity, $\mathbf{q}$ is the heat flux, and P_{ij} is the irreversible part of the momentum flux. Linearize these equations about the HCS solution (n_{hcs} constant, $T_{\text{hcs}}(t) = mv_{\text{hcs}}^2(t)/2$, and introduce the dimensionless variables

$$\delta n^* = \frac{\delta n}{n_{\text{hcs}}}, \qquad \delta T^* = \frac{\delta T}{T_{\text{hcs}}(t)}, \qquad \delta \mathbf{u}^* = \frac{\delta \mathbf{u}}{v_{\text{hcs}}(t)}. \tag{46}$$

Thus, in the long wavelength limit, neglecting all spatial gradients, these equations become

$$(\partial_s + 0)\delta n^* \approx 0, \tag{47}$$

$$(\partial_s + 1)\left(\delta T^* + \frac{\partial \ln \zeta_{\text{hcs}}}{\partial \ln n_{\text{hcs}}} \delta n^*\right) \approx 0, \tag{48}$$

$$(\partial_s - 1)\delta U_i^* \approx 0. \tag{49}$$

Therefore, it is seen that excitations obtained for the Liouville equation are the same dynamics as the long wavelength limit for the macroscopic balance equations.

More generally, it is expected that spatially inhomogeneous states for a granular gas can be described by hydrodynamics. Such hydrodynamic equations follow from the above macroscopic balance equations with constitutive relations expressing the heat flux and momentum flux in terms of the spatial gradients of n, $\mathbf{u}$, and T. The resulting linearized equations define the hydrodynamic modes for a granular system, which necessarily behave as (47) and (48) at long wavelengths. This shows that the eigenfunctions and eigenvalues of the Liouville equation found here can be interpreted as the long wavelength microscopic precursors of macroscopic hydrodynamics. This point is illustrated in more detail elsewhere.[17]

CONCLUSION

The objectives of this brief presentation have been twofold. The first was an introduction to the means to describe the dynamics of a hard sphere fluid, despite the singular forces and failure of standard Newtonian and Hamiltonian formalisms. In fact, this is a special case of a class of problems involving piecewise continuous or discontinuous potentials (e.g., the finite step potential or hard spheres with attractive square well).[4] In each case the effect of the singular force can be replaced by an associated binary collision operator to define a generator for the dynamics.

The second objective was to report some new results for the Liouville dynamics of inelastic hard spheres, an idealized model for granular fluids. The inelasticity implies a monotonic loss of energy for an isolated system. However, by a change of variables to accommodate the average decrease in the speed of each particle, the Liouville equation is given a representation that supports a stationary state. The existence of this stationary state implies certain properties of the associated generator for the dynamics. For systems with elastic collisions this implies certain invariants. For inelastic collisions the eigenfunctions identified here are no longer invariants but instead have a simple dynamics directly related to the collision cooling rate. This can be seen by choosing as the scaling function $\omega(t) \to v_{hcs}(t)$. In this case the eigenvalues (42) become

$$\lambda_n \leftrightarrow \left(\frac{1}{2}\zeta^*_{hcs}, 0, -\frac{1}{2}\zeta^*_{hcs}, -\frac{1}{2}\zeta^*_{hcs}, -\frac{1}{2}\zeta^*_{hcs}\right).$$

It was noted that the eigenvalues of the Liouville equation are the same as those of the macroscopic balance equations for average number density, energy density, and flow velocity in the long wavelength limit. Hence they are long wavelength hydrodynamic modes, whenever a closed set of hydrodynamic equations apply. This has been exploited recently to define hydrodynamic response functions for a granular fluid and to derive corresponding exact Green–Kubo expressions for the transport coefficients.[17] This identification of eigenvalues and eigenvectors of the Liouville operator is a generalization of a similar analysis at the level of the Boltzmann kinetic theory.[18,19] The connection to hydrodynamics has been established clearly in that context.

ACKNOWLEDGMENTS

This research was supported in part by Department of Energy Grant DE-FG02ER 54677.

REFERENCES

1. MCQUARRIE, D. 1973. Statistical Mechanics. HarperCollins, New York.
2. ERNST, M., J. DORFMAN, W. HOEGY & J. VAN LEEUWEN. 1969. Physica **45**: 127.
3. SENGERS, J., M. ERNST & D. GILLESPIE. 1972. J. Chem. Phys. **56**: 5583.
4. DOMARADZKI, J. 1977. Physica **86A**: 169.
5. RESIBOIS, P. & M. DE LEENER. 1977. Classical Kinetic Theory of Fluids. John Wiley, New York.
6. VAN BEIJEREN, H. & M. ERNST. 1979. J. Stat. Phys. **21**: 125.
7. MCLENNAN, J.A. 1989. Introduction to Nonequilibrium Statistical Mechanics. Prentice-Hall, New Jersey.
8. DUFTY, J. 2002. Molec. Phys. **100**: 2331.
9. DUFTY, J. & M. ERNST. 2005. Molec. Phys. **102**: 2123.
10. BREY, J.J., J.W. DUFTY & A. SANTOS. 1997. J. Stat. Phys. **87**: 1051.
11. VAN NOIJE, T.P.C. & M.H. ERNST. 2001. *In* Granular Gases. T. Poschel & S. Luding, Eds. Springer, New York.
12. POSCHEL, T. & S. LUDING, Eds. 2001. Granular Gases. Springer, New York
13. POSCHEL, T. & N. BRILLIANTOV, Eds. 2003. Granular Gases Dynamics. Springer, New York.
14. BRILLIANTOV, N. & T. POSCHEL. 2004. Kinetic Theory of Granular Gases. Oxford, New York.
15. Lutsko, J. 2004. J. Chem. Phys. **120**: 6325.
16. Lutsko, J. 2001. Phys. Rev. E. **63**: 061211.
17. BASKARAN, A., J. DUFTY & J. BREY. 2005. Linear response and hydrodynamics for a granular fluid. Unpublished.
18. DUFTY, J. & J.J. BREY. 2003. Phys. Rev. E **68**: 030302
19. DUFTY, J., J.J. BREY & M.J. RUIZ-MONTERO. 2003. *In* Granular Gases Dynamics. T. Poschel & N. Brilliantov, Eds. Springer, New York.

Nonlinear Stability of Newtonian Galaxies and Stars from a Mathematical Perspective

GERHARD REIN

Department of Mathematics, University of Bayreuth, Bayreuth, Germany

ABSTRACT: The stability of equilibrium configurations of galaxies or stars are time honored problems in astrophysics. We present mathematical results on these problems that have recently been obtained by Yan Guo and the author in the context of the Vlasov–Poisson and the Euler–Poisson models. Based on a careful analysis of the minimization properties of conserved quantities—the total energy and so called Casimir functionals—nonlinear stability results are obtained for a wide class of equilibria.

KEYWORDS: nonlinear stability; Vlasov–Poisson system; Euler–Poisson system; energy-Casimir method

INTRODUCTION

Under suitable idealization assumptions, the time evolution of a galaxy can be modeled by the Vlasov–Poisson system

$$\partial_t f + v \cdot \nabla_x f - \nabla U \cdot \nabla_v f = 0,$$

$$\Delta U = 4\pi \rho, \quad \lim_{|x| \to \infty} U(t, x) = 0,$$

$$\rho(t, x) = \int f(t, x, v) dv,$$

were $f = f(t, x, v) \geq 0$ denotes the density of the stars in phase space; $t \in \mathbb{R}$ and $x, v \in \mathbb{R}^3$, respectively, denote the time, position, and velocity; $\rho = \rho(t, x)$ is the induced spatial mass density; and $U = U(t, x)$ is the gravitational potential of the galaxy.

The problem we address is the nonlinear stability of stationary solutions of this system. Our approach automatically addresses this question for the Euler–Poisson system as well, describing a self-gravitating fluid ball, that is, a barotropic star.

By definition, a given steady state f_0 is stable if for any neighborhood N of f_0 there exists another neighborhood M of f_0 such that any solution of the system starting in M will remain in N for all time. This is the usual mathematical definition of Lyapunov stability. In the case of an infinite dimensional dynamical system, such as the Vlasov–Poisson system, the choice of the proper concept of *neighborhood* is a non-trivial part of the stability problem. We emphasize that no approach of the solution to the steady state is asserted—that would be the concept of asymptotic stability. Since the system as stated does not include dissipative effects, an approach to a

---

Address for correspondence: Gerhard Rein, Department of Mathematics, University of Bayreuth, 95440 Bayreuth, Germany. Voice: 0049-921-553287; fax: 0049-921-553293.
gerhard.rein@uni-bayreuth.de

particular steady state is not to be expected in a strict sense; in this paper we do not enter into the highly interesting questions of course graining, phase mixing, and others. The existence of global-in-time solutions of the system under consideration, at least for initial data close to the steady state, is an integral part of the stability assertion. For the Vlasov–Poisson system it is shown elsewhere[1] that every smooth, compactly supported initial datum for f launches a unique global-in-time smooth solution; a fairly short proof due to J. Schaeffer is given in Reference 2.

There is a vast astrophysics literature on the stability question. However, essentially all investigations that we are aware of proceed via linearization. This approach suffers from at least two major difficulties. First, there is no general theory for infinite dimensional dynamical systems concerning how to pass from linearized to nonlinear stability. Second, it is well known that if λ is an eigenvalue, that is, the linearized system has a solution of the form $e^{\lambda t} g(x, v)$, then the same is true for $-\lambda$. Hence, the optimal case for stability occurs if all such eigenvalues are purely imaginary, which is precisely the case where stability for the non-linear system does not follow, not even in finitely many dimensions.

We prove nonlinear stability of certain steady states by identifying them as minimizers of a conserved quantity in terms of which the neighborhoods N and M are then defined. More precisely, for a state $f = f(x, v) \geq 0$ we denote the induced spatial mass density and potential by

$$\rho_f(x) := \int f(x, v) dv, \qquad U_f(x) := -\int \frac{\rho_f(y)}{|x - y|} dy,$$

and we introduce the functionals

$$E_{\text{kin}}(f) := \frac{1}{2} \iint |v|^2 f(x, v) dv dx, \tag{1}$$

$$E_{\text{pot}}(f) := \frac{1}{2} \int U_f(x) \rho_f(x) dx = -\frac{1}{8\pi} \int |\nabla U_f(x)|^2 dx, \tag{2}$$

for the kinetic and the potential energy of the state f. The total energy

$$\mathcal{H} := E_{\text{kin}} + E_{\text{pot}}$$

is conserved along solutions of the Vlasov–Poisson system, but it is indefinite and it has no critical points, that is, the linear part in an expansion about any state f_0 with potential U_0 does not vanish,

$$\mathcal{H}(f) = \mathcal{H}(f_0) + \iint \left(\frac{1}{2}|v|^2 + U_0\right)(f - f_0) dv dx - \frac{1}{8\pi} \int |\nabla U_f - \nabla U_0|^2 dx.$$

However, according to Liouville's theorem the characteristic flow corresponding to the Vlasov equation preserves phase space volume, and hence, for any reasonable function Φ the so-called Casimir functional

$$C(f) := \iint \Phi(f(x, v)) dv dx$$

is conserved as well. If we expand the energy-Casimir functional,

$$\mathcal{H}_C := \mathcal{H} + C,$$

about an isotropic steady state

$$f_0(x, v) = \phi(E), \qquad E = E(x, v) := \frac{1}{2}|v|^2 + U_0(x),$$

we find that

$$\mathcal{H}_C(f) = \mathcal{H}_C(f_0) + \iint (E + \Phi'(f_0))(f - f_0) dv dx$$
$$- \frac{1}{8\pi} \int |\nabla U_f - \nabla U_0|^2 dx + \frac{1}{2} \iint \Phi''(f_0)(f - f_0)^2 dv dx + \dots.$$

At least formally, we can choose Φ such that f_0 is a critical point of $\mathcal{H}_C$, namely, $\Phi' = -\phi^{-1}$, provided ϕ is invertible. The essential problem is now the following: in order for the steady state to have finite total mass the function ϕ must vanish above a certain cutoff energy. For ϕ^{-1} to exist, ϕ should, thus, be decreasing, at least on its support. However, then Φ'' is positive and the quadratic part in the expansion is indefinite. Since one would like to use this quadratic part to define the concept of distance or neighborhood, the method seems to fail. This state of affairs had been observed by various authors, with the proposed conclusion that the energy-Casimir method does not work for the stellar dynamics case of the Vlasov–Poisson system (see, for example, Ref. 3). If the issue is the stability of a plasma, the sign of the source term in the Poisson equation, and hence also the one in front of the potential energy difference in the expansion above, is reversed, and up to some technicalities stability follows.[4]

The approach developed by Yan Guo and the author to overcome this difficulty is as follows. Starting with a given function Φ that defines the Casimir functional we try to minimize the energy-Casimir functional $\mathcal{H}_C$ under the constraint that only states with a prescribed total mass $M > 0$ are considered. Under suitable assumptions on Φ, a minimizer does exist despite the fact that the quadratic term in the expansion above is indefinite. One can then show that such a minimizer is a nonlinearly stable steady state. The exact statements of these results are provided in the next section—the main assumption is that Φ is strictly convex, which is equivalent to $f_0(x, v) = \phi(E)$ being strictly decreasing.

The crucial step is to prove the existence of a minimizer. Here we first construct from the energy-Casimir functional $\mathcal{H}_C$ a reduced functional $\mathcal{H}_r$, defined on the space of spatial mass densities ρ in such a way that there is a one-to-one correspondence between the minimizers of the two functionals. This reduced functional is analyzed later in this paper. The original motivation for introducing it was purely mathematical: it is defined on a simpler space, and the troublesome part of the original functional is the quadratic and negative definite potential energy, but the latter depends on the spatial mass density ρ_f and not directly on f. The detour via the reduced functional has a beautiful payoff: the minimizers of the reduced functional are stable steady states of the Euler–Poisson system with a macroscopic equation of state corresponding to the microscopic equation of state $f_0 = \phi(E)$ induced by the Casimir functional. Hence, by means of this reduction procedure we obtain a nonlinear proof of what is often referred to as Antonov's First Law (see page 305 of Ref. 5): *a spherical stellar system with $f_0 = f_0(E)$ and $df_0/dE < 0$ is stable if the barotropic star with the same equilibrium density distribution is stable.* This relation to the Euler–Poisson system (i.e., to the stability of gaseous stars) is investigated in a later section of this paper.

Subsequently, we give the main arguments leading to the existence of a minimizer for the reduced functional. Mathematically, this is the essential and non-trivial part; it can be skipped without compromising an understanding of the rest of this work.

To keep the presentation reasonably simple, we restrict ourselves mostly to spherically symmetric, isotropic steady states. However, the method has also been applied to non-isotropic steady states, to axially symmetric steady states, and to disk-like steady states. Some comments on these extensions, together with other related results and open questions are collected in the final section.

To conclude this introduction, we mention that none of the results presented here are new, although the way they are presented is new. The motivation for this paper is to collect in one place the main features of our method, and to present them in such a way that readers can appreciate the ideas involved. For details that are not so relevant to the main argument we refer to existing papers, but our presentation is aimed to be self-contained. We include almost no references to the astrophysics literature. This is *not* done out of disrespect but is due to the belief that our method is essentially the first to address the full *nonlinear* stability problem for the systems under consideration. Should these notes inspire comments or criticism from the astrophysics community, we would truly appreciate that.

NONLINEAR STABILITY FOR THE VLASOV–POISSON SYSTEM—STATEMENT OF RESULTS

We fix a Casimir functional C, that is, a function Φ such as

$$\Phi(f) = f^{1+1/k}, \quad f \geq 0, \quad \text{with } 0 < k < \frac{3}{2}. \tag{3}$$

More generally, $\Phi \in C^1([0, \infty))$ with $\Phi(0) = 0 = \Phi'(0)$, and

($\Phi 1$) Φ is strictly convex,

($\Phi 2$) $\Phi(f) \geq C f^{1+1/k}$ for $f \geq 0$ large, with $0 < k < 3/2$,

($\Phi 3$) $\Phi(f) \leq C f^{1+1/k'}$ for $f \geq 0$ small, with $0 < k' < 3/2$.

For a given constant $M > 0$ we want to minimize the energy-Casimir functional $\mathcal{H}_C$ over the constraint set

$$\mathcal{F}_M := \{ f \in L^1_+(\mathbb{R}^6) : \iint f\, dv\, dx = M, E_{\text{kin}}(f) + C(f) < \infty \},$$

were $L^1_+(\mathbb{R}^6)$ denotes the set of non-negative, integrable functions on $\mathbb{R}^6$. One can show that the potential energy is defined on this set and the two forms of E_{pot} given in (2) are equal (see Lemma 1 in Ref. 6). Due to conservation of mass, the constraint set $\mathcal{F}_M$ is invariant under solutions of the Vlasov–Poisson system.

The main step in the stability analysis is to establish the following theorem.

Theorem 1: The energy-Casimir functional $\mathcal{H}_C$ is bounded from below on $\mathcal{F}_M$ with $h_M := \inf_{\mathcal{F}_M} \mathcal{H}_C < 0$. Let $(f_j) \subset \mathcal{F}_M$ be a minimizing sequence of $\mathcal{H}_C$, that is, $\mathcal{H}_C(f_j) \to h_M$. Then, there exists a function $f_0 \in \mathcal{F}_M$, a subsequence, again denoted by (f_j) and a sequence $(a_j) \subset \mathbb{R}^3$ of shift vectors such that, for the induced gravitational fields

$$T^{a_j}\nabla U_{f_j} = \nabla U_{f_j}(\cdot + a_j) \to \nabla U_{f_0} \text{ in } L^2(\mathbb{R}^3), \quad j \to \infty.$$

The state f_0 minimizes the energy-Casimir functional, $\mathcal{H}_C(f_0) = h_M$.

We show that, to conclude stability of the state f_0, the theorem is needed in the above form; mere existence of a minimizer is not sufficient. A proof via a reduced functional is given below. The main difficulty is seen from the following sketch of the argument: to obtain a lower bound for the functional on the constraint set is easy, and by Assumption ($\Phi 2$) minimizing sequences can be seen to be bounded in $L^{1+1/k}$. A standard analysis result then implies that such a sequence has a weakly convergent subsequence, which means that for any test function g from the dual space L^{1+k} the convergence $\int f_j g \to \int f_0 g$ holds (see Sec. 2.18 of Ref. 7). The weak limit f_0 is the candidate for the minimizer, and one has to pass the limit into the energy-Casimir functional. This is easy for the kinetic energy, the latter being linear, and for the Casimir functional that is convex due to Assumption ($\Phi 1$) it relies on Mazur's lemma (see Sec. 2.13 of Ref. 7). The difficult part is the potential energy, for which one has to prove that the induced gravitational fields converge strongly in L^2. This problem we deal with on the level of the reduced functional in the section PROOF OF THE EXISTENCE OF MINIMIZERS FOR THE REDUCED PROBLEM.

Since our minimization problem is invariant under spatial translations we obtain a trivial minimizing sequence by shifting a given minimizer in space. If, for example, we shift it to spatial infinity we cannot obtain a subsequence that tends weakly to a minimizer, unless we move with the sequence. Hence, the spatial shifts in the theorem arise from the physical properties of our problem.

The Euler–Lagrange identity corresponding to our constrained variational problem implies that any minimizer is a steady state of the Vlasov–Poisson system. For the proof, see elsewhere.[8,9]

Theorem 2: Let $f_0 \in \mathcal{F}_M$ be a minimizer with potential U_0. Then

$$f_0(x, v) = \begin{cases} (\Phi')^{-1}(E_0 - E) & \text{for } E < E_0 \\ 0 & \text{else} \end{cases} \text{ where } E = \frac{1}{2}|v|^2 + U_0(x),$$

with Lagrange multiplier E_0. In particular, f_0 is a steady state of the Vlasov–Poisson system.

For example, the choice (3) yields the polytropic steady state

$$f_0(x, v) = \frac{k}{k+1} \begin{cases} (E_0 - E)^k & \text{for } E < E_0 \\ 0 & \text{else.} \end{cases}$$

It should be noted that the assumptions on Φ easily translate into assumptions on the steady state f_0 as a function of the particle energy, the main assumption being that this function is strictly decreasing on its support. Various additional properties can be derived for these minimizers/steady states, in particular they are necessarily spherically symmetric (see Thm. 3 in Ref. 6 or Thm. 2 in Ref. 9). Non-symmetric steady states are briefly considered in the last section of this paper.

To deduce our stability result we expand $\mathcal{H}_C$ about the minimizer f_0,

$$\mathcal{H}_C(f) - \mathcal{H}_C(f_0) = d(f, f_0) - \frac{1}{8\pi}\int |\nabla U_f - \nabla U_0|^2 dx, \tag{4}$$

where for $f \in \mathcal{F}_M$,

$$d(f, f_0) := \iint [\Phi(f) - \Phi(f_0) + E(f - f_0)] dv dx$$
$$\geq \iint [\Phi'(f_0) + (E - E_0)](f - f_0) dv dx \geq 0,$$

with $d(f, f_0) = 0$ iff $f = f_0$. This is due to the strict convexity of Φ and that on the support of f_0 the bracket vanishes by Theorem 2. Note also that $\iint (f - f_0) = 0$ for $f \in \mathcal{F}_M$. For suitable functions Φ, we even have $d(f, f_0) \geq c \iint |f - f_0|^2$. The point now is that, according to Theorem 1, the term with the negative sign in (4) tends to zero along any minimizing sequence. This allows us to use the sum of the two positive definite terms in the expansion as our measure of distance in the stability result. As mentioned in the introduction, initial data from the space $C_c^1(\mathbb{R}^6)$ of continuously differentiable, compactly supported functions launch smooth global-in-time solutions of the Vlasov–Poisson system that preserve all the physically conserved quantities. As above, $T^a f(x, v) := f(x + a, v)$.

Theorem 3: For any $\varepsilon > 0$ there exists a $\delta > 0$ such that for any solution $t \mapsto f(t)$ of the Vlasov–Poisson system with $f(0) \in C_c^1(\mathbb{R}^6) \cap \mathcal{F}_M$ the initial estimate

$$d(f(0), f_0) + \frac{1}{8\pi} \int |\nabla U_{f(0)} - \nabla U_0|^2 dx < \delta$$

implies that for any $t \geq 0$ there is a shift vector $a \in \mathbb{R}^3$ such that

$$d(T^a f(t), f_0) + \frac{1}{8\pi} \int |T^a \nabla U_{f(t)} - \nabla U_0|^2 dx < \varepsilon, \quad t \geq 0,$$

(provided the minimizer f_0 is unique up to shifts).

We comment on the uniqueness assumption (and why it appears in parenthesis) shortly, but first we give the proof of this result, a proof that is surprisingly simple—the difficulty resides in the proof of Theorem 1.

Proof: Assume the assertion is false. Then there exist

$$\varepsilon > 0, \ t_j > 0, \ f(0) \in C_c^1(\mathbb{R}^6) \cap \mathcal{F}_M$$

such that for $j \in \mathbb{N}$,

$$d(f_j(0), f_0) + \frac{1}{8\pi} \int |\nabla U_{f_j(0)} - \nabla U_0|^2 dx < \frac{1}{j},$$

but for any shift vector $a \in \mathbb{R}^3$,

$$d(T^a f_j(t_j), f_0) + \frac{1}{8\pi} \int |T^a \nabla U_{f_j(t_j)} - \nabla U_0|^2 dx \geq \varepsilon.$$

Since $\mathcal{H}_C$ is conserved, (4) and the assumption on the initial data implies that

$$\mathcal{H}_C(f_j(t_j)) = \mathcal{H}_C(f_j(0)) \to \mathcal{H}_C(f_0),$$

that is, $(f_j(t_j)) \subset \mathcal{F}_M$ is a minimizing sequence. Hence, by Theorem 1,

$$\int |\nabla U_{f_j(t_j)} - \nabla U_0|^2 \to 0$$

up to subsequences and shifts in x, provided that there is no other minimizer to which this sequence can converge. By (4), $d(f_j(t_j), f_0) \to 0$ as well, which is the desired contradiction. ∎

Some comments are in order. For the polytropic steady states one can show that for a given mass M the minimizer is indeed unique up to shifts, as assumed above (Thm. 3 in Ref. 6). In general, minimizers do not seem to be unique; for a numerically verified example of non-uniqueness we refer elsewhere.[9] However, minimizers always seem to be isolated up to shifts, which is sufficient for the above statement to remain true.[9] If there were a continuum of minimizers, then this set of minimizers as a whole would be stable.[10] Finally, for a closely related approach, to which we return in the last section, it is shown that the assertion of Theorem 3 holds without f_0 being unique or isolated.[11] We retained the former assumption to make the proof simple.

The spatial shifts appearing in the stability statement are again due to the spatial invariance of the system. If we perturb f_0 by giving all the particles an additional, fixed velocity, then in space the corresponding solution travels from f_0 at a linear rate in t, no matter how small the perturbation.

A nice feature of the result is that the same quantity is used to measure the deviation initially and at later times t. For infinite dimensional dynamical systems, control in a strong norm initially may be necessary to gain control in a weaker norm at later times. If our concept of distance is appropriate from a physics point of view is open to debate—it is simply what comes out of the energy-Casimir method. For the polytropic steady states our approach has been extended to yield stability with respect to the L^1-norm of f.[12] Definitely a weak point is the fact that the proof is not constructive: given a value for ε, we do not know how small the corresponding δ must be.

THE REDUCED VARIATIONAL PROBLEM

We wish to factor out the dependence on the velocity variable in our minimization problem. Starting from a given function $f = f(x,v)$ with induced spatial density $\rho_f = \rho_f(x)$ we clearly decrease $\mathcal{H}_C(f)$ by minimizing for each point x over all functions $g = g(v)$ that have as integral the value $\rho_f(x)$. This procedure does not affect the potential energy and reduces the sum of the kinetic energy and the Casimir functional into a new functional that no longer depends on f directly but only on ρ_f. More precisely, with

$$\Psi(s) := \inf\left\{\int\left(\frac{1}{2}|v|^2 g + \Phi(g)\right)dv : g \in L^1_+(\mathbb{R}^3), \int g(v)dv = s\right\} \qquad (5)$$

we have the estimate $\mathcal{H}_C(f) \geq \mathcal{H}_r(\rho_f)$, where

$$\mathcal{H}_r(\rho) := \int \Psi(\rho(x))dx + E_{\text{pot}}(\rho).$$

We now minimize $\mathcal{H}_r$ over the constraint set

$$\mathcal{R}_M := \{\rho \in L^1_+(\mathbb{R}^3) : \int \Psi(\rho) < \infty, \int \rho = M\}.$$

These constructions owe much to Reference 13. Before we analyze the reduced problem, we make sure that we can lift any information gained for the latter back to the level of the original problem:

Theorem 4: For all $f \in \mathcal{F}_M$ the estimate $\mathcal{H}_C(f) \geq \mathcal{H}_r(\rho_f)$ holds, with equality if $f = f_0$ is a minimizer. Let $\rho_0 \in \mathcal{R}_M$ minimize $\mathcal{H}_r$ and $U_0 = U_{\rho_0}$. Then

$$\rho_0 = \begin{cases} (\Psi')^{-1}(E_0 - U_0) & \text{for } U_0 < E_0 \\ 0 & \text{for } U_0 \geq E_0 \end{cases} \tag{6}$$

with Lagrange multiplier E_0, and

$$f_0 := \begin{cases} (\Phi')^{-1}(E_0 - E) & \text{for } E < E_0 \\ 0 & \text{for } E \geq E_0 \end{cases} \qquad E = E(x, v) := \frac{1}{2}|v|^2 + U_0(x), \tag{7}$$

lies in $\mathcal{F}_M$ and minimizes $\mathcal{H}_C$. If, on the other hand, $f_0 \in \mathcal{F}_M$ minimizes $\mathcal{H}_C$ then $\rho_{f_0} \in \mathcal{R}_M$ minimizes $\mathcal{H}_r$.

Equation (6) is nothing but the Euler–Lagrange identity for the reduced problem;[9] the theorem is proven in detail elsewhere.[14]

Now consider the reduced variational problem in its own right. The function Ψ defining the reduced functional is taken from the following class:

$\Psi \in C^1([0, \infty))$, $\Psi(0) = 0 = \Psi'(0)$, and

(Ψ1) Ψ is strictly convex,

(Ψ2) $\Psi(\rho) \geq C\rho^{1+1/n}$ for $\rho \geq 0$ large, with $0 < n < 3$,

(Ψ3) $\Psi(\rho) \leq C\rho^{1+1/n'}$ for $\rho \geq 0$ small, with $0 < n' < 3$,

In a later section we prove the following central result:

Theorem 5: The functional $\mathcal{H}_r$ is bounded from below on $\mathcal{R}_M$. Let $(\rho_j) \subset \mathcal{R}_M$ be a minimizing sequence of $\mathcal{H}_r$. Then there exists a sequence of shift vectors $(a_j) \subset \mathbb{R}^3$ and a subsequence, again denoted by (ρ_j), such that

$$T^{a_j}\rho_j \rightharpoonup \rho_0 \text{ weakly in } L^{1+1/n}(\mathbb{R}^3), \ j \to \infty,$$

$$T^{a_j}\nabla U_{\rho_j} \to \nabla U_0 \text{ strongly in } L^2(\mathbb{R}^3), \ j \to \infty,$$

and $\rho_0 \in \mathcal{R}_M$ is a minimizer of $\mathcal{H}_r$.

We need to translate the conditions on Ψ back into conditions on Φ. To do so we denote the Legendre transform of a function $h : \mathbb{R} \to (-\infty, \infty]$ by

$$\bar{h}(\lambda) := \sup_{r \in \mathbb{R}}(\lambda r - h(r)).$$

If Ψ arises from Φ by reduction, that is, by (5), then

$$\bar{\Psi}(\lambda) = \int \bar{\Phi}\left(\lambda - \frac{1}{2}|v|^2\right)dv.$$

This more explicit relation between Φ and Ψ is established elsewhere,[14] and it allows us to show that the assumptions on Φ imply the assumptions on Ψ if the parameters k and n are related by $n = k + 3/2$, with the same relation holding for the primed parameters.

Theorem 4 connects our two variational problems in an appropriate way to allow Theorem 1 to be derived from Theorem 5. First, $\mathcal{H}_C$ is bounded from below on $\mathcal{F}_M$

since this is true for $\mathcal{H}_r$ on $\mathcal{R}_M$. Let $(f_j) \subset \mathcal{F}_M$ be a minimizing sequence for $\mathcal{H}_C$. By Theorem 4, $(\rho_{f_j}) \subset \mathcal{R}_M$ is a minimizing sequence for $\mathcal{H}_r$. Again by Theorem 4, we can lift the minimizer ρ_0 of $\mathcal{H}_r$ obtained in Theorem 5 to a minimizer f_0 of $\mathcal{H}_C$, and the proof of Theorem 1 is complete.

Before we consider the ideas involved in the proof of Theorem 5 we reinterpret it in terms of the Euler–Poisson system.

PAYOFF OF REDUCTION—THE EULER–POISSON SYSTEM

If $\rho_0 \in \mathcal{R}_M$ minimizes the reduced functional $\mathcal{H}_r$, then ρ_0 supplemented with the velocity field $u_0 = 0$ is a steady state of the Euler–Poisson system

$$\partial_t \rho + \nabla \cdot (\rho u) = 0,$$

$$\rho \partial_t u + \rho(u \cdot \nabla)u = -\nabla p - \rho \nabla U,$$

$$\Delta U = 4\pi\rho, \quad \lim_{|x| \to \infty} U(t, x) = 0,$$

with equation of state

$$p = P(\rho) := \rho \Psi'(\rho) - \Psi(\rho).$$

This follows from the Euler–Lagrange identity (6). Here u and p denote the velocity field and the pressure of an ideal, compressible fluid with mass density ρ, and the fluid self-interacts via its induced gravitational potential U. This system is sometimes used as a simple model for a gaseous, barotropic star. The beautiful thing now is that obviously $(\rho_0, u_0 = 0)$ minimizes the energy,

$$\mathcal{H}(\rho, u) := \frac{1}{2}\int |u|^2 \rho\, dx + \int \Psi(\rho)\, dx + E_{\text{pot}}(\rho),$$

of the system, which is a conserved quantity. Expanding as before, we find that

$$\mathcal{H}(\rho, u) - \mathcal{H}(\rho_0, 0) = \frac{1}{2}\int |u|^2 \rho\, dx + d(\rho, \rho_0) - \frac{1}{8\pi}\int |\nabla U_\rho - \nabla U_0|^2\, dx,$$

where, for $\rho \in \mathcal{R}_M$,

$$d(\rho, \rho_0) := \int [\Psi(\rho) - \Psi(\rho_0) + (U_0 - E_0)(\rho - \rho_0)]\, dx \geq 0,$$

with equality iff $\rho = \rho_0$. The same proof as for the Vlasov–Poisson system implies a stability result for the Euler–Poisson system—the term with the unfavorable sign in the expansion again tends to zero along minimizing sequences (Theorem 5). However, there is an important caveat: although for the Vlasov–Poisson system we have global-in-time solutions for sufficiently nice data, and these solutions really preserve all the conserved quantities, no such result is available for the Euler–Poisson system, and we only obtain the following:

Conditional Stability Result: For every $\varepsilon > 0$ there exists a $\delta > 0$ such that for every solution $t \mapsto (\rho(t), u(t))$ with $\rho(0) \in \mathcal{R}_M$ that preserves energy and mass, the initial estimate

$$\frac{1}{2}\int |u(0)|^2 \rho(0)\, dx + d(\rho(0), \rho_0) + \frac{1}{8\pi}\int |\nabla U_{\rho(0)} - \nabla U_0|^2\, dx < \delta$$

implies that as long as the solution exists,

$$\frac{1}{2}\int |u(t)|^2 \rho(t) dx + d(\rho(t), \rho_0) + \frac{1}{8\pi}\int |\nabla U_{\rho(t)} - \nabla U_0|^2 dx < \varepsilon$$

up to shifts in x (provided the minimizer is unique up to such shifts).

The same comments as for Theorem 3 apply also in this case. Because of the above caveat we prefer not to call this a theorem, although as far as the stability analysis itself is concerned it is perfectly rigorous. The open problem is does a suitable concept of solution to the initial value problem exist.

Now that minimizers of the reduced functional are identified as steady states of the Euler–Poisson system it is instructive to reconsider the reduction procedure leading from the kinetic to the fluid dynamics picture. First, recall that for the Legendre transform

$$h'(\xi) = \eta \Leftrightarrow h(\xi) + \bar{h}(\eta) = \xi\eta \Leftrightarrow (\bar{h})'(\eta) = \xi.$$

If f_0 is a minimizer of $\mathcal{H}_C$,

$$f_0 = (\Phi')^{-1}(E_0 - E) = (\bar{\Phi})'(E_0 - E),$$

$$\rho_0 = \int f_0 dv = \int (\bar{\Phi})'\left(E_0 - U_0 - \frac{1}{2}|v|^2\right) dv,$$

$$p_0 = \frac{1}{3}\int |v|^2 f_0 dv = \int \bar{\Phi}\left(E_0 - U_0 - \frac{1}{2}|v|^2\right) dv.$$

On the other hand, if ρ_0 is a minimizer of $\mathcal{H}_r$,

$$\rho_0 = (\Psi')^{-1}(E_0 - U_0) = (\bar{\Psi})'(E_0 - U_0),$$

$$p_0 = P(\rho_0) = \rho_0 \Psi'(\rho_0) - \Psi(\rho_0) = \bar{\Psi}(\Psi'(\rho_0)) = \bar{\Psi}(E_0 - U_0).$$

For these relations on the kinetic and on the fluid level to fit, we require that

$$\bar{\Psi}(\lambda) = \int \bar{\Phi}\left(\lambda - \frac{1}{2}|v|^2\right) dv,$$

which is the relation between Φ and Ψ obtained by our reduction mechanism.

PROOF OF THE EXISTENCE OF MINIMIZERS FOR THE REDUCED PROBLEM

In this section we present the main arguments leading to the proof of Theorem 5. Constants denoted by C may only depend on M and Ψ and may change their value from line to line. For a set $S \subset \mathbb{R}^3$ the indicator function 1_S equals 1 on S and vanishes outside. By B_R we denote the ball of radius R about the origin.

Step 1: Lower bound for $\mathcal{H}_r$ and weak convergence of minimizing sequences.

We need to estimate the negative potential energy against the positive part of $\mathcal{H}_r$. The Hardy–Littlewood–Sobolev inequality,[7] tells us that

$$-E_{pot}(\rho) \leq C\|\rho\|_{6/5}^2.$$

By the restriction on n, $1 < 6/5 < 1+1/n$, and by interpolation and ($\Psi 2$),
$$-E_{\text{pot}}(\rho) \leq C\|\rho\|_1^{(5-n)/3}\|\rho\|_{1+1/n}^{(n+1)/3} \leq C + C(\int \Psi(\rho)dx)^{n/3}, \quad \rho \in \mathcal{R}_M.$$
Hence on $\mathcal{R}_M$
$$\mathcal{H}_r(\rho) \geq \int \Psi(\rho)dx - C - C(\int \Psi(\rho)dx)^{n/3}. \tag{7}$$
Since $n < 3$ this implies that $\mathcal{H}_r$ is bounded from below on $\mathcal{R}_M$,
$$h_M := \inf_{\mathcal{R}_M} \mathcal{H}_r > -\infty.$$

Let $(\rho_j) \subset \mathcal{R}_M$ be a minimizing sequence. By **(7)**, $\int \Psi(\rho_j)$ is bounded, and by ($\Psi 2$) and the fact that $\int \rho_j = M$, the minimizing sequence is bounded in $L^{1+1/n}(\mathbb{R}^3)$. Hence, we can—after extracting a subsequence—assume that it converges weakly to some function $\rho_0 \in L^{1+1/n}(\mathbb{R}^3)$, that is, for any test function $\sigma \in L^{1+n}(\mathbb{R}^3)$, $\int \rho_j \sigma \to \int \rho_0 \sigma$.[7]

As pointed out above, the main difficulty is to prove that the induced fields converge strongly in L^2; such a result is referred to as a compactness result. It is true if the sequence (ρ_j) remains concentrated.

Step 2: Concentration implies compactness.

Assume that
$$\lim_{j \to \infty} \int_{|x| \geq R} \rho_j = 0, \tag{8}$$
for some $R > 0$, that is, asymptotically the mass remains within the ball B_R. We claim that under this assumption $\nabla U_{\rho_j} \to \nabla U_{\rho_0}$ strongly in L^2.

By weak convergence, $\rho_0 \geq 0$ almost everywhere—if ρ_0 were strictly negative on some set S of positive, finite measure, the test function $\sigma = 1_S$ would yield a contradiction. Moreover, Equation **(8)** shows that ρ_0 vanishes outside the ball B_R. Again by weak convergence, $\int \rho_0 = M$. The sequence $\sigma_j := \rho_j - \rho_0$ converges weakly to 0 in $L^{1+1/n}$, $\int |\sigma_j| \leq M$, and **(8)** holds for $|\sigma_j|$ as well. We need to show that $\nabla U_{\sigma_j} \to 0$ strongly in L^2, which is equivalent to
$$I_j := \iint \frac{\sigma_j(x)\sigma_j(y)}{|x-y|} dy dx \to 0. \tag{9}$$

We choose $R > 0$ such that Equation **(8)** applies. For $\varepsilon > 0$ we split the domain of integration into three sets defined by
$$|x-y| < \varepsilon,$$
$$|x-y| \geq \varepsilon \text{ and } (|x| \geq R \text{ or } |y| \geq R),$$
$$|x-y| \geq \varepsilon \text{ and } |x| < R \text{ and } |y| < R,$$
and denote the corresponding contributions to I_j by $I_{j,1}, I_{j,2}$, and $I_{j,3}$. Since
$$\frac{2n}{n+1} + \frac{2}{n+1} = 2,$$
Young's inequality[7] implies that
$$|I_{j,1}| \leq C\|\sigma_j\|_{1+1/n}^2 \||1_{B_\varepsilon}|\cdot|^{-1}\|_{(n+1)/2} \leq C\left(\int_0^\varepsilon r^{(3-n)/2} dr\right)^{2/(n+1)} \to 0,$$

for $\varepsilon \to 0$, uniformly in j. Clearly,

$$|I_{j,2}| \leq \frac{1}{\varepsilon} M \int_{|x|>R} |\sigma_j(x)| dx \to 0$$

as $j \to \infty$, for any fixed $\varepsilon > 0$. Finally by Hölder's inequality,

$$|I_{j,3}| = \left|\int \sigma_j(x) h_j(x) dx\right| \leq \|\sigma_j\|_{1+1/n} \|h_j\|_{1+n} \leq C \|h_j\|_{1+n},$$

where in a pointwise sense,

$$h_j(x) := 1_{B_R}(x) \int_{|x-y| \geq \varepsilon} 1_{B_R}(y) \frac{1}{|x-y|} \sigma_j(y) dy \to 0$$

due to the weak convergence of σ_j and the fact that the test function against which σ_j is integrated here is in L^{1+n}. However, since $|h_j| \leq M/\varepsilon$ uniformly in j Lebesgue's dominated convergence theorem implies that $h_j \to 0$ in L^{1+n}, and the proof of **(9)** is complete. I thank my student M. Hadžić for the above rather direct compactness argument.[15]

The next two steps aim to show that minimizing sequences remain concentrated and do not split into far apart pieces or spread out uniformly in space.

Step 3: Behavior under rescaling.

For $\rho \in \mathcal{R}_M$ and $a,b > 0$ we define $\bar\rho(x) := a\rho(bx)$. Then

$$\int \bar\rho \, dx = ab^{-3} \int \rho \, dx$$
$$E_{pot}(\bar\rho) = a^2 b^{-5} E_{pot}(\rho)$$
$$\int \Psi(\bar\rho) = b^{-3} \int \Psi(a\rho) dx.$$

First we fix a bounded and compactly supported function $\rho \in \mathcal{R}_M$ and choose $a = b^3$ so that $\bar\rho \in \mathcal{R}_M$ as well. By (Ψ3) and since $3/n' > 1$,

$$\mathcal{H}_r(\bar\rho) = b^{-3} \int \Psi(b^3 \rho) dx + b E_{pot}(\rho) \leq C b^{3/n'} + b E_{pot}(\rho) < 0,$$

for b sufficiently small, and hence, $h_M < 0$ for any $M > 0$, that is, a minimizer—if one exists—is going to be a bound state.

Next we fix two masses $0 < \overline{M} \leq M$. If we take $a = 1$ and $b = (M/\overline{M})^{1/3} \geq 1$, then for $\rho \in \mathcal{R}_M$ and $\bar\rho \in \mathcal{R}_{\overline{M}}$ rescaled with these parameters,

$$\mathcal{H}_r(\bar\rho) = b^{-3} \int \Psi(\rho) dx + b^{-5} E_{pot}(\rho)$$
$$\geq b^{-5}(\int \Psi(\rho) dx + E_{pot}(\rho)) = \left(\frac{\overline{M}}{M}\right)^{5/3} \mathcal{H}_r(\rho).$$

Since for this choice of a and b the map $\rho \mapsto \bar\rho$ is one-to-one and onto between $\mathcal{R}_M$ and $\mathcal{R}_{\overline{M}}$ this estimate gives the following relation between the infima of our functional for various mass constraints,

$$h_{\overline{M}} \geq \left(\frac{\overline{M}}{M}\right)^{5/3} h_M, \quad 0 < \overline{M} \leq M. \tag{10}$$

Step 4: Spherically symmetric minimizing sequences remain concentrated.

In this step we prove Equation **(8)**, but to make matters easier we consider for a moment spherically symmetric functions $\rho \in \mathcal{R}_M$, that is, $\rho(x) = \rho(|x|)$. For any radius $R > 0$ we split ρ into the piece supported in the ball B_R and the rest, that is,

$$\rho = \rho_1 + \rho_2, \qquad \rho_1(x) = 0 \text{ for } |x| > R, \qquad \rho_2(x) = 0 \text{ for } |x| \leq R.$$

Clearly,

$$\mathcal{H}_r(\rho) = \mathcal{H}_r(\rho_1) + \mathcal{H}_r(\rho_2) - \int \frac{\rho_1(x)\rho_2(y)}{|x-y|} dx dy.$$

Due to spherical symmetry the potential energy of the interaction between the two pieces can be estimated as

$$\int \frac{\rho_1(x)\rho_2(y)}{|x-y|} dx dy = -\int U_{\rho_1} \rho_2 dx$$

$$= 4\pi \int_R^\infty \frac{4\pi}{r} \int_0^r \rho_1(s) s^2 ds\, \rho_2(r) r^2 dr \leq \frac{(M-m)m}{R},$$

where $m = \int \rho_2$ is the mass outside the radius R, which we want to make small along the minimizing sequence. We define

$$R_0 := -\frac{3M^2}{5 h_M} > 0$$

and use the scaling estimate **(10)** together with the fact that $h_M < 0$ and

$$\xi^{5/3} + (1-\xi)^{5/3} \leq 1 - \frac{5}{3}\xi(1-\xi)$$

for $0 \leq \xi \leq 1$ to conclude that

$$\mathcal{H}_r(\rho) \geq h_{M-m} + h_m - \frac{(M-m)m}{R}$$

$$\geq \left[\left(\frac{M-m}{M}\right)^{5/3} + \left(\frac{m}{M}\right)^{5/3}\right] h_M - \frac{(M-m)m}{R} \qquad (11)$$

$$\geq h_M - \left[\frac{1}{R_0} - \frac{1}{R}\right](M-m)m.$$

We claim that, if $R > R_0$, then for any spherically symmetric minimizing sequence $(\rho_j) \subset \mathcal{R}_M$ of $\mathcal{H}_r$, Equation **(8)** holds. Suppose this assertion were false so that up to a subsequence,

$$\lim_{j \to \infty} \int_{|x| \geq R} \rho_j = m > 0.$$

Choose $R_j > R$ such that

$$m_j := \int_{|x| \geq R_j} \rho_j = \frac{1}{2} \int_{|x| \geq R} \rho_j.$$

By **(11)**,

$$\mathcal{H}_r(\rho_j) \geq h_M + \left[\frac{1}{R_0} - \frac{1}{R_j}\right](M - m_j)m_j \geq h_M + \left[\frac{1}{R_0} - \frac{1}{R}\right](M-m_j)m_j,$$

and letting $j \to \infty$ leads to a contradiction.

Step 5: Removing the symmetry assumption.

The restriction to spherical symmetry means that stability would hold only against spherically symmetric perturbations. Fortunately, the restriction can be removed using a general result due to Burchard and Guo. To explain this we define for a given function $\rho \in L^1_+(\mathbb{R}^3)$ its spherically symmetric decreasing rearrangement ρ^* as the unique spherically symmetric, decreasing function with the property that for every $\tau \geq 0$ the sup-level-sets $\{x \in \mathbb{R}^3 : \rho(x) > \tau\}$ and $\{x \in \mathbb{R}^3 : \rho^*(x) > \tau\}$ have the same volume; the latter set is, of course, then a ball about the origin whose radius is determined by the volume of the former. The important point is that for any monotone function Ψ the integral $\int \Psi(\rho) dx$ does not change under such a rearrangement, whereas the potential energy can only decrease, and it does not decrease if and only if ρ is already spherically symmetric (with respect to some center of symmetry) and decreasing. These facts can be found in Chapter 3 of Reference 7. In particular, a minimizer must *a posteriori* be spherically symmetric.

Now let $(\rho_j) \subset \mathcal{R}_M$ be a, not necessarily spherically symmetric, minimizing sequence. Obviously, the sequence of spherically symmetric decreasing rearrangements (ρ_j^*) is again minimizing. Hence, by the previous steps, up to a subsequence (ρ_j^*) converges weakly to a minimizer $\rho_0 = \rho_0^*$ and

$$\nabla U_{\rho_j^*} \to \nabla U_{\rho_0} \text{ in } L^2, \quad \text{hence} \quad \int \Psi(\rho_j^*) \to \int \Psi(\rho_0).$$

Moreover,

$$E_{pot}(\rho_j) = \mathcal{H}_r(\rho_j) - \int \Psi(\rho_j) = \mathcal{H}_r(\rho_j) - \int \Psi(\rho_j^*)$$
$$\to \mathcal{H}_r(\rho_0) - \int \Psi(\rho_0) = E_{pot}(\rho_0).$$

In this situation the result of Burchard and Guo[16] says that up to translations in space

$$\nabla U_{\rho_j} \to \nabla U_{\rho_0} \text{ in } L^2. \tag{12}$$

The proof of this general result is by no means easy, and it is possible to obtain stability against general perturbations without resorting to it.[6,8,14] However, since this general result may be useful for other problems of this nature we wanted to mention and exploit it here.

Step 6: Proof of Theorem 5.

Given a minimizing sequence (ρ_j) we already know that up to a subsequence it converges weakly in $L^{1+1/n}$ to a non-negative limit ρ_0 of mass M. The functional $\rho \mapsto \int \Psi(\rho) dx$ is convex by Assumption (Ψ1), so that by Mazur's Lemma[7] and Fatou's Lemma,[7]

$$\int \Psi(\rho_0) dx \leq \limsup_{j \to \infty} \int \Psi(\rho_j) dx, \tag{13}$$

in particular, $\rho_0 \in \mathcal{R}_M$. Together with Equation (12) this implies that

$$\mathcal{H}_r(\rho_0) \leq \limsup_{j \to \infty} \mathcal{H}_r(\rho_j) = h_M,$$

so that ρ_0 is a minimizer of $\mathcal{H}_r$, and the proof of Theorem 5 is complete.

RELATED RESULTS, CONCLUDING REMARKS, AND OPEN PROBLEMS

In this last section we touch upon a number of questions that may come to mind.

The Threshold k = 3/2

First we ask what happens if we choose the exponent k in Assumptions ($\Phi 2$) and ($\Phi 3$) to be larger than 3/2. This question is answered by the following observations. For the Vlasov–Poisson system the Casimir functional C is preserved. Hence, we can pursue an alternative approach, namely to minimize the energy

$$\mathcal{H} = E_{\text{kin}} + E_{\text{pot}}$$

under the mass-Casimir constraint

$$\iint f\,dv\,dx + C(f) = M.$$

As is shown elsewhere,[8,17] this works, provided $\Phi \in C^1([0,\infty))$ with $\Phi(0) = 0 = \Phi'(0)$ satisfies the assumptions ($\Phi 1$) (strict convexity) and

($\Phi 2'$) $\Phi(f) \geq C f^{1 + 1/k}$ for $f \geq 0$ large, with $0 < k < 7/2$;

($\Phi 3$) is no longer needed. However, reduction, which combined the kinetic energy and the Casimir functional into the new functional $\int \Psi(\rho)$, can no longer work because the former two functionals now appear in different places in the variational problem. Moreover, for polytropes one can show that the energy-Casimir functional changes its sign from negative to positive at $n = 3$ (i.e., $k = 3/2$). If a perturbation leads to an initial datum with positive energy $\mathcal{H}(f(0)) > 0$ then

$$\sup\{|x| : (x,v) \in \text{supp}\,f(t)\} \geq Ct, \quad t \to \infty.$$

The analogous result holds for the Euler–Poisson system, except that for a minimizer the energy in the Euler–Poisson picture equals the energy-Casimir functional in the kinetic Vlasov–Poisson picture. Hence, the fact that $\mathcal{H}_C$ (for the polytropes) changes sign at $k = 3/2$ does not signify any stability change on the kinetic level, but it does destroy stability on the fluid level.

Are All Relevant Isotropic Models Covered?

To answer this question it is useful to translate the conditions on the function Φ determining the Casimir functional into conditions on the dependence $f_0(x,v) = \phi(E_0 - E)$ of the resulting steady state on the particle energy. Since $(\Phi')^{-1} = \phi$, Φ is strictly convex as required iff ϕ strictly increases; if $\phi(\eta) \leq C\eta^k$ for η large and $\phi(\eta) \geq C\eta^{k'}$ for $\eta > 0$ small, then Φ satisfies the growth conditions ($\Phi 2$) and ($\Phi 3$) provided $0 < k, k' < 3/2$. To satisfy the alternative growth condition ($\Phi 2'$) we only need to require $\phi(\eta) \leq C\eta^k$ for η large with $0 < k < 7/2$. This means that all those polytropic steady states that are compactly supported and decrease as a function of the local energy are covered by our results—either by the approach via reduction with the bonus of the stability of the corresponding fluid model or, when this is no longer possible, by the alternative approach mentioned above. The alternative approach also covers the limiting case $f_0(x,v) = (E_0 - E)^{7/2}$, the so-called Plummer sphere.[8] The minimization property of the latter has also been investigated,[18] without deducing stability.

An important example from astrophysics that is not yet covered is the King model where $f_0(x, v) = \exp(E_0 - E) - 1$ on its support. It leads to a Casimir function Φ growing as $f \ln f$ that is too slow. Possible extensions of the method to include the King model are currently under investigation.

Does the Method Apply to Non-Isotropic Steady States?

In the presence of spherical or axial symmetry (with respect to the x_3-axis) angular momentum quantities, such as

$$L := |x \times v|^2 \text{ or } L_3 := x_1 v_2 - x_2 v_1 = (x \times v)_3,$$

are conserved along particle trajectories. If we let the function Φ depend also on L or L_3 the corresponding variational problem has a solution that is a steady state depending on the additional invariant. The dependence on L,[10,19,20] and axially symmetric steady states depending on L_3,[21] are treated elsewhere. The main assumption again is the strict monotonicity of the dependence on the local energy. There is however one important difference to the isotropic case: since a Casimir where Φ also depends on angular momentum is preserved by time dependent solutions only if they have the corresponding symmetry, we obtain stability only with respect to either spherically symmetric or axially symmetric perturbations. The question of stability of non-isotropic steady states against non-symmetric perturbations is under investigation.

The method has also been applied to disk-like steady states.[22] Here we need to assume that the perturbations live only in the plane of the disk. An extension of these results in the spirit of the later developments reported in this paper or in Reference 8 is under investigation. Stability against perturbations perpendicular to the disk is another challenging open problem in this area.

ACKNOWLEDGMENTS

This paper is an expanded version of my presentation at the workshop Nonlinear Dynamics in Astronomy and Physics, held in November 2004 at the University of Florida. I truly appreciated the kind invitation to this inspiring event, as well as the feedback I received there. The results reported here originate from my collaboration with Y. Guo of Brown University, whom I thank for many stimulating discussions.

REFERENCES

1. PFAFFELMOSER, K. 1992. Global classical solutions of the Vlasov–Poisson system in three dimensions for general initial data. J. Diff. Eqns. **95**: 281–303.
2. REIN, G. 1997. Selfgravitating systems in Newtonian theory—the Vlasov–Poisson system. Banach Center Publications **41**(1): 179–194.
3. KANDRUP, H.E. 1990. Geometric approach to secular and nonlinear stability for spherical star clusters. Astrophys. J. **351**: 104–113.
4. REIN, G. 1994. Nonlinear stability for the Vlasov–Poisson system—the energy-Casimir method. Math. Meth. Appl. Sci. **17**: 1129–1140.
5. BINNEY, J. & S. TREMAINE. 1987. Galactic Dynamics. Princeton University Press, Princeton.
6. REIN, G. 2002. Stability of spherically symmetric steady states in galactic dynamics against general perturbations. Arch. Rational Mech. Anal. **161**: 27–42.

7. LIEB, E.H. & M. LOSS. 1996. Analysis. American Math. Soc., Providence. 29.
8. GUO, Y. & G. REIN. 2001. Isotropic steady states in galactic dynamics. Commun. Math. Phys. **219:** 607–629.
9. REIN, G. 2003. Nonlinear stability of gaseous stars. Arch. Rational Mech. Anal. **168:** 115–130.
10. GUO, Y. & G. REIN. 1999. Stable steady states in stellar dynamics. Arch. Rational Mech. Anal. **147:** 225–243.
11. SCHAEFFER, J. 2004. Steady states in galactic dynamics. Arch. Rational Mech. Anal. **172:** 1–19.
12. SÁNCHEZ, O. & J. SOLER. 2005. Orbital stability for polytropic galaxies. Preprint.
13. WOLANSKY, G. 1999. On nonlinear stability of polytropic galaxies. Ann. Inst. Henri Poincaré **16:** 15–48.
14. REIN, G. 2002. Reduction and a concentration-compactness principle for energy-Casimir functionals. SIAM J. Math. Anal. **33:** 896–912.
15. HADŽIĆ, M. 2005. Private communication.
16. BURCHARD, A. & Y. GUO. 2004. Compactness via symmetrization. J. Funct. Anal. **214:** 40–73.
17. GUO, Y. 2000. On the generalized Antonov's stability criterion. Contem. Math. **263:** 85–107.
18. ALY, J.J. 1989. On the lowest energy state of a collisionless selfgravitating system under phase space volume constraints. Month. Not. Roy. Astro. Soc. **241:** 15–27.
19. GUO, Y. 1999. Variational method in polytropic galaxies. Arch. Rational Mech. Anal. **150:** 209–224.
20. GUO, Y. & G. REIN. 1999. Existence and stability of Camm type steady states in galactic dynamics. Indiana Univ. Math. J. **48:** 1237–1255.
21. GUO, Y. & G. REIN. 2003. Stable models of elliptical galaxies. Month. Not. Roy. Astr. Soc. **344:** 1396–1406.
22. REIN, G. 1999. Flat steady states in stellar dynamics—existence and stability. Commun. Math. Phys. **205:** 229–247.

Chaos in Orbits Due to Disk Crossings

C. HUNTER

*Department of Mathematics, Florida State University,
Tallahassee Florida, USA*

> ABSTRACT: We study orbits of halo stars in simple models of galaxies with disks and halos to see if the cumulative effects of the sudden changes in acceleration that occur at disk crossings can induce chaos. We find that they can, although not in all orbits and not in all potentials. Most of the orbits that become chaotic stay relatively close to the disk and range widely in the radial direction. Heavier disks and increased halo flattening both enhance the extent of the chaos. A limited range of experiments with a three-component model of the Milky Way with an added central bulge finds that many chaotic disk-crossing orbits can be expected in the central regions, and that prolateness of the halo is much more effective than oblateness in generating chaos.
>
> KEYWORDS: celestial mechanics; stellar dynamics; galaxies; kinematics and dynamics; galaxy; halo

INTRODUCTION

Many galaxies have both a flattened disk-like component and much less flattened non-disk components. Normal spirals with their bulges and halos are one example. S0s, which bridge the gap between ellipticals and normal spirals in the Hubble tuning fork diagram,[1] are another. More recently it has been recognized that disks can be found in less extreme elliptical galaxies; Bender[2] and Nieto et al.[3] found that many ellipticals can be classified as disk-like, and that their "diskiness" is consistent with a disk plus bulge model.[4] The stars that populate the disk can remain forever in the disk, but the gravitational attraction of the disk causes the orbits of non-disk stars to cross back and forth through the disk. As a star crosses the disk, it experiences an abrupt change in the gravitational force field perpendicular to the disk. This causes an abrupt change in the acceleration of the star, although not in its velocity, with the result that the curvature of the orbit changes abruptly. FIGURE 1 illustrates this phenomenon for a razor-thin disk, for which the change is discontinuous.

This paper examines when the cumulative effect of repeated disk crossings leads to chaos. In his efforts to get galactic theorists to recognize the importance of chaos, I recall Henry Kandrup asking me for examples of chaos in axisymmetric systems to help make his point that there is no essential link between triaxiality and chaos. Although I had not thought about the present topic when he first asked, it does provide clear examples to reinforce his point. FIGURE 2 shows a surface of section for one simple example of a model galaxy with a disk and halo. It shows that chaotic

---

Address for correspondence: C. Hunter, Department of Mathematics, Florida State University, Tallahassee FL 32306-4510, USA.
  hunter@math.fsu.edu

orbits occupy a significant portion of phase space and that there are other parts of phase space in which the orbits are quite regular.

The idea that chaos can be induced by disk crossings is not new, but it has not attracted much attention. One notable exception is that of Ostriker, Spitzer, and Chevalier,[5] who discussed the effect that the compressive gravitational shocks, caused by passage through the disk, have on the internal structure of globular clusters. Although we study the simpler matter of individual orbits rather than an entire globular cluster, Ostriker *et al.* point out that the disruptive effect they study is due to changes in the relative motion of two nearby points of a cluster caused by disk shocking. Orbits of nearby points in a stable cluster are nearby when viewed on the

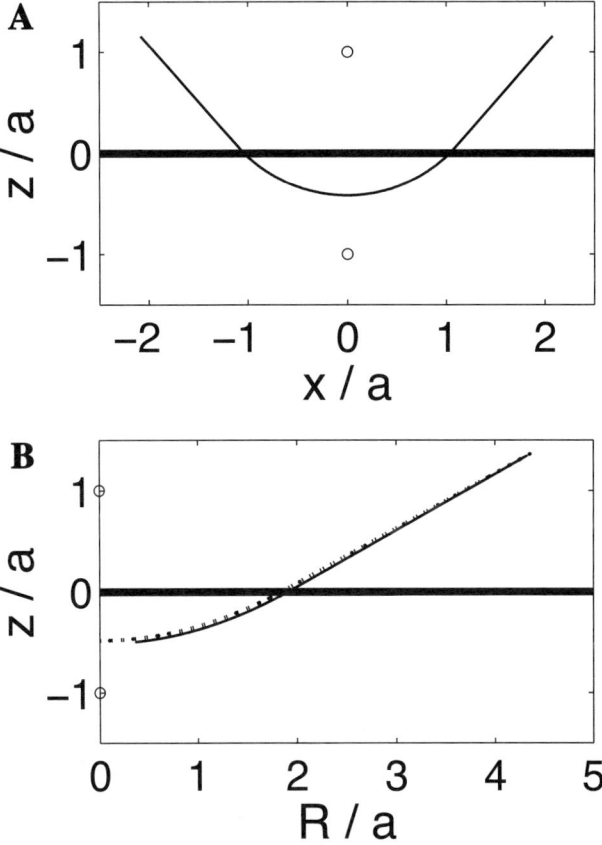

FIGURE 1. (**A**) A banana periodic polar orbit for the Kuzmin disk, with two exactly linear segments. This orbit is possible only if $E > -GM/2a\sqrt{2}$. It is drawn for the slightly larger value $E = -GM/3a$ when the curved segment is almost circular. The *open circles* mark the centers of force. (**B**) The *full curve* shows the saucer orbit for the logarithmic Kuzmin-like potential (**6**) for $R_c = 3a$ and $L_z/L_c(E) = 0.2$. For comparison, the *dotted curve* shows half of the $L_z = 0$ polar orbit for the same potential and R_c.

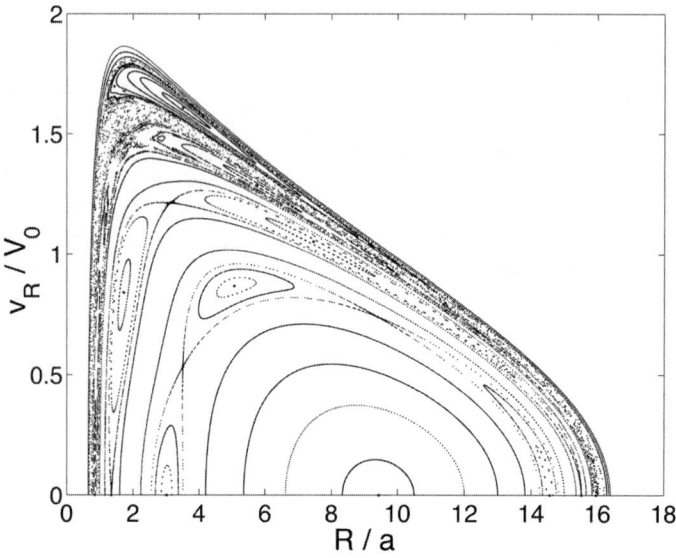

FIGURE 2. Surface of section for the logarithmic Kuzmin-like potential (6) for $R_c/a = 10$ and $L_z/L_c(E) = 0.15$.

galactic scale. They grow far apart if and when the cluster disrupts. Hence Ostriker *et al.*, like us, are basically concerned with circumstances in which small changes in initial conditions lead to greatly different outcomes. This is well-known to be a hallmark of chaos.[6]

The issue of the extent to which orbits in our Galaxy are chaotic was also studied in the 1970s by Martinet and coauthors (see Ref. 7 and earlier work cited therein). They integrated orbits in the Schmidt models of the Galaxy, which have highly flattened disk components, and found extensive regions in which orbits are chaotic. Their surfaces of section are cruder than those we display, which is not in the least surprising because they worked in the era of punched cards and with computers that are primitive by today's standards, and we have not attempted any detailed comparisons with their work.

The present work grew out a study[8] of the Kuzmin-like potentials introduced by Tohline and Voyages.[9] Kuzmin-like potentials provide a simple set of model galaxies with disk and non-disk components. We discuss them and the orbits they support in the next section. We then compare and contrast two other simple classes of models, those of Miyamoto and Nagai[10] and the scalefree logarithmic models of Monet, Richstone, and Schechter[11] and Toomre.[12] The scalefree models have razor-thin disks, whereas those of Miyamoto and Nagai have disks with a non-zero thickenings. Because of their simplicity, Miyamoto and Nagai disks have often been used to represent the disk in more realistic models of galaxies.[13,14] Helmi and White[15] gave an analytic description of how fossils of accreted satellite galaxies remain coherent and detectable in the Milky Way provided that they inhabit regions of phase space in which the motion is essentially integrable. We delineate where this is so in a more realistic three-component model of the Galaxy with disk, halo, and bulge. We offer our conclusions in the final section.

KUZMIN-LIKE MODELS

The Kuzmin Disk

To understand Kuzmin-like potentials, we first consider the Kuzmin disk.[16] It is a flat and razor-thin density distribution with the remarkable property that the gravitational potential on one of its sides is that due to an image mass on the opposite side, as illustrated in Figure 2-5 of Binney and Tremaine.[17] Its gravitational potential is

$$\Phi = \frac{-GM}{\sqrt{R^2 + (a + |z|)^2}}, \tag{1}$$

where $z = 0$ is the plane of the disk, R measures radial distance from the z-axis of symmetry, M is the mass of each image source, and a is their distance from the plane of the disk. The equipotentials are spherical, those in $z > 0$ are centered on the image mass at $z = -a$ below the disk, and those in $z < 0$ are centered on the image mass at $z = a$ above the disk.

Kuzmin-Like Potentials

The Kuzmin-like potentials of Tohline and Voyages[9] are defined by the more general formula

$$\Phi = \Phi_K(\xi), \qquad \xi = \sqrt{R^2 + (a + |z|)^2}. \tag{2}$$

The Kuzmin disk is simply the special case of the point mass potential $\Phi_K = -GM/\xi$. The more general Kuzmin-like potentials also have spherical equipotential surfaces centered on opposite sides of the $z = 0$ plane. However, they are produced by a combination of a thin disk *and* a volume density. The Poisson equation shows that the general Kuzmin-like potential (2) is generated by the density

$$\rho = \frac{1}{4\pi G}\left[\Phi_K''(\xi) + \frac{2(1 + a\delta(z))}{\xi}\Phi_K'(\xi)\right]. \tag{3}$$

The Dirac delta function $\delta(z)$ arises from a discontinuity in the derivative of $|z|$, and gives the thin disk. Kuzmin-like potentials, therefore, arise from a volume density that is stratified on the equipotentials, and a surface density $\Sigma = a\,\Phi_K'(\xi)/2\pi G\xi$ on the plane $z = 0$. It is this surface density that causes the z-component of force to be discontinuous. The volume density vanishes for the special case of the Kuzmin disk, when there is only a surface density

$$\Sigma(R) = \frac{aM}{2\pi(a^2 + R^2)^{3/2}}. \tag{4}$$

The freedom to choose the spherical potential $\Phi_K(\xi)$ allows there to be a considerable variety among Kuzmin-like potentials for modeling galaxies with both disks and halos. However, a choice of $\Phi_K(\xi)$ fixes both the volume and the surface density.

The masses of the parts of the disk and the whole that are interior to the equipotential $\xi = \xi_0$ are:[8]

$$M_{\text{disk}}(\xi_0) = \frac{a}{G}[\Phi_K(\xi_0) - \Phi_K(a)],$$

$$M_{\text{total}}(\xi_0) = \frac{\xi_0}{G}(\xi_0 - a)\,\Phi'_K(\xi_0).$$

(5)

Generally the disk mass becomes a progressively smaller part of the whole at large distances as ξ_0 increases. For example, if we choose $\Phi_K(\xi)$ to be a logarithmic potential so as to obtain a rotation curve that becomes flat at large distances, then

$$\Phi_K(\xi) = V_0^2 \ln \xi, \quad \rho = \frac{V_0^2}{4\pi G \xi^2} + \frac{aV_0^2 \delta(z)}{2\pi G(R^2 + a^2)}.$$

(6)

Both disk and halo masses are now infinite, although the disk mass grows logarithmically with increasing ξ_0, whereas the total mass has a faster linear growth. An intermediate case between this, and that of the Kuzmin disk for which all of the mass is in the disk, is given by

$$\Phi_K(\xi) = -\frac{GM_D}{\sqrt{a\xi}}.$$

(7)

Now the disk has finite total mass M_D, whereas the halo has infinite mass.

Dynamics in Kuzmin-Like Potentials

Motion in a spherical potential is super-integrable. It conserves the energy E and the angular momentum vector about the center of force. The motion is confined to the plane through the center of force and perpendicular to the angular momentum vector. Disk-crossing orbits in a Kuzmin-like potential pass continually back and forth between the regions $z > 0$ and $z < 0$, and hence from one spherical potential field to the other. The angular momenta that are conserved in the two regions are both described by the vector

$$\mathbf{J} = [\mathbf{r} + a\,\text{sgn}(z)\mathbf{k}] \times \mathbf{v},$$

(8)

where $\mathbf{r}$ is the position relative to the origin, $\mathbf{v}$ is velocity, and $\mathbf{k}$ is the unit vector in the z-direction. This vector changes discontinuously as z changes sign unless $\mathbf{v}$ is then perpendicular to the disk. The only component of $\mathbf{J}$ that does not change as the disk is crossed is the z-component J_z. That is because it is also the z-component L_z of the angular momentum $\mathbf{L} = \mathbf{r} \times \mathbf{v}$ about the origin midway between the two centers of force, and is a constant of the motion everywhere, as in any axisymmetric potential. Since it is known that energy and angular momenta are the only constants of motion in general spherical potentials and the other two components of $\mathbf{J}$ are not conserved in Kuzmin-like potentials, it follows that, contrary to what is stated in Reference 9, Kuzmin-like potentials are not generally integrable. The Kuzmin disk is exceptional. It is integrable because its potential is Stäckel, and so has a third integral. That third integral is a linear combination of $\mathbf{J}^2$ and the z-component of the Laplace–Runge–Lenz vector, the fifth integral of a Keplerian potential.[18]

Orbits in Kuzmin-Like Potentials

Orbits must be computed numerically. We use the method described in Reference 8 to compute surfaces of section (SOS) by integrating orbits from one crossing to the next, with no interpolation needed to find when and where crossings occur. The SOS show periodic orbits, regular regions associated with stable periodic orbits, and stochastic regions. One well-known family of periodic orbits which appears in many surfaces of section, and whose dynamics are particularly simple in a Kuzmin-like potential, is that associated with the period-1 periodic orbit of banana/saucer type.[19] We label a periodic orbit by the number of times that it crosses $z = 0$ with $\dot{z} > 0$ during a complete period of its motion in both R and z. The banana periodic orbit, shown in the left panel of FIGURE 1, is a polar orbit with zero component L_z of angular momentum about the axis of symmetry. It moves in a straight line directly toward or away from the center of force while on one side of the disk. Its two linear segments are connected by a curved segment on the other side of the disk when it is controlled by the other center of force. This banana orbit is modified to a saucer orbit when $L_z \neq 0$. Angular momentum then bars the orbit from approaching the axis of symmetry, and reflects it before it reaches the z-axis.[19] The right half of FIGURE 1 shows how little the orbit is changed otherwise; its abrupt change in curvature at the disk crossing is still evident, and its upper segment is close to straight.

Figure 2 displays the $z = 0$ SOS for orbits in the logarithmic Kuzmin-like potential (6) for a specific pair of values of the energy E and L_z. We identify the energy by the value of R_c, the radius of a circular orbit of that energy. This gives a typical distance of the orbit from the center of the model, and is more easily comprehensible than a numerical value of E. The value of E can be computed from it using

$$E = \Phi_K(\xi_c) + \frac{R_c^2}{2\xi_c} \Phi_K'(\xi_c), \qquad \xi_c = \sqrt{R_c^2 + a^2}. \tag{9}$$

We express L_z as a fraction of the angular momentum

$$L_c(E) = R_c^2 \sqrt{\frac{\Phi_K'(\xi_c)}{2\xi_c}}, \tag{10}$$

of the circular orbit of radius R_c, that is, as a fraction of its maximum possible value.

Small values of $L_z/L_c(E)$, such as we have in our SOS, allow orbits that range widely, and have a range of geometric shapes. They include orbits that are flat and remain close to the plane of the disk at all times. In fact, the outer boundary of each SOS is formed by the orbit that lies always in the disk; a small value of $L_z/L_c(E)$ allows its pericenter to be a small fraction of R_c. Generally orbits become progressively flatter from the center point of an SOS to its outside. The center point shows the single crossing of a symmetric period-1 thin tube orbit. This orbit travels far above and below the plane of the disk. It is closest to the z-axis at its extremities, and furthest when crossing the disk. We plot only the upper, $v_R > 0$, half of each SOS, so as to show more detail; the lower, $v_R < 0$, half is its mirror image reflected in the R-axis.

The disk contains 25% of the matter inside the equipotential $\xi = \xi_c$ when $R_c/a = 10$, and the angular momentum for FIGURE 2 is 15% of its maximum. The figure shows a large regular region of short axis tubes surrounding the thin tube orbit represented by the central fixed point at $R/a = 9.44$ and $v_R = 0$. That orbit rises to distances $z/a = \pm 9.76$ above and below the disk, at which points $R/a = 1.64$. It remains always far from the center. Two four-island chains in the regular region surround pairs of period-2 orbits. Those for the inner chain are of the symmetric spaceship type shown by full curves in Figure 7(b) of Reference 8. The reason for four islands, but period 2, is that the orbits can be traversed in either direction. Those for the outer chain, for which the stability properties are reversed, are a pair of unsymmetric twisted fish form, shown by dashed curves in Figure 7(b) of Reference 8. Orbits cross the disk closer and closer to its center as we move outward through the SOS, as well as remaining progressively closer to the disk. They eventually become primarily chaotic when they approach closer to the z-axis than $R/a \approx 1$. The two families of regular orbits within the chaotic sea are a three-island chain associated with period-3 pretzel orbits[19] at $R/a = 5.76$ and $v_R/V_0 = 1.29$, and the outer two-island chain associated with the period-1 saucer orbit at $R/a = 3.52$ and $v_R/V_0 = 1.61$. The latter is flatter than that shown in FIGURE 2B. Adjacent to outer boundary of the SOS is a narrow strip of regular orbits that remain very close to the plane of the disk at all times.

Figure 11 of Reference 8 displays the SOS for the same potential and energy but the lower angular momentum $L_z/L_c(E) = 0.1$. It is not greatly different, but its chaotic region is larger and now envelops the outer four-island chain. The extent of the chaotic region shrinks as $L_z/L_c(E)$ increases, and the four-island chains disappear before $L_z/L_c(E) = 0.2$. The reason for their disappearance is that they merge and annihilate at $L_z/L_c(E) = 0.184$ in the manner discussed by Terzić.[20] That same merger is close to happening in FIGURE 11 (see below) for a model of the Milky Way.

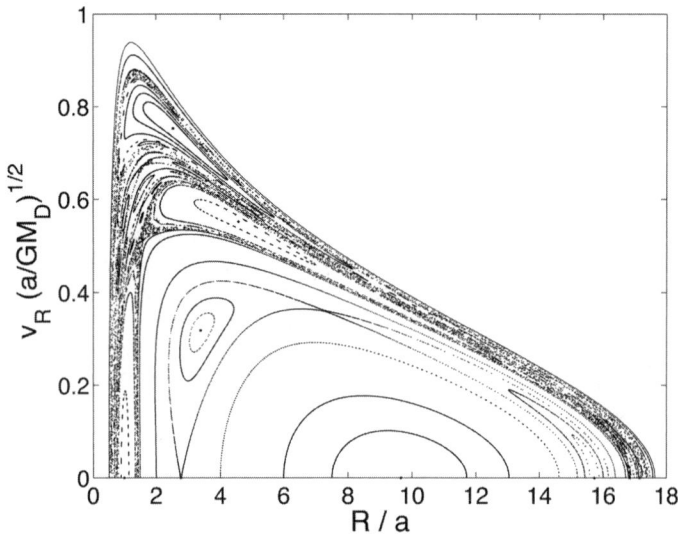

FIGURE 3. Surface of section for the power-law Kuzmin-like potential (7) for $R_c/a = 10$ and $L_z/L_c(E) = 0.15$.

The extent of the chaotic region decreases at lower energies, but it is at least equally important in FIGURE 3 for the power law potential (7) as it is in FIGURE 2. FIGURE 3 is drawn for the same R_c and relative angular momentum as in FIGURE 2, although now the disk has only 18% of the matter inside $\xi = \xi_c$. It lacks the period-2 orbits of FIGURE 2, but has a three-island chain associated with period-3 reflected fish orbits[19] in its regular region. The chaotic region surrounds prominent islands of stability associated with the saucer orbits and a second three-island periodic chain, as well as many smaller higher order resonances. The stable orbit at the center of the outer three-island chain is again of pretzel type, like the unstable period-3 orbit of the inner chain. The two three-island chains also merge and annihilate at higher angular momenta, as in Terzić.[20]

OTHER SIMPLE MODELS

Miyamoto-Nagai Models

Miyamoto and Nagai proposed the potential

$$\Phi_{MN} = -\frac{GM}{\xi_b}, \qquad \xi_b = \sqrt{R^2 + (a + \sqrt{z^2 + b^2})^2}, \qquad (11)$$

with two length scales a and b for modelling galaxies. It has been used widely. Binney and Tremaine[17] comment that the flattened isophotes obtained for $b/a = 0.2$ are qualitatively similar to the light distributions of disk galaxies. Johnson, Spergel, and Hernquist[13] use a Miyamoto–Nagai potential with $b/a = 0.04$ in their model of our Galaxy, whereas Patsis et al.[14] use one with $b/a = 1/3$ for their explanation of boxy profiles in normal spirals.

The Miyamoto-Nagai potential links the Kuzmin disk, its $b = 0$ case, to the Plummer sphere,[21] which is its $a = 0$ case. Both of these limiting cases are integrable, although their third integrals differ. We have looked for, but failed to find, evidence of chaos at intermediate cases. FIGURE 4 is typical. It is for $a = b$, which is halfway between the two integrable extremes, and for a low relative angular momentum to allow a wide range of orbits. The SOS is dominated by thin tubes, interrupted only by two slender island chains around the period-1 saucer and period-3 pretzel orbits.

Alar Toomre (private communication) proposed a combination of a Kuzmin disk of mass δM and a Plummer sphere of mass $(1 - \delta)M$ with the same length scale as an alternative link between the Kuzmin disk and the Plummer sphere. We computed some of its SOS for various values of δ, which we do not show because we found them also to be highly regular.

Thickened Kuzmin-Like Potentials

Non-zero values of the length scale b in the Miyamoto–Nagai variable ξ_b of Equation (11) thicken the Kuzmin disk. The Kuzmin-like potentials can similarly be thickened by replacing ξ in the Kuzmin-like potential of Equation (2) by ξ_b to obtain

$$\Phi = \Phi_K(\xi_b). \qquad (12)$$

FIGURE 5 shows the SOS that is obtained when the logarithmic Kuzmin-like potential with the same E and L_z as in FIGURE 2 is thickened with $b/a = 1/3$. Overlapping the

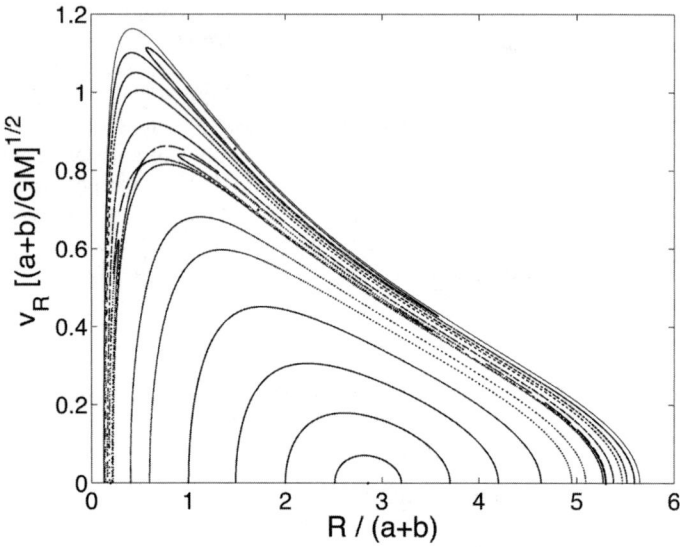

FIGURE 4. Surface of section for the Miyamoto an Nagai potential (11) for $R_c/(a+b)$ = 3 and $L_z/L_c(E) = 0.1$.

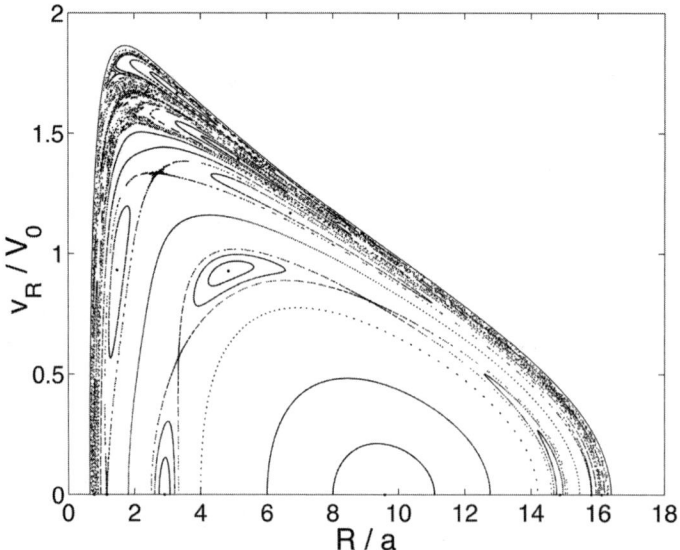

FIGURE 5. Surface of section for the thickened logarithmic Kuzmin-like potential (6) for $R_c/(a+b) = 10$, $L_z/L_c(E) = 0.2$, and $b/a = 1/3$.

two figures shows that the quite substantial thickening has enlarged the regular central region only a little at the expense of the chaotic region, the islands within the chaotic sea are somewhat diminished, but the basic qualitative structure is unchanged. FIGURE 5 provides evidence, in addition to that in Reference 8, that the chaos we found to be induced by disk crossings is not due to the use of a razor-thin disk, because it persists when the disk is thickened significantly.

Scalefree Logarithmic models

The last class of simple models that we consider are the scalefree logarithmic models of Monet et al.[11] and Toomre.[12] They have potentials and densities of the form

$$\Phi = V_0^2\left[\ln\left(\frac{r}{a}\right) + P(\theta)\right], \qquad \rho = \rho_0 S(\theta)\left(\frac{a}{r}\right)^2, \qquad (13)$$

where θ is the colatitude measured from the axis of symmetry and ρ_0 is the average density on the sphere $r = a$. These models can incorporate a razor-thin disk in the equatorial plane by including a singular component $2\beta\delta(\theta - \pi/2)$ in the angular density function $S(\theta)$. The disk then contains a fraction β of the mass within any finite radius, because of the scalefree property. The total mass of both disk and halo are now infinite.

The Toomre models are more general and are constructed so as to have two-integral distribution functions

$$f(E, L_z) \propto L_z^{2n} \exp\left[-\frac{(2n+2)E}{V_0^2}\right]. \qquad (14)$$

Monet et al. consider only the $n = 0$ case, for which the distribution function depends only on the energy because, as Toomre notes, the density is then constant on the equipotentials. The volume densities (3) of Kuzmin-like models are also constant on equipotentials, and so their two-integral distribution functions, which we do not give, also depend only on energy. The $n = 0$ scalefree models are not Kuzmin-like. Monet et al. introduced their models to investigate the effect of massive disks on bulge isophotes. The mass models (13) become flatter and the density increasingly concentrated toward the equatorial plane with increasing n. Toomre's Figure 1 shows this for the diskless $\beta = 0$ case, and the same happens when there is a disk.

Neither Monet et al., nor Toomre discuss orbits in their models. Scalefree models have the advantage that orbits at each energy are scaled versions of those at other energies.[22] SOS are therefore needed for just one value of E. We take that value to be $V_0^2/2$, corresponding to $R_c/a = 1$. That still leaves three parameters, the angular momentum L_z, the relative mass β of the disk, and the index n, to be explored. Our three SOS, all with the same $L_z/L_c(E) = 0.2$, study the effects of varying halo flattening and disk mass. FIGURE 6 is for the otherwise spherical $n = 0$ halo flattened by a 20% disk. Its SOS is simpler than that for a logarithmic Kuzmin-like potential, although it too is for a logarithmic potential. The regular and chaotic regions are now equally large. The chaotic region is interrupted only by a prominent family of saucer orbits, and a smaller family of period-3 pretzel orbits. The slender three-island chain surrounding the saucer orbit is associated with a periodic orbit that has a small period-3 oscillation about the saucer. Increasing the disk mass enlarges the chaotic

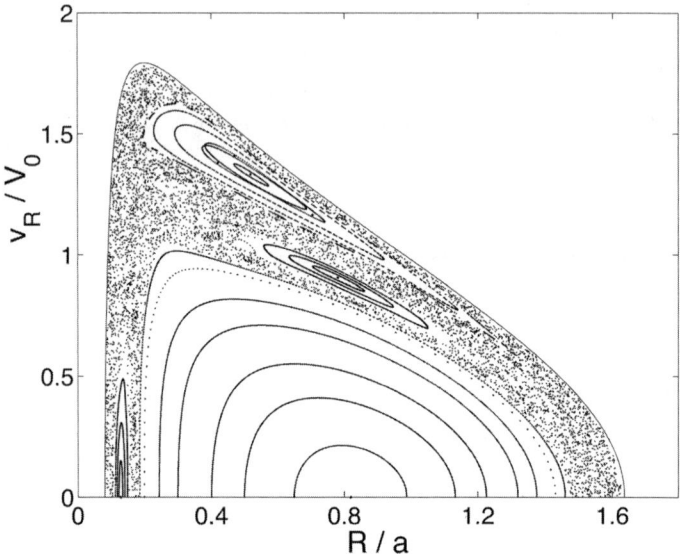

FIGURE 6. Surface of section for the $n = 0$ scalefree logarithmic potential (13) for a 20% disk ($\beta = 0.2$) and $L_z/L_c(E) = 0.2$.

region, whereas decreasing the disk diminishes the chaotic region. Chaos of course disappears when there is no disk and the halo becomes spherical.

FIGURE 7 shows delay plots for two orbits which lie on the lower edge of the saucer orbit region of FIGURE 6. Delay plots are used widely for analyzing time series generated by dynamical systems,[6] and can show more clearly than trajectories in (R, z)-space which orbits are chaotic. FIGURE 7 plots the azimuthal component of angular momentum $L_\phi = -R\dot{z}$ at one crossing of the SOS with $\dot{z} > 0$, versus its value at its previous crossing. The role of L_ϕ here is simply that of a scalar property of the orbit. It varies continuously along the orbit, unlike the **J** components used in the delay plots in Reference 8, which remain constant in $z > 0$ and change only because of disk crossings. Nevertheless, both kinds of delay plots have the same character. Those for regular orbits form closed loops, whereas those for chaotic orbits show a widespread scatter. FIGURE 7 shows how sudden the transition can be. FIGURE 7A shows the closed loop characteristic of an orbit of the saucer family and FIGURE 7B, for a close starting point in the SOS, shows nearly random scattering with little evidence of the previous structure.

FIGURE 8 shows the SOS for the flatter $n = 1$ halo, also with a 20% disk. The regular core has shrunk, and the general extent of the chaotic region has grown, but it now embeds a much larger regular island. That island has the usual saucer orbit at its center and it is now surrounded by islands of period-4 orbits that wobble around the saucer orbit. The pretzel family seen in FIGURE 6 has disappeared, but other small islands have appeared, including a fish orbit at $R/a = 0.27$ and $v_R/V_0 = 1.67$.

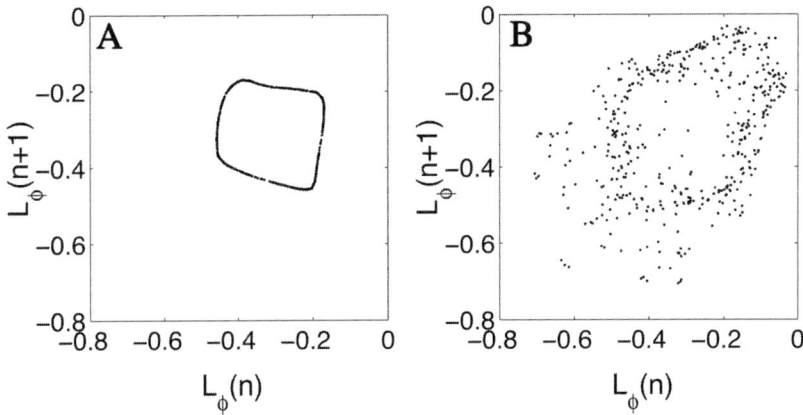

FIGURE 7. Delay plots of L_ϕ, the angular momentum perpendicular to the axis of symmetry, when the SOS is crossed with $\dot{z} > 0$. The two orbits start in the SOS of FIGURE 6 from **(A)** $R/a = 0.4$, $v_R/V_0 = 1.314$ and **(B)** $R/a = 0.4$, $v_R/V_0 = 1.312$.

FIGURE 9, also for $n = 1$ but now with no disk, shows the extent to which the features seen in FIGURE 8 are due to halo flattening alone. It shows that the disk is responsible for the large size of the chaotic region. With no disk, chaos is confined to narrow strips around the regular regions and the neighborhoods of some unstable

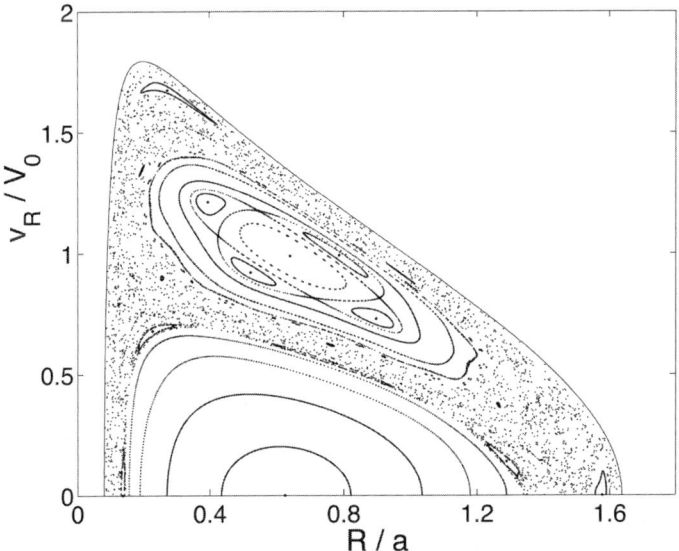

FIGURE 8. Surface of section for the $n = 1$ scalefree logarithmic potential **(13)** for a 20% disk ($\beta = 0.2$), and $L_z/L_c(E) = 0.2$.

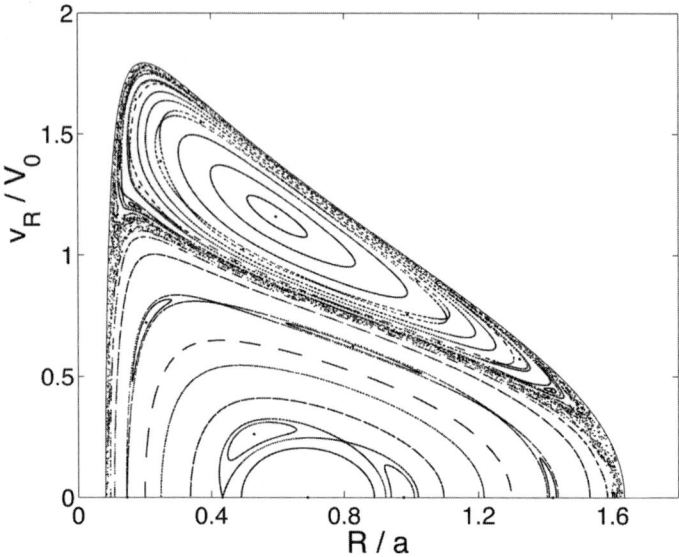

FIGURE 9. Surface of section for the $n = 1$ scalefree logarithmic potential **(13)** with no disk ($\beta = 0$) and $L_z/L_c(E) = 0.2$.

fixed periodic orbits. The absence of the disk has allowed a large variety of periodic orbits and island chains to appear, including a period-3 reflected fish orbit near the center and two four-island chains now surround the saucer orbit.

MODELS OF THE GALAXY

We investigate orbits in the three-component model of the Galaxy introduced by Johnston et al.,[13] and extended by Helmi[23,24] to allow for halo flattening, with potential

$$\Phi = \Phi_{MN} - \frac{GM_{bulge}}{r+c} + V^2_{halo} \ln\left(d^2 + R^2 + \frac{z^2}{q^2}\right). \quad (15)$$

The rotation curves obtained from its components and their combinations are shown in FIGURE 10. The mass of the bulge is 0.34 times that of the disk and has a length scale $c = 0.7$ kpc, versus $a + b = 6.76$ kpc for the disk and $d = 12$ kpc for the halo. The centrally concentrated bulge dominates the potential near the center, and thereby distinguishes the potential **(15)** from those we considered previously.

The SOS in FIGURES 11 through 13 are all for an energy for which $R_c = 30$ kpc, and a relative angular momentum of $L_z/L_c(E) = 0.15$ as previously. The reason for this choice of R_c is that it corresponds approximately to that of four of the six orbits studied in Helmi and White,[15] although they had only a disk and spherical halo, and no bulge. When we too ignored the bulge, we obtained a purely regular SOS with no prominent resonances. FIGURE 11 shows the effect of adding the bulge. The SOS is

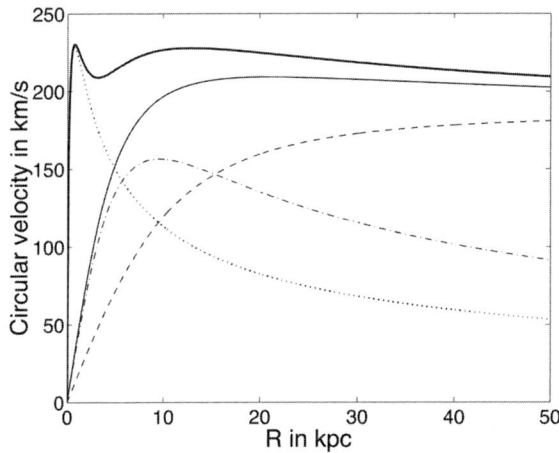

FIGURE 10. Rotational velocities of the components of models of the Galaxy. The *dot–dashed*, *dashed*, and *dotted curves* are those due to the disk, halo, and bulge, respectively. The *thin full curve* is that for the disk plus halo model used in Reference 15. The *thick full curve* is that for the full three-component model.

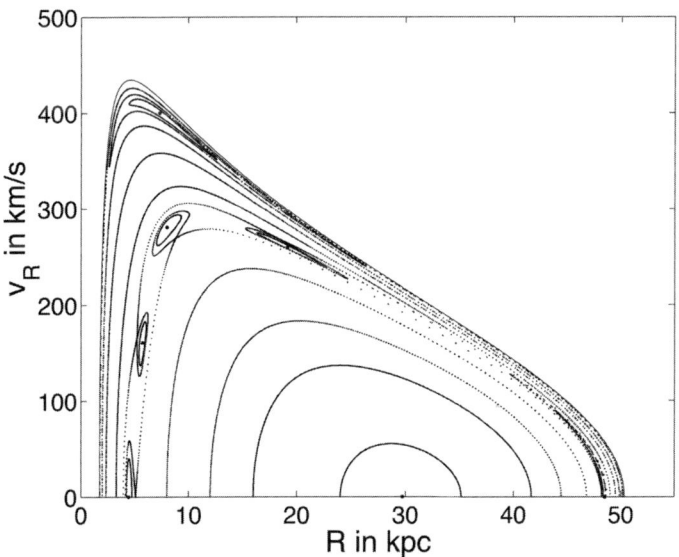

FIGURE 11. Surface of section for the Galactic potential (15) with a spherical halo ($q = 1$), $R_c = 30$ kpc, and $L_z/L_c(E) = 0.15$.

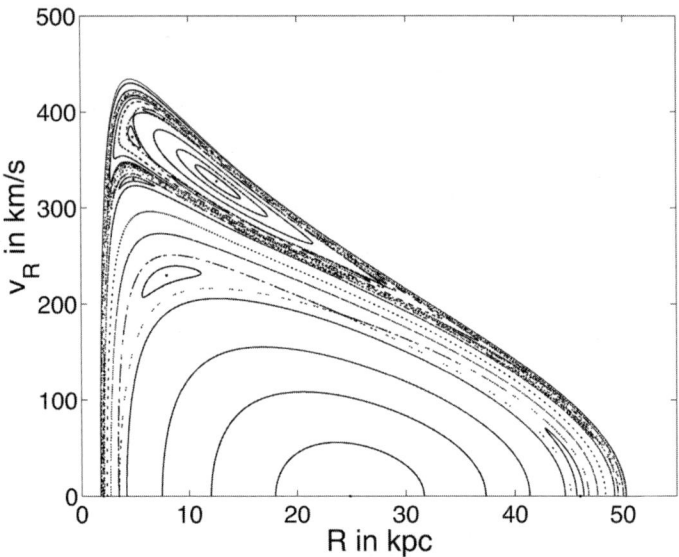

FIGURE 12. As FIGURE 11, but with an oblate spheroidal halo with $q = 0.8$.

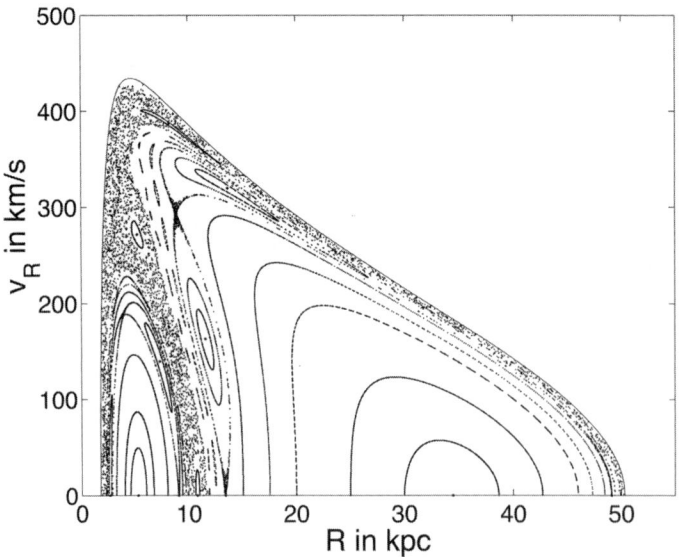

FIGURE 13. As FIGURE 11, but with a prolate spheroidal halo with $q = 1.25$.

still regular, but has three periodic orbits in addition to the ever-present thin tube; a saucer orbit near its edge, and a pair of period-2 orbits, like those of FIGURE 2, which are about to merge and disappear.

The consequences of making the halo spheroidal are illustrated in the next two figures. FIGURE 12 shows that oblateness enlarges the saucer family of orbits, and introduces more periodic orbits, including a central period-3 reflected fish orbit like those in FIGURES 3 and 9, but that there is not much chaos at $L_z/L_c(E) = 0.15$. More chaos develops at lower angular momenta when resonances grow and overlap. FIGURE 13 shows that prolateness makes a much bigger difference. It introduces new periodic orbits, and replaces old ones. The left hand regular region is centered on a thin inner long-axis tube.[18] The three-island chain around it is associated with a period-3 orbit that wobbles around that inner tube and is of the same type as that shown in Figure 7(c) of Evans.[25] The orbits associated with the right hand regular region are similar to those found in oblate potentials, with a thin outer long axis tube at its center, and period-2 twisted fish further out. There are no longer any unsymmetric period-1 saucer orbits; they have been replaced by the symmetric period-1 inner tube that crosses the SOS at $R = 5.23$ kpc, but then travels out to $R = 27$ kpc, and to distances $z = \pm 48$ kpc above and below the Galactic plane. The chaotic region is no longer confined to orbits in the outer part of the SOS which remain close to the Galactic plane. The band of chaotic orbits that lie between the two regular islands range to distances greater than 50 kpc above and below the Galactic plane.

FIGURE 14 is for a much lower energy, of the order of that of near circular orbits in the solar neighborhood, although with much lower angular momenta. Its SOS, which is for a spherical halo, is the simplest of all. It has a large chaotic region that

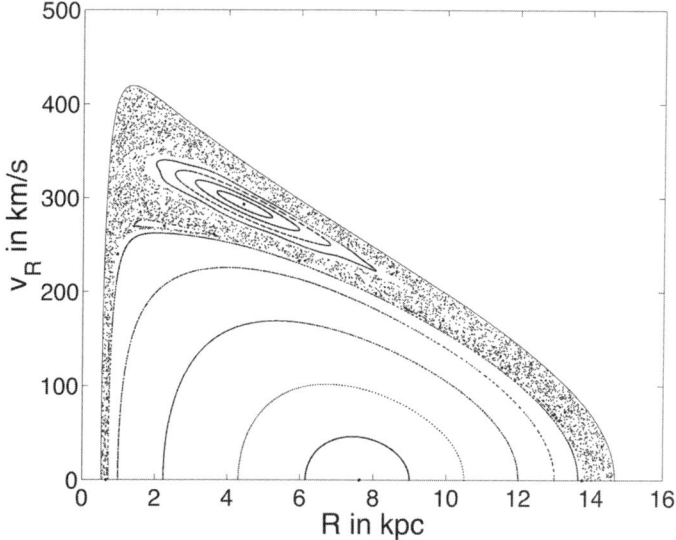

FIGURE 14. As FIGURE 11, but at the lower energy for which $R_c = 9$ kpc.

becomes larger at lower energies, and hence, lower R_c. FIGURE 14 is somewhat like FIGURE 6 for the scalefree disk, but it shows no evidence of a three-island chain. Some differences are to be expected because the spherical bulge dominates the halo for the central regions of FIGURE 14. The singular halo of the scalefree model of FIGURE 6 is equally important everywhere, and it is not spherical because the disk gives it oblateness.

CONCLUSIONS

We have studied orbital behavior in a variety of galactic models with disks and halos. Crossing the disk causes an abrupt change in acceleration, but not in the velocity of an orbit. Hence, disk crossing is a milder perturbation than the impulses, and resulting velocity changes, that Wisdom[26] used to explain the Kirkwood gaps in the asteroid belt. Nevertheless, we find that repeated disk crossings can allow some orbits to become chaotic and that this effect is not sensitive to disk thickness. The orbits that become chaotic are orbits whose pericenters and apocenters are far apart, and whose trajectories are confined relatively closely to the plane of the disk. The Kuzmin disk is exceptional in that every one of its disk-crossing orbits is regular. The Miyamoto–Nagai disks appear to inherit a near-integrability from the Kuzmin disk and the Plummer sphere, which they connect. All our other examples, with the single exception of FIGURE 11 for a Miyamoto–Nagai disk in a spherical potential, have both purely regular regions, and chaotic regions that contain islands of regularity. There are many more islands than our SOS show, because more, smaller, and higher period islands can always be found, given sufficient persistence. However, we believe that our SOS do not omit any islands of significant size.

We show SOS with low values of $L_z/L_c(E)$ because only then can orbits range widely in R or travel far from the plane of the disk. Chaotic regions diminish as $L_z/L_c(E)$ increases and orbits range less widely. The scalefree models considered already enable us to vary the mass of the disk relative to that of the halo, and the interrelated flattening of the halo. Increased flattening increases the number of noticeable resonances, whereas increasing the disk mass enhances their overlap and the generation of chaos. Our previous study[8] of Kuzmin-like potentials, in which there was no freedom to adjust the relative mass of the disk, showed more chaos at higher energies.

In restricting our study to the orbits of test particles, our calculations are simpler than those of Helmi and White[15] and Helmi[23,24] who also modeled the orbiting satellite galaxy. Helmi and White found that their satellites soon disrupted so that its stars behaved as test particles for most of their evolution, as do our simple orbits. Aguilar, Hut, and Ostriker[27] modeled orbiting globular clusters, and found that the clusters that are most vulnerable to destruction by bulge and disk shocking are those on highly radial orbits that come closer and faster to the central regions of the Galaxy. That is consistent with our finding of chaotic regions in the outer parts of our SOS.

Our investigations of orbits in the model (15) of the Milky Way are of very limited scope. When the halo is spherical like the bulge, then only the disk perturbs the integrability of the potential. Chaos cannot then be expected to arise unless an orbit approaches closely enough for the perturbing effect of the disk to be large. The disk enhances resonances at the energy used in FIGURE 11, but no chaos is evident. The less energetic orbits in FIGURE 14 show considerable chaos. This is largely the result of the interaction of the bulge with the Miyamoto–Nagai disk because the halo is much less important closer in.

It is well known that spheroidal halos introduce resonances and periodic orbits,[25] and hence, more scope for a disk to generate chaos. Comparing FIGURES 12 and 13 shows how much more effective a prolate halo is than an oblate one for generating chaos. The orbits used by Helmi[23,24] in her modeling of the Sagittarius dwarf have pericenters in the range 10.1 to 14.7 kpc, and apocenters in the range 59.2 to 52.3 kpc as q varies from 0.8 to 1.25. These ranges require somewhat larger values of R_c than the value $R_c = 30$ kpc we have used, but the large size of the central regular region in FIGURE 12 for an oblate $q = 0.8$ halo leaves little doubt that an orbit with $R_{peri} > 10$ kpc will lie safely inside it and be regular. Although the chaotic region of the prolate $q = 1.25$ halo in FIGURE 13 extends as far out as $R = 11.7$ kpc, the wide ranging orbits that pass through there have much smaller pericentric distances because they come closer than 3 kpc of the center of the Galaxy elsewhere in their orbits. Hence, orbits similar to those considered by Helmi are unlikely to be chaotic even if the Galactic halo is prolate. Of course, the fact that the debris from the Sagittarius remnant is identifiable is direct evidence that its orbit is regular and not chaotic.

ACKNOWLEDGMENTS

This work was supported by the National Science Foundation grant DMS-0104751. The author declares that he has no competing financial interest.

REFERENCES

1. HUBBLE, E.P. 1936. The Realm of the Nebulæ. Yale University Press, New Haven.
2. BENDER, R. 1988. Velocity anisotropies and isophote shapes in elliptical galaxies. Astron. Astrophys. **193:** L7–10.
3. NIETO, J.-L., M. CAPACCIOLI & E.V. HELD. 1988. More isotropic oblate rotators in elliptical galaxies. Astron. Astrophys. **195:** L1–4.
4. SCORZA, C. & R. BENDER. 1995. The internal structure of disky elliptical galaxies. Astron. Astrophys. **293:** 20–43.
5. OSTRIKER, J.P., L. SPITZER & R.A. CHEVALIER. 1972. On the evolution of globular clusters. Astrophys. J. **176:** L51–56.
6. ALLIGOOD, K.T., T.D. SAUER & J.A. YORKE. 1996. Chaos: an Introduction to Dynamical Systems. Springer, New York.
7. MARTINET, L. & F. MAYER. 1975. Galactic orbits and integrals of motion for stars of old Galactic populations. Astron. Astrophys. **44:** 45–57.
8. HUNTER, C. 2003. Disk-crossing orbits. *In* Galaxies and Chaos. Lecture Notes on Physics, 626. G. Contopoulos & N. Voglis, Eds.: 137–153. Springer, New York.
9. TOHLINE, J.E. & K. VOYAGES. 2001. Integrals of motion in Kuzmin-like potentials. Astrophys. J. **555:** 524–531.
10. MIYAMOTO, M. & R. NAGAI. 1975. Three-dimensional models for the distribution of mass in galaxies. Publ. Astron. Soc. Japan **27:** 533–543.
11. MONET, D.G., D.O. RICHSTONE & P.L. SCHECHTER. 1981. The effects of massive disks on bulge isophotes. Astrophys. J. **245:** 454–458.
12. TOOMRE, A. 1982. Some flattened isothermal models of galaxies. Astrophys. J. **259:** 535–543.

13. JOHNSTON, K.V., D.N. SPERGEL & L. HERNQUIST. 1995. The disruption of the Sagittarius dwarf galaxy. Astrophys. J. **451:** 598–606.
14. PATSIS, P.A., E. ATHANASSOULA, P. GROSBØL & CH. SKOKOS. 2002. Edge-on boxy pro les in non-barred galaxies. Mon. Not. R. Astron. Soc. **335:** 1049–1053.
15. HELMI, A. & S.D.M. WHITE. 1999. Building up the stellar halo of the Galaxy. Mon. Not. R. Astron. Soc. **307:** 495–517.
16. KUZMIN, G.G. 1956. Model of the steady Galaxy allowing of the triaxial distribution of velocities. Astron. Zh. **33:** 27–45.
17. BINNEY, J. & S. TREMAINE. 1987. Galactic Dynamics. Princeton University Press, Princeton.
18. DE ZEEUW, P.T. 1985. Elliptical galaxies with separable potentials. Mon. Not. R. Astron. Soc. **216:** 273–334.
19. LEES, J.F. & M. SCHWARZSCHILD. 1992. The orbital structure of galactic halos. Astrophys. J. **384:** 491–501.
20. TERZIĆ, B. 1998. Irregular period-tripling bifurcations in axisymmetric scalefree potentials. Ann. New York Acad. Sci. **867:** 85–92.
21. PLUMMER, H.C. 1911. On the problem of distribution in globular star clusters. Mon. Not. R. Astron. Soc. **71:** 460–470.
22. RICHSTONE, D.O. 1980. Scale-free axisymmetric galaxy models with little angular momentum. Astrophys. J. **238:** 103–109.
23. HELMI, A. 2004. Is the dark halo of our Galaxy spherical? Mon. Not. R. Astron. Soc. **351:** 643–648.
24. HELMI, A. 2004. Velocity trends in the debris of Sagittarius and the shape of the dark matter halo of our Galaxy. Astrophys. J. **610:** L97–100.
25. EVANS, N.W. 1994. The power law galaxies. Mon. Not. R. Astron. Soc. **267:** 333–360.
26. WISDOM, J. 1982. The origin of the Kirkwood gaps: a mapping for asteroidal motion near the 3/1 commensurability. Astron. J. **87:** 577–593.
27. AGUILAR, L., P. HUT & J.P. OSTRIKER. 1988. On the evolution of globular cluster systems. I. Present characteristics and rate of destruction in our Galaxy. Astrophys. J. **335:** 720–747.

Systems with Escapes

G. CONTOPOULOS AND M. HARSOULA

Research Center of Astronomy, Academy of Athens, Athens, Greece

> ABSTRACT: There are two types of escapes in conservative dynamical systems with two degrees of freedom: escapes to infinity and escapes to certain singular points at a finite distance. In both cases the areas on a surface of section are not preserved. We consider the basins of escape to infinity in simple Hamiltonian systems. The initial conditions of orbits escaping after 1, 2, ... intersections with a surface of section form in general spiral fractal sets. Then we consider the sets of escapes into two fixed black holes for various values of the energy. The forms of these sets depend on the unstable periodic orbits and their asymptotic curves. We find the characteristics of the simple periodic orbits and their changes for various values of the energy.
>
> KEYWORDS: escapes; black holes; periodic orbits; asymptotic curves; fractals; Lyapunov characteristic numbers

INTRODUCTION

I was happy to attend the symposium devoted to the memory of Henry Kandrup. Henry Kandrup was one of the leaders in modern dynamical astronomy. He played an important role in many difficult fields, such as the N-body problem, and he emphasized the role of order and chaos in dynamical systems. I had many lively discussions with Henry on many scientific topics during my extended visits to the University of Florida.

This paper on escapes in dynamical systems started as a collaboration with Kandrup. We published five papers together (with additional collaborators)[1–5] dealing with the fractal properties of escapes and with possible universal scaling laws.

The subject of escapes has attracted much interest in recent years; for example, in galactic dynamics,[6] in the restricted three-body problem,[7,8] in the general three-body problem,[9] and in mappings.[10,11] The particular problem of chaotic scattering has been discussed extensively.[12–22] In 1993, a special issue of the journal *Chaos* was devoted to chaotic scattering.[23]

More recently we have attacked the problem of escapes in more detail.[24] We have separated the basins of escape into independent zones according to the time required for escape. Most of these zones have a spiral form, which we could explain theoretically. We review this work here and provide additional results. In particular we study the change of the escape rates as a function of time.

We argue that our results are generic for dynamical systems with escapes. What is also important is that the areas on a surface of section are not preserved, although

---

Address for correspondence: George Contopoulos, Research Center of Astronomy, Academy of Athens, Athens, Greece.
gcontop@cc.uoa.gr

such systems are conservative. In fact, infinity acts as an attractor. In these cases the volumes in phase space are conserved.

We then consider conservative systems with attractors at a finite distance. Such is the system of two fixed black holes in general relativity. In such a system many orbits escape into the two black holes, therefore, we have not only nonconservation of the areas on a surface of section, but also nonconservation of the volumes in phase space.

The sets of orbits falling into the two black holes are fractal[25,26] and this is a clear indication of chaos. Additional work on fractals and chaos in the case of two or more black holes, has been done by Dettmann et al.[27,28] and by Yurtsever.[29] More recently we have studied in detail the asymptotic curves from the unstable periodic orbits and their homoclinic and heteroclinic intersections.[30,31] The corresponding orbits are doubly asymptotic orbits, asymptotically approaching the various unstable periodic orbits.

In this paper, we extend this work for various masses of the black holes and for various energies. We calculate the main periodic orbits, the asymptotic curves of the unstable orbits and the basins of attraction of the two black holes.

ESCAPES TO INFINITY

When the orbits in a conservative system with two degrees of freedom extend to infinity we have escapes. In general, there are no Poincaré surfaces of section. In fact, the escaping orbits, in general, have only a finite number of intersections with a surface of section.

We consider three types of Hamiltonians (see FIGURE 1)

$$H = \frac{1}{2}(\dot{x}^2 + \dot{y}^2 + \omega_1^2 x^2 + \omega_2^2 y^2) - \varepsilon x y^2 = h, \tag{1}$$

$$H = \frac{1}{2}(\dot{x}^2 + \dot{y}^2 + x^2 + y^2) - \varepsilon x^2 y^2 = h, \text{ and} \tag{2}$$

$$H = \frac{1}{2}(\dot{x}^2 + \dot{y}^2 + x^2 + y^2) - \frac{x^3}{3} + \varepsilon x y^2 = h. \tag{3}$$

In each case the curves of zero velocity (CZV) are the equipotentials $H(\dot{x} = \dot{y} = 0) = h$. If h is larger than a critical value $h = h_{esc}$ (escape energy) these curves are open and allow particles to escape. If $h < h_{esc}$ there are no escapes and the surface $y = 0$ in phase space is a Poincaré surface of section. However, for $h > h_{esc}$ the escaping orbits do not intersect this surface beyond a certain distance and the surface of section is no longer a Poincaré surface of section. As a consequence, the areas on this surface are not conserved, although the volumes in phase space are preserved.

The openings of the CZV for $h > h_{esc}$ are two in the first case, four in the second case, and three in the third case. In all cases, at every opening there is an unstable periodic orbit, called Lyapunov orbit (FIG. 1) and any orbit crossing a Lyapunov orbit outward escapes from the system and never returns close to the origin.[32]

If $h < h_{esc}$ we distinguish the orbits on the surface of section as regular and chaotic. In FIGURE 2 we show the case of the Hamiltonian (1) with $\omega_1^2 = 1.6$, $\omega_2^2 = 0.9$, $\varepsilon = 0.08$, and $h = 25.2$. In this case $h_{esc} = 25.31$. The successive inter-

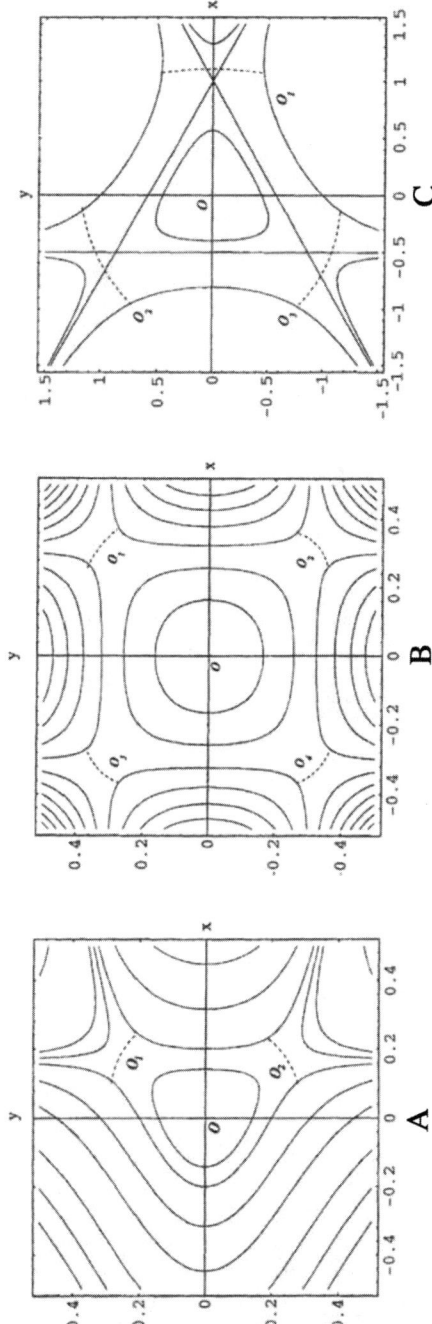

FIGURE 1. Equipotentials, curves of zero velocity (CZV), for the systems (**1**), (**2**), and (**3**). Some Lyapunov orbits are indicated by O_i.[3]

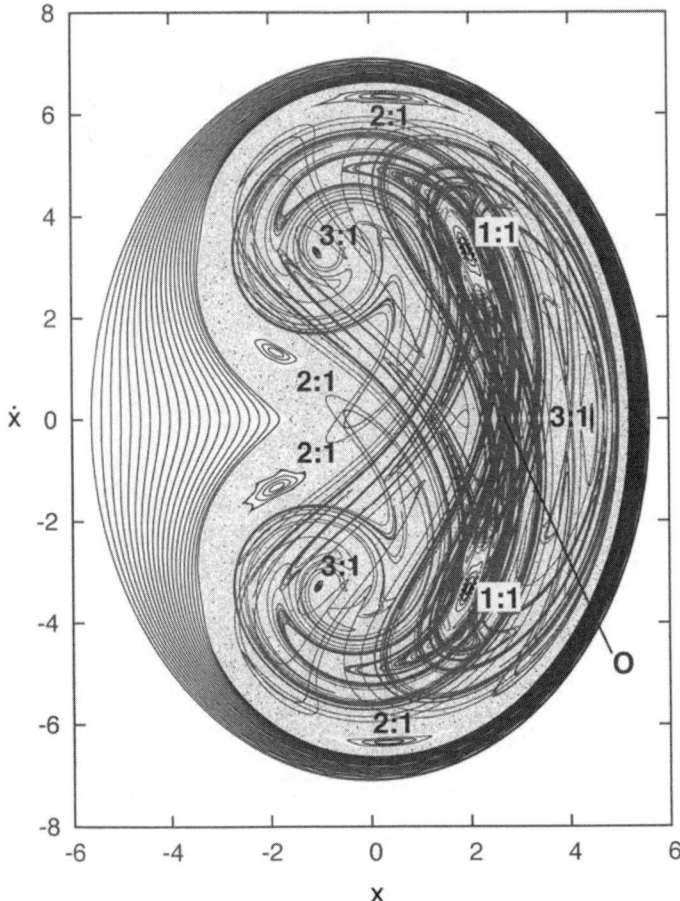

FIGURE 2. A Poincaré surface of section (i.e., $y = 0$, $\dot{y} > 0$) of the Hamiltonian **(1)** for $\omega_1^2 = 1.6$, $\omega_2^2 = 0.9$, $\varepsilon = 0.08$, and $h = 25.2$. There are invariant curves near the boundary and in the islands of stability 1:1, 2:1, 3:1. The scattered points represent a chaotic orbit. We also show long arcs of the asymptotic curves from the unstable periodic orbit O.

sections of the regular orbits lie on invariant curves, whereas they are scattered in an irregular way in the case of chaotic orbits.

The regular invariant curves are of two types: (1) curves near the boundary of the phase space ($\dot{x}^2 + \omega_1^2 x^2 = 2h$) surrounding the chaotic domain, and (2) islands of stability corresponding to particular resonances. In FIGURE 2 ($h = 25.2$) we indicate two islands of stability of type 1:1, two pairs of islands of type 2:1, and three small islands around a stable orbit of type 3:1. There are also higher order islands, both in the chaotic domain and in the regular domain. The two types of invariant curves are not essentially different. In fact, the invariant curves close to the boundary can be considered to form an island around the periodic orbit $y = \dot{y} = 0$.

For smaller values of h ($h < 22.16$) the central orbit (O) is stable and it is surrounded by closed invariant curves. However, for $h = 22.16$, this orbit becomes unstable and generates, by bifurcation, two stable orbits 1:1 above and below it. In FIGURE 2 we have drawn large parts of the asymptotic curves from the orbit O. These asymptotic curves form a homoclinic tangle that generates a large degree of chaos. The asymptotic curves cover most of the chaotic domain of FIGURE 2.

This chaotic domain contains additional unstable periodic orbits (e.g., of types 2:1 and 3:1) with their corresponding homoclinic tangles. However, for this value of h the various homoclinic tangles intersect and form a large heteroclinic tangle (for the terminology and details about tangles see elsewhere[33]). In FIGURE 2 we see a large chaotic domain that we call a *chaotic sea*. The scattered points belong to a single chaotic orbit.

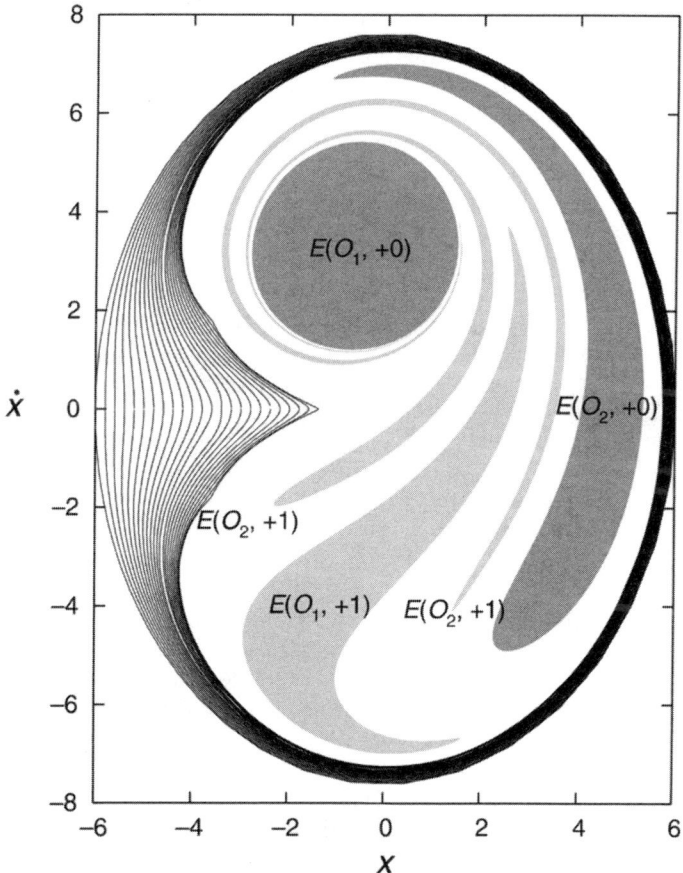

FIGURE 3. The escape regions $E(O_1,+0)$, $E(O_2,+0)$, $E(O_1,+1)$, and $E(O_2,+1)$ on the surface of section $(x, \dot{x})$ of the Hamiltonian (1) for $\omega_1^2 = 1.6$, $\omega_2^2 = 0.9$, $\varepsilon = 0.08$, and $h = 29$. Near the boundary there are several invariant curves.[24]

Close to the outer boundaries of the islands there are sticky chaotic orbits (confined chaotic orbits[5]) that stay for a long time in rings around the islands, but eventually they become unconfined and fill the large chaotic sea. An example of such a sticky orbit is shown around one pair of 2:1 islands in FIGURE 2. This orbit remains in the sticky zone for 12,000 iterations before emerging into the large chaotic sea.

The sticky orbits are limited by cantori, that is, invariant Cantor sets, surrounding the islands with an infinity of holes that produce a temporary obstacle to the orbits inside them.[33–35] Therefore, the density of points around the islands, due to chaotic orbits starting in these regions, is larger than average. However, subsequently these orbits pass through the holes of the cantori and finally the density inside and outside these cantori is equalized.

When the energy h becomes larger than the escape energy $h_{esc} = 25.31$ the chaotic orbits of the large chaotic sea escape from the system to infinity. However, the escapes are not immediate. In FIGURE 3 we mark the domains of initial conditions of orbits escaping through the Lyapunov orbits O_1 and O_2 without any further intersection with the surface of section (domains marked $E(O_1,+0)$ and $E(O_2,+0)$, respectively), and the domains of orbits that escape after one intersection ($E(O_1,+1)$ and $E(O_2,+1)$). There are more domains that escape after n iterations ($E(O_i,+n)$ with $i = 1,2,...,n \geq 2$). These domains fill the empty space between the invariant curves close to the boundary and the domains $E(O_i,+n)$ ($i = 1, 2, n = 0, 1$). There are also domains $E(O_i,-n)$ ($i = 1, 2$), $n \geq 0$, that escape in the backward time direction. The domains $E(O_1,-n)$ and $E(O_2,-n)$ are symmetric to $E(O_2,+n)$ and $E(O_1,+n)$, respectively, with respect to the axis $\dot{x} = 0$.

The totality of the escape domains form a region very similar to the large chaotic sea that existed for $h < h_{esc}$. In fact, the chaotic domain of FIGURE 2 is very similar to the total escape domain of FIGURE 3, because the change of the energy is rather small. However, if h becomes much larger the escape domains change and they become larger (see FIGURE 4). The image of the domain $E(O_1,+0)$ after one iteration backward in time is the domain $E(O_1,+1)$. In fact, any point inside $E(O_1,+1)$ has only one iterate (forward in time) before escaping, therefore, after one iteration it has to be in a corresponding region $E(O_1,+0)$ of points that escape immediately afterward. Since the domain $E(O_1,+0)$ does not intersect the domains $E(O_1,-0)$ or $E(O_2,-0)$, there are no escapes of the points of $E(O_1,+0)$ between $t = 0$ and $t = -1$. As a consequence, the area of $E(O_1,+1)$ is equal to the area of $E(O_1,+0)$.

On the other hand, the domain $E(O_2,+0)$ intersects its symmetric domain $E(O_1,-0)$, therefore, the common part CP = $E(O_2,+0) \cap E(O_1,-0)$ (see FIGURE 5) of these two domains escapes between $t = 0$ and $t = -1$. As a consequence, the total area of $E(O_2,+1)$ is smaller than the area of $E(O_2,+0)$ by the amount CP. The domain $E(O_2,+1)$ is split into two parts, corresponding to the upper and lower parts of $E(O_2,+0)$ (i.e., parts that are outside $E(O_1,-0)$).

In FIGURES 3 and 5 it can be seen that the two subdomains $E(O_2,+1)$ spiral around $E(O_1,+0)$. This can be explained as follows. The boundaries of $E(O_1,+0)$ and $E(O_1,-0)$ are the intersections of the stable and unstable manifolds of the Lyapunov orbit O_1 with the surface of section. For example, orbits from the boundary of $E(O_1,+0)$ asymptotically approach the orbit O_1, and orbits inside this boundary escape to infinity by crossing O_1. Similarly, orbits from the boundary of $E(O_1,-0)$ asymptotically approach O_1 in the backward time direction.

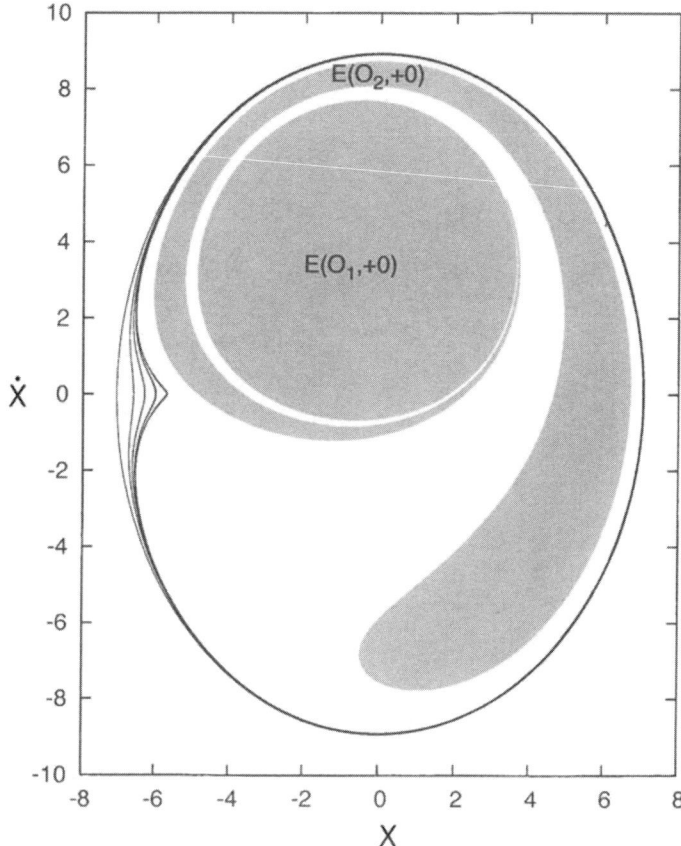

FIGURE 4. The escape regions $E(O_i,+0)$ ($i = 1, 2$) for $h = 40$.

Orbits along a small segment δ perpendicular to the boundary of the region CP (which is a part of the boundary of $E(O_1,-0)$, FIG. 5) are close to the unstable manifold of O_1. The orbits starting in the part of δ inside the boundary of CP escape directly to infinity without any further intersection with the surface of section. On the other hand, the orbits starting in the part of δ outside the boundary of CP approach O_1, but later they deviate from it, remaining close to the stable manifold of O_1 (always backward in time), and reach the surface of section close but outside the boundary of $E(O_1,+0)$. However, the points of δ are known to be mapped after one iteration backward in time into a line inside the domain $E(O_2,+1)$, approaching the boundary of $E(O_1,+0)$. Thus, they produce the thick line inside the gray region $E(O_2,+1)$ in FIGURE 5. As the outer points of δ approach the boundary of CP the corresponding orbits make more and more oscillations close to O_1 and deviate from it after longer and longer times. As a consequence, nearby points along δ form orbits that deviate further and further from each other as they move away from O_1 toward

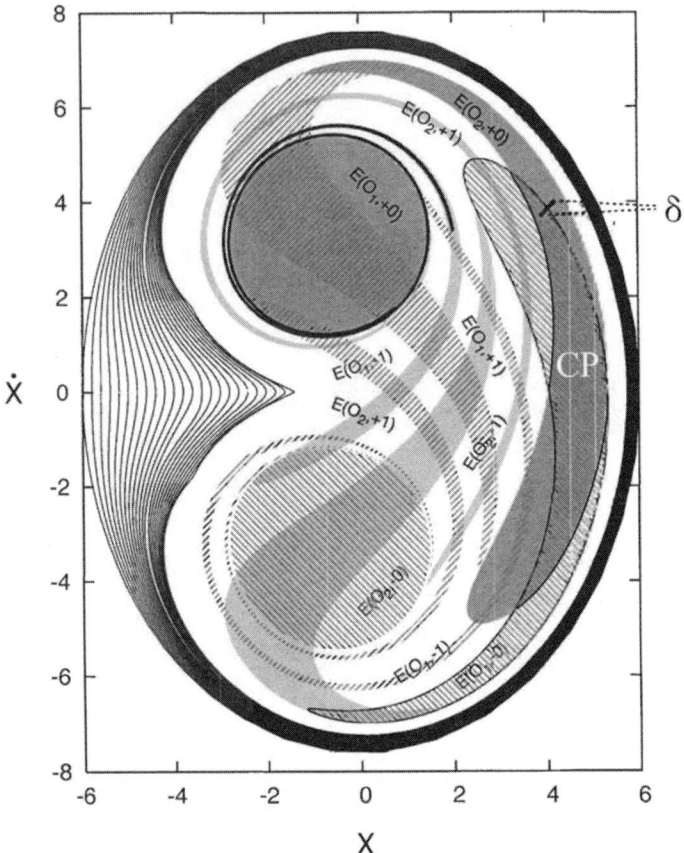

FIGURE 5. The escape regions $E(O_i, \pm n)$ ($i = 1, 2, n = 0, 1$) for $h = 29$. The common region $E(O_2,+0) \cap E(O_1,-0)$ is denoted by *CP*. A small segment δ in $E(O_2,+0)$ reaching the boundary of $E(O_1,-0)$ is mapped backward in time into an infinite spiral inside $E(O_2,+1)$ (*dark line*).

the surface of section close to the boundary of $E(O_1,+0)$. Thus, the image of the outer part of δ becomes infinitely long, surrounding the boundary of $E(O_1,+0)$ infinitely many times before reaching it asymptotically as $t \to -\infty$.

The escapes of points through the Lyapunov orbits follow a pattern. In FIGURE 6 we show the logarithm of the proportion of escapes $\Delta N/(N_0 - N_{non})$ as a function of n (the number of iterations, an integer time). ΔN is the number of escapes between $t = n$ and $t = n + 1$, N_0 is the total number of available orbits, and N_{non} is the number of nonescaping orbits. We note that initially we have a large number of escapes, then an almost linear decrease in the number of escapes, and later a slower decrease of escapes.

During the large linearly decreasing part of each curve of FIGURE 6 we have

$$\ln\left(\frac{\Delta N}{N_0 - N_{\text{non}}}\right) = a + \lambda t, \quad (4)$$

hence,

$$\Delta N = A e^{\lambda t}, \quad (5)$$

with $\lambda < 0$, that is, an exponential decrease in the number of escapes. After the transition to a smaller escape rate we have

$$\Delta N = A' e^{\lambda' t}, \quad (6)$$

where $\lambda < \lambda' < 0$. Note that the transition to the lower escape rate $|\lambda| \to |\lambda'|$ occurs at smaller times for larger values of the energy ($h \geq 26.5$). As a consequence, the curves of FIGURE 6 intersect themselves; therefore, the rate of escapes at a fixed time is not monotonic in h. For example, at $t = 200$ the escapes ΔN decrease as h increases from 25.5 to 26.5, but then increase as h goes to $h = 27$ and $h = 28.5$.

Similar results appear for other values of ω_1 and ω_2 and for other Hamiltonians, such as those given by Equations (2) and (3). In our previous papers[4,5] we noticed an interesting similarity between the systems (1), (2), and (3), namely, that the escape probability P in all cases is given by a power law $(\varepsilon - \varepsilon_i)^a$ (for fixed h), where ε_i is a critical perturbation larger than the escape perturbation ε_{esc}, and a has approximately the same value, $a = 0.5$.

In the case of the Hamiltonian (1) with $\omega_1 = \omega_2 = 1$, $\varepsilon = 1$, and $h = 0.15$, the escape regions are shown in FIGURE 7. This figure is qualitatively similar to FIGURE 3, but

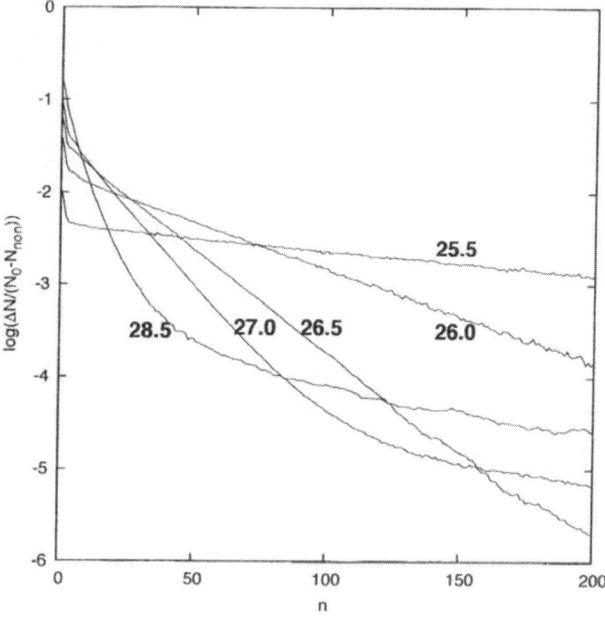

FIGURE 6. Logarithm of the proportion of escapes between times n and $n + 1$ as a function of n for various values of the energy h.

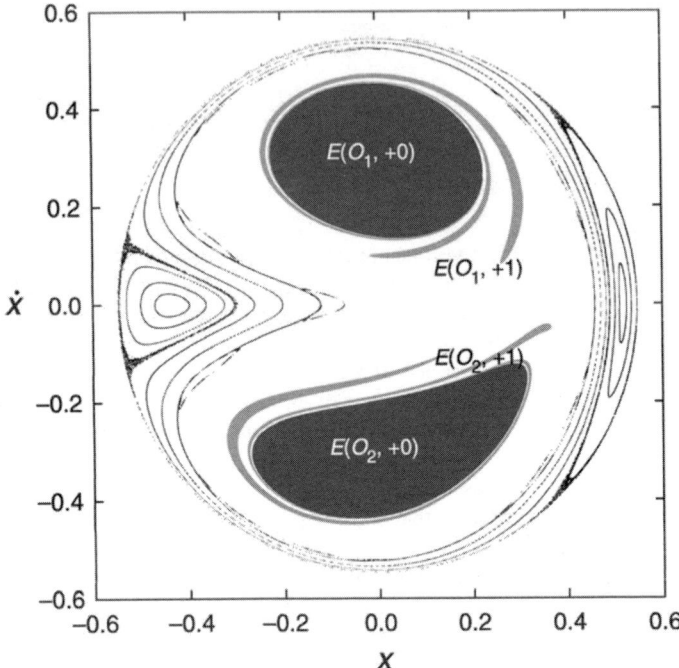

FIGURE 7. The escape regions $E(O_i, n)$ ($i = 1, 2, n = 0, 1$) in the Hamiltonian **(1)** for $\omega_1 = \omega_2 = 1$, $\varepsilon = 1$, and $h = 0.15$.[24]

there are certain differences: (1) Although the domain $E(O_1,+0)$ of FIGURE 7 is in about the same position as in FIGURE 3, the domain $E(O_2,+0)$ is not on the right but is below the center of the figure. (2) The boundary $(\dot{x}^2 + x^2 = 2h)$, which is the periodic orbit $y = \dot{y} = 0$, in this case is unstable and close to it there are small chaotic regions independent of the large chaotic sea that surrounds the domains of escape. On the left and on the right there are two large islands of stability, and further away there are invariant curves surrounding the large chaotic sea. As the energy increases the escape domain covers most of the phase space, reaching the boundary.[24]

ESCAPES INTO TWO BLACK HOLES

In the case of two fixed black holes, a large set of orbits escape into these black holes. This is a genuine relativistic effect that does not occur in Newtonian mechanics. In fact the Newtonian problem of two fixed centers is integrable and the orbits can be given explicitly.[36,37] In this case only a set of orbits with measure zero reaches the two centers (black holes). In particular, in the case of one fixed center, only exactly radial orbits reach the center. However, in the relativistic case the black holes act as attractors; they attract all orbits that come close to them. In this case, not only the areas on a surface of section are not conserved, but also the volumes in phase space are not conserved.

The motions of particles in the case of two black holes are governed by the Lagrangrian[38]

$$2L = V^{-2}\dot{t}^2 - V^2(\dot{x}^2 + \dot{y}^2 + \dot{z}^2) = 1, \tag{7}$$

where

$$V = -\left[1 + \frac{M_1}{r_1} + \frac{M_2}{r_2}\right] \tag{8}$$

is the total potential. M_1 and M_2 are the masses of the two black holes, which we assume to be situated along the axis z, at $z = \pm 1$; r_1 and r_2 are the distances of a particle with infinitesimal (but nonzero) mass from M_1 and M_2. The dots mean derivatives with respect to the line element $s = \tau$ (proper time). Without loss of generality, we can take $M_2 = 1$ and various values of M_1. We consider orbits on a "meridian" plane (x, z), that is, a plane passing through M_1 and M_2. Furthermore, we consider an elliptic energy $E < 1$, that is, the orbits are not allowed to escape to infinity. The energy is given by

$$E^2 = \dot{x}^2 + \dot{z}^2 - V^{-2}. \tag{9}$$

We consider now the intersections of the orbits by a surface of section $x = 0$ ($\dot{x} > 0$). These intersections are inside the limiting curve,[31]

$$\dot{z}^2 = E^2 - V^{-2}, \tag{10}$$

where

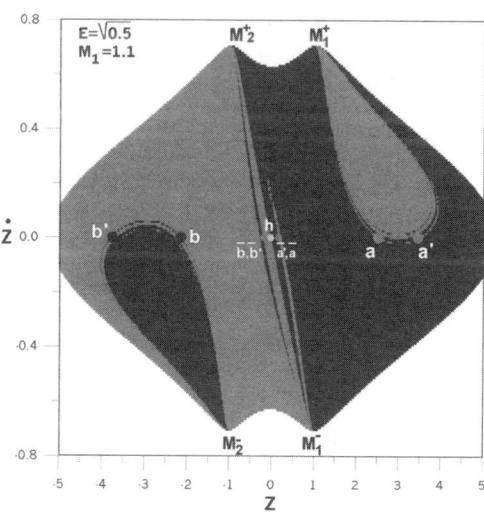

FIGURE 8. The basins of attraction of two fixed black holes for $E = \sqrt{0.5}$, $M_1 = 1.1$ ($M_2 = 1$). The periodic orbits are a and a' (clockwise around M_1, black), b and b' (counterclockwise around M_2, gray), $\bar{a}$, $\bar{a}'$, $\bar{b}$, and $\bar{b}'$ (the same as a, a', b, and b' in opposite directions), and h (like a hyperbolic arc).

$$V = -\left[1 + \frac{M_1}{|z-1|} + \frac{M_2}{|z+1|}\right]. \tag{11}$$

The curve (10) passes through the black holes M_1 and M_2 at the points M_1^- ($z=1$, $\dot{z}=-E$), M_1^+ ($z=1$, $\dot{z}=+E$), M_2^- ($z=-1$, $\dot{z}=-E$), and M_2^+ ($z=-1$, $\dot{z}=+E$) (see FIGURES 8–10).

The limiting curves (10) extend along the z-axis ($\dot{z}=0$) from

$$z_{\min} = -\mu - \left[(\mu-1)^2 + \frac{2M_2}{\varepsilon}\right]^{1/2} \tag{12}$$

to

$$z_{\max} = \mu + \left[(\mu+1)^2 - \frac{2M_2}{\varepsilon}\right]^{1/2}, \tag{13}$$

where

$$\varepsilon = \frac{1}{E} - 1 \tag{14}$$

and[31]

$$\mu = \frac{M_1 + M_2}{2\varepsilon}. \tag{15}$$

The upper curve (10) (i.e., $\dot{z}>0$) has a minimum $\dot{z}$ for

$$z = \frac{\sqrt{M_2} - \sqrt{M_1}}{\sqrt{M_2} + \sqrt{M_1}}. \tag{16}$$

The minimum value of $\dot{z}^2$ is

$$\dot{z}^2 = E^2 - 4[2 + M_1 + M_2 + 2\sqrt{M_1 M_2}]^{-2}, \tag{17}$$

and this is negative if

$$E < E_0 = \frac{2}{2 + M_1 + M_2 + 2\sqrt{M_1 M_2}}. \tag{18}$$

For such values of E the limiting curve (10) is split into two separate curves (FIG. 10A), one closing around M_2 ($z=-1$, $\dot{z}=0$) and the other closing around M_1 ($z=1$, $\dot{z}=0$). The inequality (18) for given E and $M_2 = 1$ is written

$$M_1 + 2\sqrt{M_1} + 3 - \frac{2}{E} < 0, \tag{19}$$

therefore, $\sqrt{M_1}$ must be positive and lie between the two roots

$$\sqrt{M_1} = -1 \pm \left(\frac{2}{E} - 2\right)^{1/2}. \tag{20}$$

One root is positive if

$$E < \frac{2}{3}. \tag{21}$$

Therefore, the limiting curve is split if E satisfies condition (21) and

$$M_1 < M_{10} = \left\{-1 + \left(\frac{2}{E} - 2\right)^{1/2}\right\}^2. \tag{22}$$

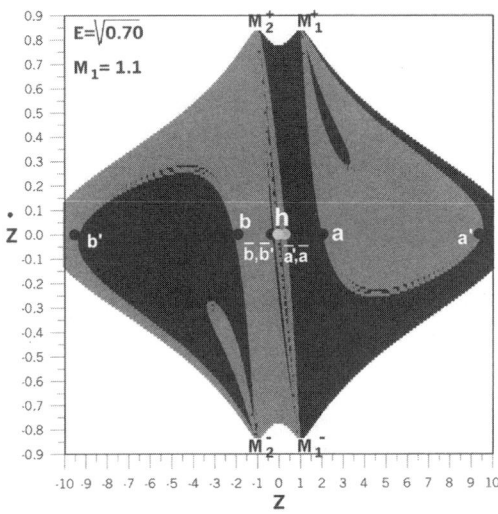

FIGURE 9. The basins of attraction for $E = \sqrt{0.7}$, $M_1 = 1.1$ ($M_2 = 1$) on the surfact of section $(z, \dot{z})$, that is $x = 0$ and $\dot{x} > 0$. The periodic orbits are a and a' (clockwise around M_1, *black*), b and b' (counterclockwise around M_2, *gray*), $\bar{a}$, $\bar{a}'$, $\bar{b}$, and $\bar{b}'$ (the same as a, a', b, and b' in opposite directions), and h (like a hyperbolic arc).

In the cases $M_2 = 1$ and $E = \sqrt{0.5}$ or $E = \sqrt{0.7}$ the limiting curve cannot be split for any value of M_1. For $M_1 = 1.1$ the limiting curves are shown in FIGURES 8 and 9. In the case $E = E = \sqrt{0.38}$ ($< 2/3$) the limiting curves are split if $M_1 < M_{10} = 0.013$

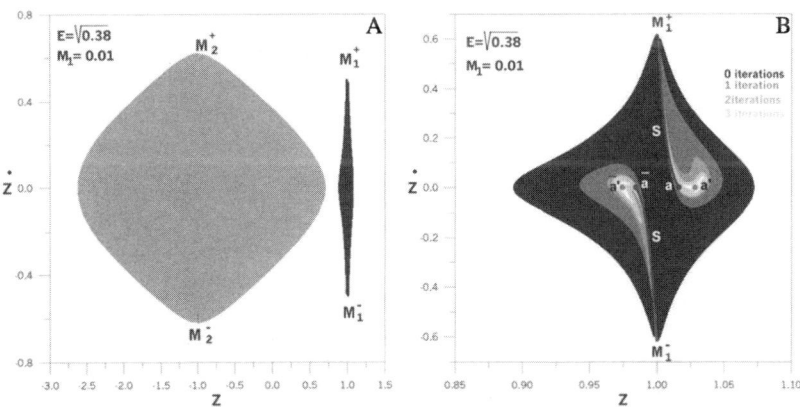

FIGURE 10. **A.** The basins of attraction for $E = \sqrt{0.38}$, $M_1 = 0.01$ ($M_2 = 1$): M_1, *black*; M_2, *gray*. **B.** The basin of attraction around $z = 1$ in detail. We mark the periodic orbits a' and $\bar{a}'$ (stable) and a and $\bar{a}$ (unstable), and the stable asymptotic curves S from a and $\bar{a}$. The various shades indicate regions of escape after 0 (*black*), 1, 2, and 3 iterations.

($M_2 = 1$). Such is the case shown in FIGURE 10, where $M_1 = 0.01$. The points z', where this curve intersects the axis $\dot{z} = 0$ between M_1 and M_2 are given by the equation

$$\frac{M_1}{1-z'} + \frac{M_2}{1+z'} = \varepsilon. \qquad (23)$$

These roots are

$$z' = \frac{M_2 - M_1}{2\varepsilon} \pm \left[\left(\frac{M_2 - M_1}{2} + 1\right)^2 - \frac{2M_2}{\varepsilon}\right]^{1/2}, \qquad (24)$$

and they are both between $z = -1$ and $z = +1$. The quantity under the square root is positive if E satisfies inequality (18).

Inside the limiting curves (10) most orbits escape to the black holes M_1 and M_2. In FIGURES 8–10 we show the basins of escape of the black holes M_1 (black) and M_2 (gray) for $M_1 = 1.1$, $M_2 = 1$, and $E = \sqrt{0.5}$ (FIG. 8), for $M_1 = 1.1$, $M_2 = 1$, and $E = \sqrt{0.7}$ (FIG. 9), and for $M_1 = 0.01$, $M_2 = 1$, and $E = \sqrt{0.38}$ (FIG. 10). For increasing M_1 the proportion of orbits falling into M_1 increases, whereas for decreasing M_1 this proportion decreases and tends to zero as $M_1 \to 0$.

The largest part of the escaping particles escape very fast without any intersection of their orbits with the surface of section (except for M_1 or M_2). However, there are also sets of orbits (of finite measure) that escape after 1, 2, 3, …, n intersections. In the case $E = \sqrt{0.5}$, $M_1 = 0$, and $M_2 = 1$, all orbits escape to M_2 immediately, that is, without any intersection with the surface of section except M_2.

The sets of escaping orbits into M_1 (type I) and M_2 (type II) shown in FIGURES 8 and 9 consist of large compact regions and of thin filaments. The thin filaments have finite thickness but there are infinite filaments in certain regions of the surface of section. In fact, these filaments form fractal sets that provide a clear indication of chaos.

The structure of the filaments is not easily seen in FIGURES 8 and 9 because of poor resolution. A much better picture is obtained if we consider the intersections of these filaments by the asymptotic curves of the main periodic orbits a and a' clockwise around M_1 (a closer to M_1), b and b' counterclockwise around M_2 (b closer to M_2), and h (like an arc of a hyperbola between the orbits a' and b'). We also have the orbits $\bar{a}$, $\bar{a}'$, $\bar{b}$, and $\bar{b}'$, which are the same as a, a', b, and b' but in opposite directions. The positions of these orbits for $E = \sqrt{0.5}$ and various values of M_1 (with $M_2 = 1$) are shown in FIGURE 11.

In FIGURE 12 we show an unstable asymptotic curve of the periodic orbit a for $E = \sqrt{0.5}$ and $M_1 = 1.1$ ($M_2 = 1$). This asymptotic curve starts from a to the right [arc (1)] and reaches the black hole M_1 at the point M_1^- (i.e., with $z = 1$ and $\dot{z} = -E$). Then there is a gap after which a second arc (2) starts from M_1^+ (i.e., $z = 1$ and $\dot{z} = E$) to M_2^- ($z = -1$ and $\dot{z} = -E$). Then there is a second gap followed by arc (3) from M_2^+ back to M_2^+, and so on. (For details see Contopoulos and Harsoula[30]).

Along the asymptotic curve we mark numbers referring to the fourth iterates of initial points very close to a, along the asymptotic curve, at distances $m \cdot 10^{-8}$. The points with $0 < m < 207$ are mapped (at their fourth iteration) along the arc (1) that reaches the point M_1^-. Then there is a gap for $208 < m < 279$. The orbits starting at these values of $m \cdot 10^{-8}$ do not have a fourth intersection with the surface of section other than the black hole M_1 itself. The points $280 < m < 340$ form the second arc, and so on. All of these orbits are asymptotic to the periodic orbit a, that is, they tend

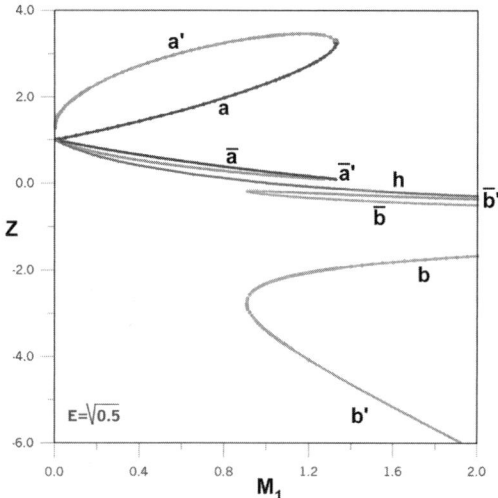

FIGURE 11. Characteristics of the orbits a', a, $\bar{a}$, $\bar{a}'$, h, $\bar{b}'$, $\bar{b}$, b, and b' for $E = \sqrt{0.5}$. These orbits intersect the z-axis perpendicularly; z is given as a function of M_1 ($M_2 = 1$).

asymptotically to the orbit a as $t \to -\infty$. This asymptotic curve cannot intersect itself, or other unstable asymptotic curves, but it does intersect the stable asymptotic curve from a at an infinity of homoclinic points, and the stable asymptotic curves of other

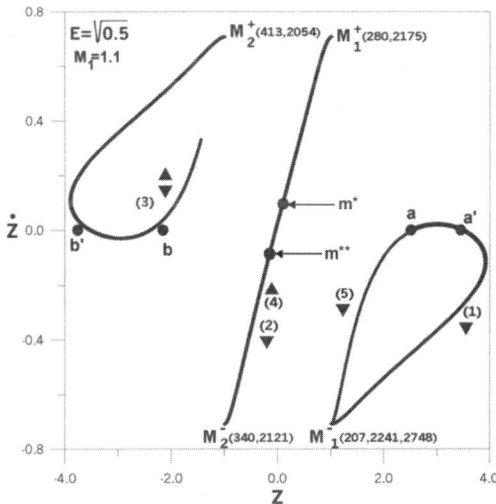

FIGURE 12. Successive arcs, (1)–(5), of the unstable asymptotic curve from the periodic orbit a for $E = \sqrt{0.5}$ and $M_1 = 1.1$ ($M_2 = 1$).

unstable periodic orbits at an infinity of heteroclinic points. The existence of these homoclinic and heteroclinic points proves the chaotic character of the problem of two fixed centers in general relativity.

The homoclinic and heteroclinic points can be taken as initial conditions for doubly asymptotic orbits, either homoclinic (asymptotic to the same periodic orbit a as $t \to \infty$) or heteroclinic (asymptotic to another periodic orbit as $t \to \infty$). These homoclinic and heteroclinic orbits are studied in detail elsewhere.[30,31] The most important result of that study was that there are infinite intervals Δm of orbits of types I and II (that is, falling on the black hole M_1 or M_2) limited by homoclinic, or heteroclinic points. Between two such intervals there are infinitely many more intervals of type I and II. This fact establishes the fractal structure of sets of type I and II.

In particular there are infinitely many intervals I and II between $m = m^* = 307.643390$ and $m = m^{**} = 314.384627$ along the arc (2) of the asymptotic curve from a (FIG. 12). The orbit with $m = m^*$ is heteroclinic between a and $\bar{a}$, that is, the same orbit described in opposite direction (because of that, it is considered a different orbit). This orbit after infinitely many clockwise rotations around a reaches a point on the curve of zero velocity (CZV) on the left (see FIGURE 13A) and returns exactly along its previous path reaching the orbit $\bar{a}$ after infinitely many counterclockwise rotations. Similarly, the orbit of FIGURE 13B ($m = m^{**}$) connects the orbits a and $\bar{b}$. In this case, the orbit makes an oscillation to the left, but does not reach the CZV.

Between these two heteroclinic orbits there are infinitely many more homoclinic and heteroclinic orbits connecting a with a, $\bar{a}$, a', $\bar{a}'$, h, b, b', $\bar{b}$, and $\bar{b}'$. One way to find such orbits is by calculating the stable asymptotic curves of these periodic orbits and determine their intersections with the unstable asymptotic curve of the orbit a (arc (2)). By comparing FIGURES 8 and 12, we can see that the points $\bar{a}$, $\bar{a}'$, h, $\bar{b}$, and $\bar{b}'$ are close to arc (2) of the orbit a. Their stable asymptotic curves are almost orthogonal to the arc (2).

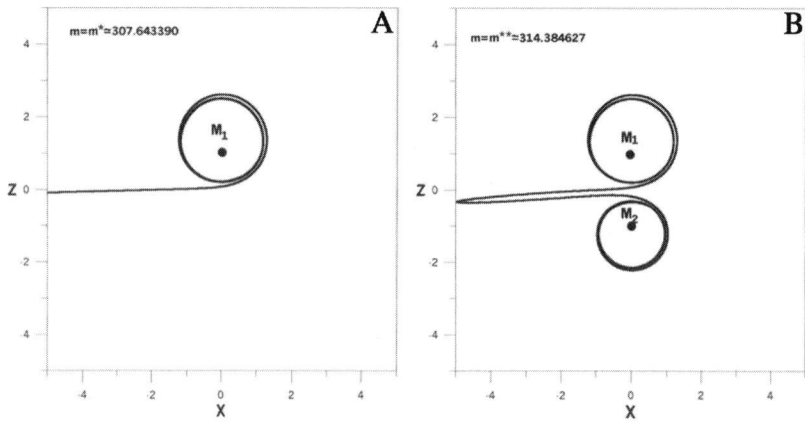

FIGURE 13. Heteroclinic orbits: (A) between a and $\bar{a}$ for $m = m^*$ and (B) between a and $\bar{b}$ for $m = m^{**}$.

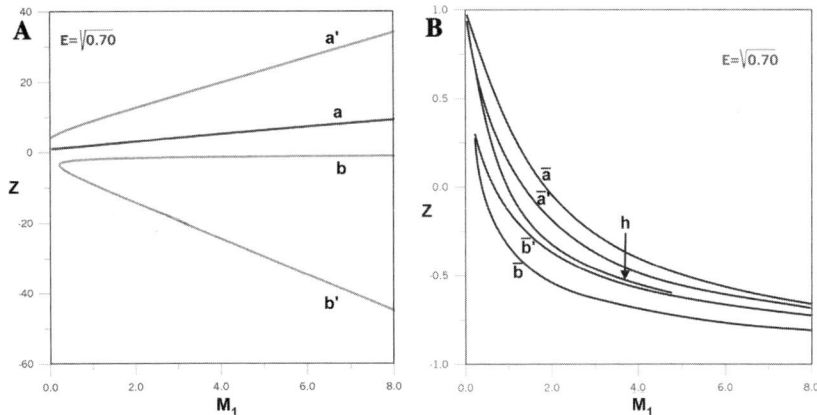

FIGURE 14. Characteristics for $E = \sqrt{0.7}$ of the families **(A)** a', a, b, and b', and **(B)** $\bar{a}$, $\bar{a}'$, h, $\bar{b}'$, and $\bar{b}$ in detail.

The orbits above the point m^* on arc (2) (FIG. 12) are of type I (i.e., they all fall on the black hole M_1) thus, they belong to the black region of FIGURE 8. Similarly the orbits below the point m^{**} on arc (2) (FIG. 12) are of type II (i.e., they fall on the black hole M_2) and they belong to the gray region of FIGURE 8. Additional regions along the unstable asymptotic curve from a that contain homoclinic and heteroclinic orbits are close to the points a, a', b, and b' (FIG. 8).

The existence of escaping orbits in the systems of two fixed black holes has as a consequence the reduction of the areas on a surface of section and of the volumes in phase space. The escapes are either fast, for example, without any intersection with the surface of section other than the black holes M_1 and M_2, or slow, after some intersections with the surface of section (see, e.g., FIG. 10B).

In the case $E = \sqrt{0.5}$ and $M_1 = 1.1$ ($M_2 = 1$) the proportion p of particles along the asymptotic curve that remains after n iterations (n intersections with the surface of section) is given by the approximate formula

$$p = Ae^{-bn}, \tag{25}$$

where $A = 1,120$ and $b = 2.17$ (valid for $n \geq 4$). Therefore, the decrease of the number of particles along the asymptotic curve is exponential. The rate of escapes $b = 2.17$ is related to the eigenvalue of the periodic orbit a, which is $\lambda \approx 32.33$ for $M_1 = 1.1$. For larger values of M_1 the value of λ decreases and b also decreases.

At a maximum value $M_1 = M_{1\max} = 1.32576$, λ reaches the value $\lambda = 1$. At this value of M_1 the orbits a and a' join and they disappear for larger M_1 (FIG. 11). As M_1 increases up to $M_{1\max}$ the escape rate b decreases and tends to zero. Thus, escapes of particles into the black holes M_1 and M_2 become slower. Furthermore, for M_1 close to $M_{1\max}$ (but smaller) the orbit a' is stable and some orbits never escape to the black holes M_1 and M_2.

All of the above examples refer to the particular value, $E = \sqrt{0.5}$. In FIGURE 14 we show the characteristics (z as a function of M_2) for the orbits a, a', $\bar{a}$, $\bar{a}'$, h, $\bar{b}'$, $\bar{b}$, b, and b' in the case $E = \sqrt{0.7}$. The topology of these characteristics is similar to that of the case $E = \sqrt{0.5}$ (FIG. 11). However, the families b and b', and also $\bar{b}$ and $\bar{b}'$, join and disappear much closer to $M_1 = 0$ than in the case $E = \sqrt{0.5}$; namely, at $M_1 \approx 0.2$, whereas for $E = \sqrt{0.5}$ the minimum is $M_1 = 0.90706$. Furthermore, the families a, a', $\bar{a}$, and $\bar{a}'$ join and disappear in the case $E = \sqrt{0.7}$ at a much larger M_1 value ($M_1 \approx 50$).

In both cases, $E = \sqrt{0.5}$ and $E = \sqrt{0.7}$ the characteristics of the family a' reach the axis $M_1 = 0$ at points above $z = 1$, whereas the characteristics of the families a, $\bar{a}$, $\bar{a}'$, and h reach the point $z = 1$ for $M_1 = 0$. In the case $E = \sqrt{0.85}$ (see FIGURE 15) the families b, b', $\bar{b}$, and $\bar{b}'$ have also reached the axis $M_1 = 0$. For $M_1 = 0$ the characteristics of the families $\bar{b}'$ and $\bar{b}$ reach the point $z = 1$, whereas b and b' reach points below $z = -1$. The orbits for $M_1 = 0$ are described in the APPENDIX.

Two new families of periodic orbits, c' and c, are families of almost elliptical orbits surrounding both black holes M_1 and M_2. An example of these families is shown in FIGURE 15 for $E = \sqrt{0.85}$. The families c' and c are described clockwise. The same orbits described counterclockwise are called $\bar{c}'$ and $\bar{c}$. The families c' and c exist for E larger than $\sqrt{27/32} = 0.84375$.[39] The family c' is stable and exists up to $E = 1$, whereas the family c is unstable and exists all the way to $E = \infty$.

In FIGURE 15 it can be sees that the families c' and c exist from $M_1 = 0$ up to a maximum $M_{1\max} \approx 0.05$ and then from $M_{1\min} = 1.07$ up to $M_1 = \infty$. Thus, there is a gap in these families from $M_{1\max}$ to $M_{1\min}$ ($> M_{1\max}$). For somewhat larger values of E ($1 > E > \sqrt{0.855}$) there is no such gap and the families c' and c exist all the way from $M_1 = 0$ to $M_1 = \infty$.

The unstable asymptotic curve U of the unstable periodic orbit c comprises several arcs (see FIGURE 16) separated by gaps, as is the case in FIGURE 12. On the other hand, the unstable asymptotic curve UU makes a small loop around the stable point c' (see insert of FIGURE 17) and then it oscillates very close to U, following the same arcs as the curve U. The stable asymptotic curves S and SS are symmetric to U and UU with respect to the axis $\dot{z} = 0$, but they are not drawn in FIGURE 16.

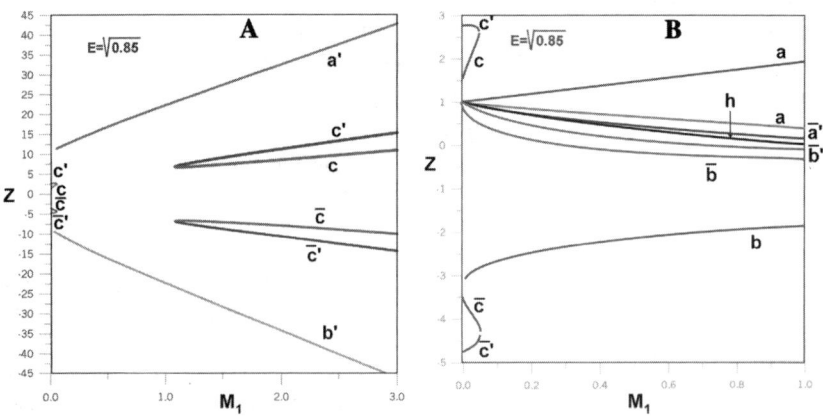

FIGURE 15. Characteristics for $E = \sqrt{0.85}$ of the families **(A)** a', c', c, $\bar{c}$, $\bar{c}'$, and b', and **(B)** c', c, a, $\bar{a}$, $\bar{a}'$, h, b', $\bar{b}'$, $\bar{b}$, b, $\bar{c}$, and $\bar{c}'$ in detail.

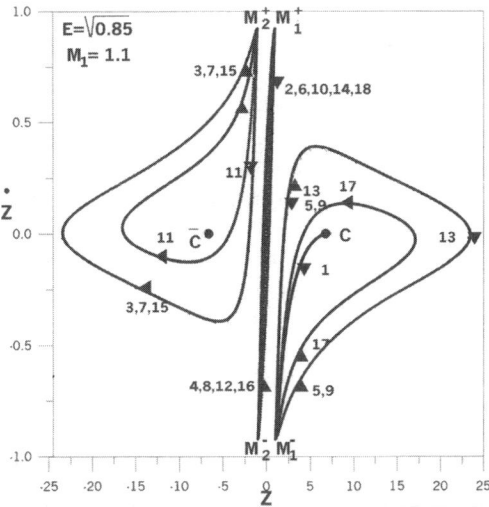

FIGURE 16. The successive arcs of the unstable asymptotic curve of the unstable periodic orbit c, for $E = \sqrt{0.85}$ and $M_1 = 1.1$.

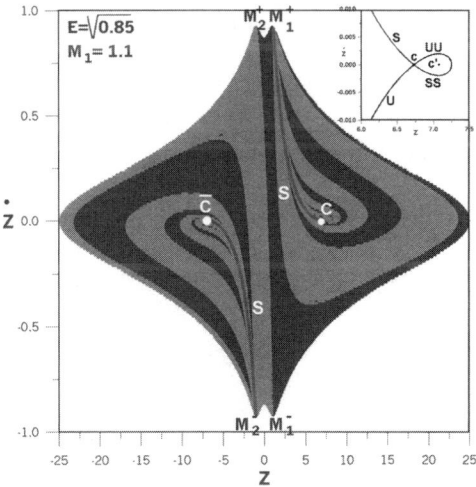

FIGURE 17. Escape domains for $E = \sqrt{0.85}$ and $M_1 = 1.1$ ($M_2 = 1$), and the first arcs S of the stable asymptotic curves from the unstable orbits c and $\bar{c}$. In the **insert** we show the region close to the orbits c and c', together with the unstable and stable asymptotic curves of c (U, UU and S, SS).

FIGURE 17 shows the basins of attraction in the case $E = \sqrt{0.85}$, $M_1 = 1.1$, and $M_2 = 1$. The black regions are attracted by the black hole M_1 and the gray regions are attracted by the black hole M_2. We also indicate the unstable orbits c and $\bar{c}$ (the same orbit as c, but in opposite direction) and the first arcs of the stable asymptotic curves S from the points c and $\bar{c}$. The boundaries of the black and gray regions are very close to various arcs of the stable asymptotic curve S (compare FIG. 17 with FIG. 16, inverted around the axis $\dot{z} = 0$). In particular, the orbits close to the arcs S fall into the black hole M_2. There are also small stable regions around the stable orbits c' and $\bar{c}'$ (inside the loops of the unstable curves UU and SS from c and $\bar{c}$) that never escape. The stable region around c' is shown in detail in the insert of FIGURE 17.

For the case $E = \sqrt{0.38}$ the characteristics of the families a', a, $\bar{a}$, $\bar{a}'$, h, $\bar{b}'$, $\bar{b}$, b, and b' are shown in FIGURE 18. The families a', a, $\bar{a}$, and $\bar{a}'$ exist only for small values of M_1, much smaller than for the case $E = \sqrt{0.5}$ (FIG. 11). If we decrease E further, the families a', a, $\bar{a}$, and $\bar{a}'$ disappear for $E < \sqrt{0.3755}$. The characteristic of the family h does not reach the axis $M_1 = 0$, but it terminates at the value $M_1 = M_{10} = 0.013$ (Eq. (22)), where the limiting curve reaches the axis $\dot{z} = 0$.[40] For this value of M_1 the curve h degenerates to a point and for smaller M_1 it does not exist at all. In general using Equations (16) and (22) we find that the limiting point for $M_1 = M_{10}$ is at

$$z = \frac{2}{\left(\frac{2}{E} - 2\right)^{1/2}} - 1. \tag{26}$$

This limiting point exists for $E \leq 2/3$. For $E = 2/3$, $z = 1$, and for $E < 1/3$, $z < 0$, reaching the point $z = -1$ for $E = 0$. The families a, $\bar{a}$, h, b, and $\bar{b}$ are always unstable, whereas the families a', $\bar{a}'$, b', and $\bar{b}'$ are stable near their maxima and minima. In particular, for $M_1 = 0.01$, the orbits a' and $\bar{a}'$ are stable (FIG. 10B).

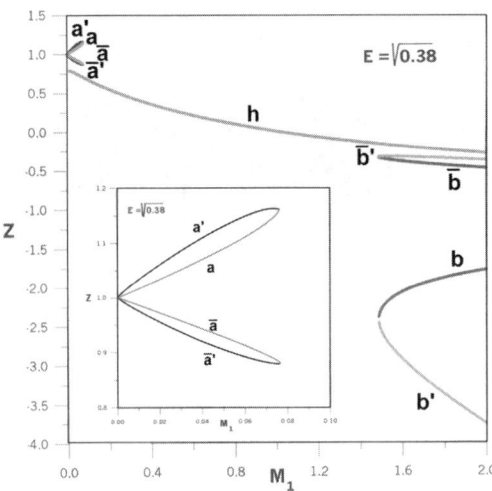

FIGURE 18. Characteristics for $E = \sqrt{0.38}$ of the families a', a, $\bar{a}$, $\bar{a}'$, h, $\bar{b}'$, $\bar{b}$, b, and b'. In the **insert** we show the families a', a, $\bar{a}$, and $\bar{a}'$ in more detail.

In the cases where the basins of attraction are separated, as in FIGURE 10A, we do not see the fractal structure of orbits escaping into M_1 and M_2 that we see in FIGURES 8 and 9. However, in FIGURE 10B we see regions around a' and $\bar{a}'$ that never escape. These regions are limited by the asymptotic curves of the orbits a and $\bar{a}$ (not shown in this figure) that intersect at infinite homoclinic points. There are also probably heteroclinic intersections with asymptotic curves of higher order periodic orbits.

In FIGURE 10B we see fast and slow escapes into M_1, that is, escapes after 0 iterations (black regions) and escapes after 1, 2, 3, ... iterations (lighter shades). The regions of escape after a longer time are congested, close to the stable asymptotic curves S from the unstable orbits a and $\bar{a}$. In particular, orbits starting exactly on the asymptotic curves S asymptotically approach the periodic orbits a and $\bar{a}$, and never escape to the black hole M_1.

For the limiting case $M_1 = 0$ ($M_2 = 1$, $E = \sqrt{0.85}$) there is only one black hole (M_2) and most orbits fall into it (FIG. 19). In this case also there are two stable regions around the stable periodic orbits c' and $\bar{c}'$ that never escape into the black hole. The stable regions are surrounded by the asymptotic curves SS and UU from c and $\bar{c}$. The curves SS and UU are very close to each other, but in fact they intersect at an infinity of homoclinic points. In FIGURE 19 we use different shades of gray for orbits that escape into M_2 after 0, 1, 2, and 3 iterations. The darker regions are closer to the asymptotic curves S. If an orbit starts very close to S it stays in its neighborhood for a long time before falling into M_2.

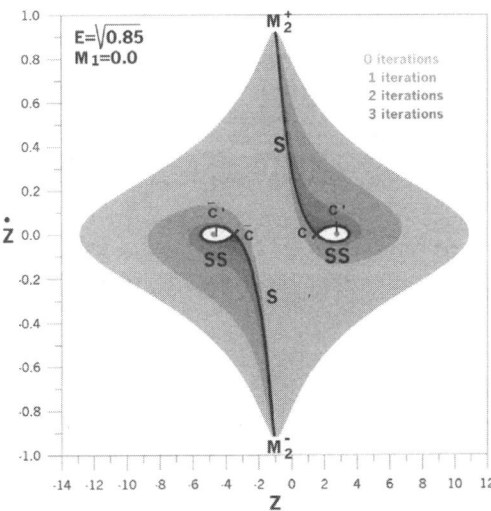

FIGURE 19. Domains of escape after 0, 1, 2, and 3 iterations and nonescaping domains around the points c' and $\bar{c}'$ for $M_1 = 0$ and $E = \sqrt{0.85}$. We also indicate the stable asymptotic curves S and SS of the unstable circular orbits c and $\bar{c}$. The unstable asymptotic curves U, UU are symmetric to S, SS with respect to the axis $\dot{z} = 0$.

SOME SIMPLE CASES

The existence of point attractors at a finite distance depends on the form of the forces close to these points. This can be seen in the very simple case of a point mass μ with potential

$$V = -\frac{\mu}{r^2 + a}, \tag{27}$$

for values of a close to zero. In this case the Hamiltonian is

$$H = \frac{1}{2}\left(\dot{r}^2 + \frac{J^2}{r^2} - \frac{\mu}{r^2 + a}\right) = h, \tag{28}$$

with energy h and angular momentum

$$J = r^2\dot{\theta}. \tag{29}$$

Hence, we derive

$$\dot{r}^2 = 2h + \frac{1}{r^2}(2\mu r^{-a} - J^2). \tag{30}$$

If $a > 0$, the value of $\dot{r}^2$ close to the center is positive for all values of h and J. Therefore, all the orbits moving inward fall into the center $r = 0$. If $a = 0$, the same happens if $J^2 < 2\mu$ (i.e., for small values of J^2). Such is the case of a Schwarzschild black hole. On the other hand, if $a < 0$, the maximum of $\dot{r}^2$ occurs for

$$4hr - 2\mu a r^{-1-a} = 0, \tag{31}$$

that is,

$$r = r_m = \left(\frac{2h}{\mu\alpha}\right)^{\frac{1}{2+\alpha}} \tag{32}$$

and this is real and positive for $h < 0$. If we insert this value in Equation (30) we find that $\dot{r}^2$ is positive if

$$J^2 < 2hr_m^2 + 2\mu r_m^{-a}, \tag{33}$$

whereas for $r \to 0$ and $r \to \infty$ the value of $\dot{r}^2$ is negative. Therefore, if $h < 0$ and J^2 satisfies (33), we have real orbits between a minimum and a maximum r.

In conclusion, the center $r = 0$ acts as an attractor only for $a \geq 0$ (in the case $a = 0$ only if J^2 is sufficiently small). If $a < 0$, the orbits do not reach the center $r = 0$ unless $J = 0$.

LYAPUNOV CHARACTERISTIC NUMBERS AND DYNAMICAL SPECTRA

One way to establish the chaotic character of a dynamical system is by calculating, for a number of orbits, the Lyapunov characteristic number

$$\text{LCN} = \lim_{\tau \to \infty} \chi, \tag{34}$$

where

$$\chi = \frac{\ln(\xi/\xi_0)}{\tau}, \tag{35}$$

and ξ_0 and ξ are infinitesimal deviations from an orbit at proper times τ_0 and τ. If LCN is positive the system is chaotic, whereas if it zero it is ordered. In the present case we take

$$\xi = \left[\Delta x^2 + \Delta z^2 + \frac{1}{E^2}(\Delta \dot{x}^2 + \Delta \dot{z}^2)\right]^{1/2}, \tag{36}$$

where Δx, Δz, $\Delta \dot{x}$, and $\Delta \dot{z}$ are small quantities and ξ is rescaled to unity at each step.[34]

Instead of calculating orbits for very long times τ we may calculate the spectrum of the short time Lyapunov characteristic numbers or *stretching numbers*

$$a_i = \frac{\ln(\xi_{i+1}/\xi_i)}{\Delta \tau}. \tag{37}$$

In the case of maps, we take successive iterates (i and $i + 1$) of the map,[41,42] and $\Delta \tau = 1$. In the case of continuous orbits we can take $\Delta \tau$ to be very small (e.g., equal to the integration step).[43] In the present case, we take $\Delta \tau = 10^{-6}$.

The distribution of the stretching numbers a_i forms the dynamical spectrum of the system. Namely, we calculate the function

$$S(a) = \frac{\Delta N}{N \Delta a}, \tag{38}$$

where ΔN is the number of values between a and $a + \Delta a$, and $\Delta a = 10^{-3}$.

The Lyapunov characteristic number is the average value of a_i. In a system with a connected and compact chaotic domain the spectrum of stretching numbers is invariant.[42] In general, convergence to an approximately invariant spectrum is rather fast and one can easily determine if the average value of a (i.e., the LCN) is positive.

The above considerations are applicable to systems without escapes. On the other hand, in cases with escapes to infinity the orbits far from the central part of the system are like hyperbolæ and the values of ξ increase only linearly with τ. Therefore, LCN = 0. However, in many cases the stretching numbers for short times $\Delta \tau$ are usually positive and only in the limit $\tau \to \infty$ do we find $\chi \to 0$. This phenomenon

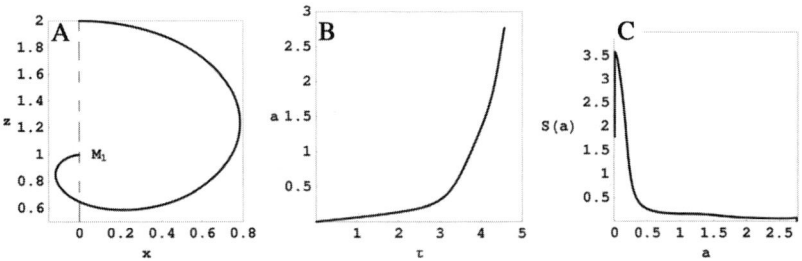

FIGURE 20. (A) An orbit falling on the black hole M_1 ($M_1 = 1.1$, $M_2 = 1$, $E = \sqrt{0.5}$, and initial conditions $z_0 = 2$, $\dot{z}_0 = x_0 = 0$. (B) Corresponding variation in the stretching numbers a in proper time τ. (C) Spectrum for the stretching numbers.

appears in the case of *chaotic scattering*. Such is the case for the Mixmaster universe[44] and in the case considered already in the first part of this paper.

The cases with escapes to attractors at a finite distance present a different problem. Namely, the orbits reach the attractor at a finite time τ; therefore, the limit $\tau \to \infty$ in **(34)** does not exist. However, in these cases we find the spectra of stretching numbers up to the time of collision with the attractor and, thus, derive a finite time Lyapunov characteristic number (LCN_f).

In FIGURES 20A and 21A we show two orbits, falling on the black holes M_1 and M_2, respectively. FIGURES 20B and 21B show the variation in the stretching number a as a function of the (proper) time τ. (We note here that the value of a depends on variations of nearby orbits in the four-dimensional phase space). Finally, in FIGURES 20C and 21C we show the corresponding dynamical spectra. It is obvious that the average values of a are positive in both cases; namely, $\text{LCN}_f = 0.47$ in the first case and $\text{LCN}_f = 0.01$ in the second case. Therefore, these two orbits can be called chaotic.

In the case of FIGURE 20B it can be seen that the value of a increases from zero along the orbit. The rate of increase is small initially and this is why we have large values of $S(a)$ close to $a = 0$. On the other hand, the second orbit (FIG. 21A) is more complicated. The value of a in this case changes in a rather complicated way (FIG. 21B) and the maxima of $S(a)$ in the corresponding spectrum (FIG. 21C) are marked by letters A, B, C, D, and E that represent maxima or minima of a in FIGURE 21B. In these regions the variation in a is relatively small.

The two spectra shown in FIGURES 20C and 21C are quite different. The spectra of other orbits are also different. Therefore, we do not have the invariance of the chaotic spectra that we find in cases without escapes. This is easily understood, because the orbits do not explore the entire available phase space before falling on one of the two black holes. Furthermore, we cannot even define a dynamical spectrum if we take only the intersections of the orbits with the surface of section $(z, \dot{z})$, because the number of such intersections is in general small (for most initial conditions it is zero).

Therefore, the dynamical spectra of the orbits can be defined only along the orbits and they give a positive finite time Lyapunov characteristic number. However, these spectra depend on the initial conditions of the orbits and they are not invariant.

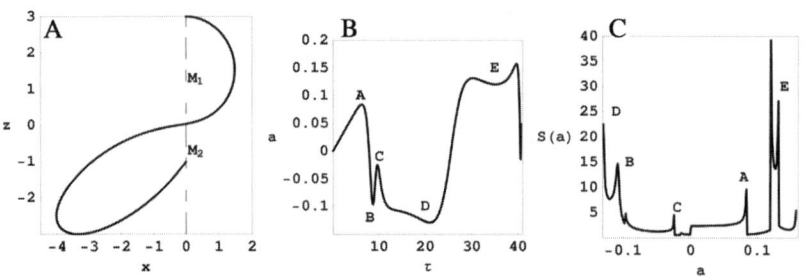

FIGURE 21. As FIGURE 20 for $z_0 = 3$. The orbit falls on the black hole M_2.

CONCLUSIONS

The following conclusions are derived from the present study.

1. In the case of systems with escapes, infinity acts as an attractor. In these cases the volumes in phase space are preserved but the areas on a surface of section are not preserved.
2. The domains of escape on a surface of section (sets of initial conditions of orbits escaping after 0, 1, 2, ... intersections with the surface of section), in general, have spiral form. This form is explained theoretically.
3. The domains of escape are of fractal form. The measure of escaping orbits that have not yet escaped after time t decreases exponentially with t, but at different rates for small and large t (slower for larger t).
4. In the case of escapes into singular points at a finite distance the volumes in phase space and the areas on a surface of section are not preserved.
5. The domains of escape into two (fixed) black holes form fractal sets near the asymptotic curves of the unstable periodic orbits.
6. For given masses M_1 and M_2 of the black holes, the available area on a surface of section is separated into two parts if the energy is small enough.
7. The simple periodic orbits around the two black holes are of four types: (1) orbits surrounding the black hole M_1 ($a, a', \bar{a}$, and $\bar{a}'$), (2) orbits surrounding the black hole M_2 ($b, b', \bar{b}$, and $\bar{b}'$), (3) an orbit like an arc of hyperbola between M_1 and M_2 (h), and (4) orbits surrounding both black holes ($c, c', \bar{c}$, and $\bar{c}'$). We have considered the characteristics of these families (z as a function of M_1 for $M_2 = 1$ and $x = \dot{z} = 0$) for various values of the energy E. The families of type c (i.e., c, c', $\bar{c}$, and $\bar{c}'$) exist only for sufficiently large E (but $E < 1$). The families of type a exist only for small M_1 if E is relatively small, but extend to large M_1 for large E. The families of type b exist only for large M_1 if E is small, but reach $M_1 = 0$ for large E. The family h exists for all values of E larger than a limiting value E_0 (Eq. (**18**)).
8. The families $a, \bar{a}, b, \bar{b}, h, c$, and $\bar{c}$ are always unstable. The families c' and $\bar{c}'$ are stable, whereas the families $a', \bar{a}', b'$, and $\bar{b}'$ are only stable close to the maxima and minima of their characteristics.
9. The asymptotic curves of the unstable families intersect at infinitely many homoclinic and heteroclinic points. The corresponding orbits are asymptotic to the same or different unstable periodic orbits. The existence of infinite homoclinic and heteroclinic orbits proves the chaotic character of the problem of two fixed black holes.
10. The existence of point attractors at a finite distance depends on the form of the force (or of the potential) near these points. If the potential is of the form $V \propto r^{-2-a}$ these points are attractors if $a > 0$. If $a < 0$ only the orbits with zero angular momentum fall on them.
11. The Lyapunov characteristic numbers (LCN) of orbits escaping to infinity are, in general, zero. However, the short time Lyapunov characteristic numbers (stretching numbers) are mostly positive. These are cases of chaotic scattering.

12. In the cases of escapes to attractors at a finite distance, the orbits escape after a finite time, thus we cannot even define a Lyapunov characteristic number. However, we can define short time LCNs (stretching numbers) and finite time Lyapunov characteristic numbers up to the escape time. This is positive for chaotic orbits, but the spectra of the stretching numbers in this case depend on the initial conditions of an orbit and they are not invariant.

ACKNOWLEDGMENTS

This research was supported by the Research Committee of the Academy of Athens (Program 200/557). We thank Dr. C. Efstathiou for calculating FIGURES 2–7 and Mr. G. Lukes-Gerakopoulos (supported by IKY) for calculating FIGURES 20 and 21.

REFERENCES

1. CONTOPOULOS, G., H.E. KANDRUP & D. KAUFMANN. 1993. Physica D **64:** 310.
2. SIOPIS, C.V., G. CONTOPOULOS & H.E. KANDRUP. 1995. Ann. N.Y. Acad. Sci. **751:** 205.
3. SIOPIS, C.V., H.E. KANDRUP, G. CONTOPOULOS & R. DVORAK. 1995. Ann. N.Y. Acad. Sci. **773:** 189.
4. SIOPIS, C.V., H.E. KANDRUP, G. CONTOPOULOS & R. DVORAK. 1997. Cel. Mech. Dyn. Astron. **65:** 57.
5. KANDRUP, H.E., C. SIOPIS, G. CONTOPOULOS & R. DVORAK. 1999. Chaos **9:** 381.
6. CONTOPOULOS, G. 1988. In The Few-Body Problem. M.J. Valtonen, Ed.: 265. Kluwer, Dordrecht.
7. BENET, L., D. TRAUTMANN & T.H. SELIGMAN. 1997. Cel. Mech. Dyn. Astron. **66:** 203.
8. BENET, L., T.H. SELIGMAN & D. TRAUTMANN. 1999. Cel. Mech. Dyn. Astron. **71:** 167.
9. ANOSOVA, J.P. 1986. Astroph.Space Sci. **124:** 217.
10. BARTLETT, J.H. 1978. Cel. Mech. **17:** 3.
11. BARTLETT, J.H. 1982. Cel. Mech. **28:** 295.
12. JUNG, C. & H.J. SCHOLZ. 1987. J. Phys. A **20:** 3607.
13. JUNG, C. & H.J. SCHOLZ. 1988. J. Phys. A **21:** 2301.
14. BLEHER, S., C. GREBOGI, E. OTT & R. BROWN. 1988. Phys. Rev. A **38:** 930.
15. BLEHER, S., C. GREBOGI & E. OTT. 1990. Physica D **46:** 87.
16. ECKHARDT, B. 1988. Physica D **33:** 87.
17. HÉNON, M. 1989. La Recherche **20:** 490.
18. GASPARD, P. & S.A. RICE. 1989. J. Chem. Phys. **90:** 2225.
19. DING, M., C. GREGOBI, E. OTT & J.A. YORKE. 1990. Phys. Rev. A **42:** 7025.
20. DING, M., C. GREGOBI, E. OTT & J.A. YORKE. 1991. Phys. Lett. A **153:** 21.
21. LAU, Y.-T., J.M. FINN & E. OTT. 1991. Phys. Rev. Lett. **66:** 978.
22. CHRISTIANSEN, F. & P. GRASSBERGER. 1993. Phys. Lett. A **181:** 47.
23. OTT, E. & T. TEL, Eds. 1993. Chaos **3:** 417.
24. CONTOPOULOS, G. & K. EFSTATHIOU. 2004. Cel. Mech. Dyn. Astron. **88:** 163.
25. CONTOPOULOS, G. 1990. Proc. Roy. Soc. Lond. A **431:** 183.
26. CONTOPOULOS, G. 1991. Proc. Roy. Soc. Lond. A **435:** 551.
27. DETTMANN, C.P., N.E. FRANKEL & N.J. CORNISH. 1994. Phys. Rev. D **50:** R618.
28. DETTMANN, C.P., N.E. FRANKEL & N.J. CORNISH. 1995. Fractals **3:** 161.
29. YURTSEVER, U. 1995. Phys. Rev. D **52:** 3176.
30. CONTOPOULOS, G. & M. HARSOULA. 2004. J. Math. Phys. **45:** 4932.
31. CONTOPOULOS, G. & M. HARSOULA. 2005. Cel. Mech. Dyn. Astron. In press.
32. CHURCHILL, R., G. PECELLI & D. ROD. 1975. J. Diff. Eqn. **17:** 329.
33. CONTOPOULOS, G. 2004. Order and Chaos in Dynamical Astronomy, revised edit. Springer, Berlin, Heidelberg, New York.

34. AURBRY, S. 1978. *In* Solitons and Condensed Matter Physics. A.R. Bishop & T. Schneider, Eds.: 264. Springer, New York.
35. PERCIVAL, I.C. 1979. *In* Nonlinear Dynamics and the Beam–Beam Interaction. M. Month & J.C. Herrera, Eds.: 302. Amer. Inst. Physics, New York.
36. CHARLIER, C.L. 1902. Die Mechanik des Himmels. von Veit, Leipsig.
37. DEPRIT, A. 1960. *In* Mathématiques du Xème sciècle. Univ. Louvain **1**: 45.
38. CHANDRASEKHAR, S. 1989. Proc. Roy. Soc. Lond. A **421**: 227.
39. CONTOPOULOS, G. 2005. Int. J. Bif. Chaos. To appear.
40. CONTOPOULOS, G. & H. PAPADAKI. 1993. Cel. Mech. Dyn. Astron. **55**: 47.
41. FROESCHLÉ, C., CH. FROESCHLÉ & E. LOHINGER. 1993. Cel. Mech. Dyn. Astron. **56**: 307.
42. VOGLIS, N. & G. CONTOPOULOS. 1994. J. Phys. A **27**: 4899.
43. SMITH, H. & G. CONTOPOULOS. 1996. Astron. Astrophys. **314**: 795.
44. CONTOPOULOS, G., N. VOGLIS & C. EFTHYMIOPOULOS. 1999. Cel. Mech. Dyn. Astron. **73**: 1.

[See Appendix on next page.]

APPENDIX:
PERIODIC ORBITS IN THE LIMIT $M_1 \to 0$

As M_1 tends to zero, the periodic orbit a (see FIGURE A1) shrinks to a point at $z = 1$. Thus, the characteristics a and $\bar{a}$ reach the point $z = 1$. The orbit h, for small M_1, becomes like an elongated arc of a hyperbola (FIG. A1) and as M_1 tends to zero, it tends to a straight line from $z = 1$ to the curve of zero velocity at $z_{max} = (2E-1)/(1-E)$.[30] If $E = \sqrt{0.5}$, we have $z_{max} = \sqrt{2}$.

The orbit a' for small M_1 is an elongated ellipse (FIG. A1), with its uppermost point above $z = 1$ and its lowest point is below, but very close to, the point $z = 1$. At the limit $M_1 = 0$, the value of z becomes $z = z_{max} = (2E-1)/(1-E)$ for the orbit a' and $z = 1$ for the orbit $\bar{a}'$. Therefore, in the limit $M_1 = 0$ the orbit a' becomes a straight line from $z = 1$ to z_{max}—that is, it coincides with the orbit h.

If E decreases, the value of z_{max} also decreases and, for $E = 2/3$, it reaches the value $z_{max} = 1$, therefore, the limiting orbit a' for $M_1 = 0$ is the point $z = 1$. For $E < 2/3$ and M_1 very small, the orbit a' is a very small closed curve around $z = 1$ and, when $M_1 \to 0$, this tends again to the point $z = 1$. Thus, the characteristics of a' and $\bar{a}'$ reach the point $z = 1$, together with a and $\bar{a}$ (FIG. 18 insert).

When $E^2 > 27/32 = 0.84275$ we have also circular periodic orbits of types c', c, $\bar{c}$, and $\bar{c}'$. The orbits c' and $\bar{c}'$ are stable whereas the orbits c and $\bar{c}$ are unstable (FIGS. 15 and 17). The radius of the stable orbit increases from $R = 3$ ($z = 2$ for c' and $z = -4$ for $\bar{c}'$) to $R \to \infty$ as E^2 increases from 27/32 to ∞, whereas the radius of the unstable orbit decreases from $R = 3$ ($z = 2$ for c and $z = -4$ for $\bar{c}$) to $R = 1$ ($z = 0$ for c and $z = -2$ for $\bar{c}$) as E^2 increases from 27/32 to ∞.[31]

Since the orbits c' and $\bar{c}'$ (i.e., the same orbit as c' but described in the opposite direction) are stable, they are surrounded by quasiperiodic orbits that form closed

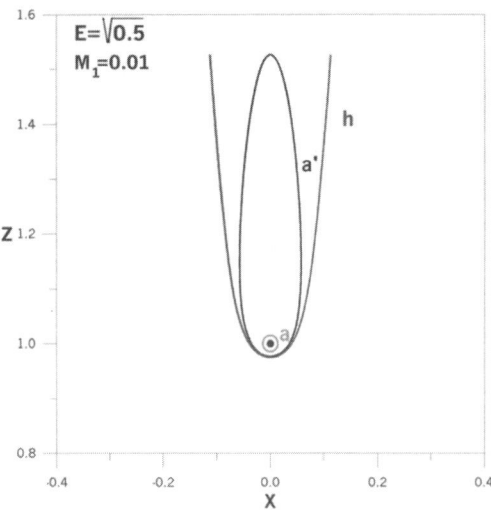

FIGURE A1. The periodic orbits a, a', and h close to $M_1 = 0$ ($E = \sqrt{0.5}$, $M_1 = 0.01$, $M_2 = 1$).

invariant curves around c' and $\bar{c}'$ (FIG. 19). The rotation numbers of these curves are functions of their distances from the point c' or $\bar{c}'$. When these rotation numbers are rational (n/m) we have periodic orbits of multiplicity m. Therefore, we have an infinity of periodic orbits for such values of the energy E.

In the case $E = \sqrt{0.85}$ the rotation number varies from about 0.35 near the central orbit c' to about 0 at the point c (i.e., at the curves SS and UU that limit the island around c'). The simplest periodic orbits around c' are those with rotation number 1/3, one stable and one unstable. These orbits close after three revolutions around M_2.

In the case $E = \sqrt{0.9}$, the rotation number varies from about 0.69 near the central orbit c' to about 0 at the point c. The simplest periodic orbits around c' have rotation number 1/2, one stable and one unstable. There orbits close after two revolutions around M_2.

On Bars and Haloes

Their Interaction and Their Orbital Structure

E. ATHANASSOULA

Observatoire de Marseille Provence, Marseille, France

ABSTRACT: A live halo plays an active role in the formation and evolution of bars by participating in the angular momentum redistribution that drives the dynamical evolution. Angular momentum is emitted mainly by near-resonant material in the bar region and is absorbed mainly by near-resonant material in the halo and in the outer disc. This exchange determines the strength of the bar, the decrease in its pattern speed, as well as its morphology. Thus, contrary to previous beliefs, a halo can help the bar grow, so that bars growing in galaxies with responsive massive haloes can become stronger than bars growing in disc dominated galaxies. During the evolution the halo does not stay axisymmetric. It forms a bar that is shorter and fatter than the disc bar and stays so throughout the simulation, although its length grows considerably with time. I discuss the orbital structure in the disc and the halo and compare it with periodic orbits in analytical barred galaxy potentials. A central mass concentration (e.g., a central black hole or a central disc) weakens a bar and increases its pattern speed. The effect of the central mass concentration depends strongly on the model, being less strong in models with a massive concentrated halo and a strong bar.

KEYWORDS: barred galaxies; dynamical evolution; resonances; bar; halo; peanuts; bulges; orbits; periodic orbits; chaos

INTRODUCTION

Bars are elongated structures seen in the central parts of a large fraction of disc galaxies. Their morphology, photometric, and kinematic properties have been widely studied. Unfortunately, there is no review of observation results that is both complete and up to date, but the reader can consult earlier reviews or reviews covering sub-topics.[1–5]

Bars are very common features. Eskridge *et al.*,[6] using a statistically well-defined sample of 186 disc galaxies from the Ohio State University bright spiral galaxy survey, find that 56% are strongly barred in the H band, whereas another 16% are weakly barred. Grosbøl, Patsis, and Pompei,[7] using a smaller sample of 53 spirals observed in the K band, find that about 75% of them have bars or ovals.

In this paper I discuss a number of results I have obtained recently on the formation and the dynamical evolution of bars. In particular, I discuss the effect of angular

Address for correspondence: E. Athanassoula, Observatoire de Marseille Provence, 2, place le Verrier, 13248 Marseille cedex 04, France.
 lia@oamp.fr

momentum exchange within the galaxy, the role of the halo, the orbital structure in barred galaxies, and the effect of a central mass concentration on the evolution.

ORBITAL STRUCTURE IN BARRED GALAXIES

The first step toward understanding the dynamics of a given structure is to understand its main families of periodic orbits. Indeed, if a periodic orbit is stable it can trap around it regular orbits of similar orientation and morphology. On the other hand, unstable periodic orbits introduce chaos. A growing body of evidence shows that chaotic orbits can indeed be a considerable fraction of the total and that they contribute significantly to the morphology and to the kinematics of bars.[8–16]

Periodic Orbits

The basic families of periodic orbits were first calculated in two-dimensional models. Because of restricted computing power, the earlier studies were limited.[17,18] The first extensive study was done by Contopoulos and Papayannopoulos,[19] who used a very simplified potential to calculate orbits both inside and outside corotation (CR). They called the main family x_1 and showed that within CR it is elongated along the bar. They also presented the banana-like Lagrangian orbits, around the Lagrangian points L_4 and L_5, the family x_2 of central orbits elongated perpendicular to the bar, and the retrograde family x_4. Athanassoula et al.[9] extended this work using a much more realistic bar potential, namely the Ferrers potential.[20] Unfortunately their nomenclature is different from that of Contopoulos and Papayannopoulos,[19] the main family being called B, the perpendicular one A, and the retrograde family R. They confirmed that the B (x_1) family is the backbone of the bar. They used surfaces of section to show that most regular quasiperiodic orbits are trapped around family B (x_1) or around the main retrograde family R (x_4). They also showed that more massive and/or more eccentric bars introduce more chaos.

FIGURE 1[21] shows a sequence of orbits of the x_1 family. Following them in order of increasing Jacobi constant, we see a morphology sequence, first discussed by Athanassoula;[22] namely, the orbits first become cuspy at the apocenter, where, for yet larger energies, they acquire two loops, one at each apocenter. At yet higher energies these two loops disappear and the orbits become oval-like and then rectangle-like. At the largest energies the orbits form four loops, one at each of the four corners of the rectangular shape.

More than half of the orbits displayed in FIGURE 1 close after one revolution around the center and two radial oscillations. Thus, they are resonant 1:2 orbits and are often referred to as such. At higher energies, however, the orbits close after one rotation and four (or more) radial oscillations. Such orbits are often considered as members of the x_1 family, but can also be called 1:4, 1:6, or in general, 1:n, orbits.

Three-dimensional studies showed that the orbital structure is much richer and more complicated. Pfenniger[23] initiated such studies and showed that there are several families of three-dimensional orbits, bifurcated at the vertical resonances of the main planar family. This work was supplemented and extended in a series of four papers.[21,24–26] I briefly recall results from these papers that are relevant to the subjects discussed here. I also follow their notation.

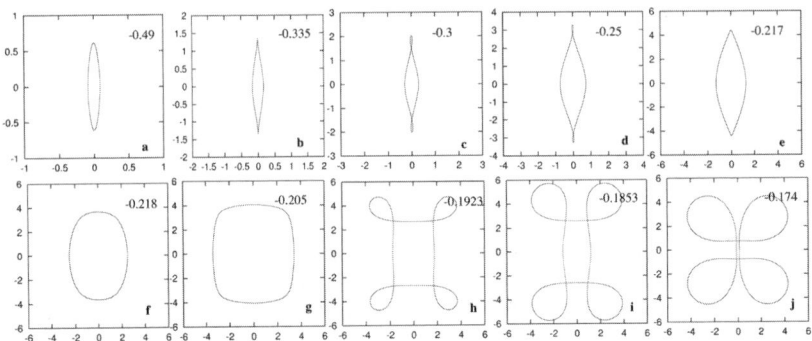

FIGURE 1. Examples of x_1 orbits. The bar is oriented along the y axis and has a semi-major axis equal to six length units. The Jacobi constant, given in the upper right corner of each panel, increases from *left* to *right* and from *top* to *bottom*. (Reproduced from Ref. 21 with permission from the Blackwell Publ.)

The backbone of three-dimensional bars is the x_1 tree, that is, the x_1 family plus a tree of two-dimensional and three-dimensional families bifurcating from it.[21,26] These three-dimensional families are called x1vn, where n is the order with which they are bifurcated in the fiducial model. Thus, the family that bifurcates at the lowest energy is x1v1, followed by x1v2, then x1v3, and so forth. Characteristic orbits from these families in the fiducial model of Skokos, Patsis, and Athanassoula[21] are given in FIGURE 2. The first four orbits (from families x1v1, x1v3, x1v4, and x1v5) are stable; the last (from family x1v2) is unstable. A comparison of FIGURES 1 and 2 clearly shows the morphologic similarity between orbits of the x_1 family in the two-dimensional problem and the (x,y) projection of the x1vn orbits. On the other hand, the x1vn orbits extend well out of the equatorial plane. Thus, it is the trapping around the periodic orbits of the x_1 tree that defines the shape of the bar in the equatorial plane, as well as its thickness perpendicular to it.

Skokos, Patsis, and Athanassoula[21] found also three-dimensional families of banana-like orbits around the L_4 and L_5 Lagrangian points. Considerable sections of the families are stable. More surprising, they also found *stable* periodic orbits around the unstable Lagrangian points L_1 and L_2. The family is planar, starts unstable, but turns stable at larger energy values.

Chaos and How To Measure It

Galaxies are systems that contain both order and chaos, that is, they have both ordered and chaotic orbits. For two-dimensional systems for which there is a rotating frame of reference in which the energy is an integral of the motion, the most straightforward way of distinguishing between the two is to use surfaces of section. Even idealized galaxies, however, often do not fulfill these conditions. Thus, many methods for measuring chaos have been proposed (see Ref. 14 for a review).

Here, I use a method proposed by the person this volume honours, Henry Kandrup, and his collaborators.[27] By Fourier transforming a quantity related to the orbital coordinates, for example, the cylindrical radius R, or the complex quantity $x + iy$,

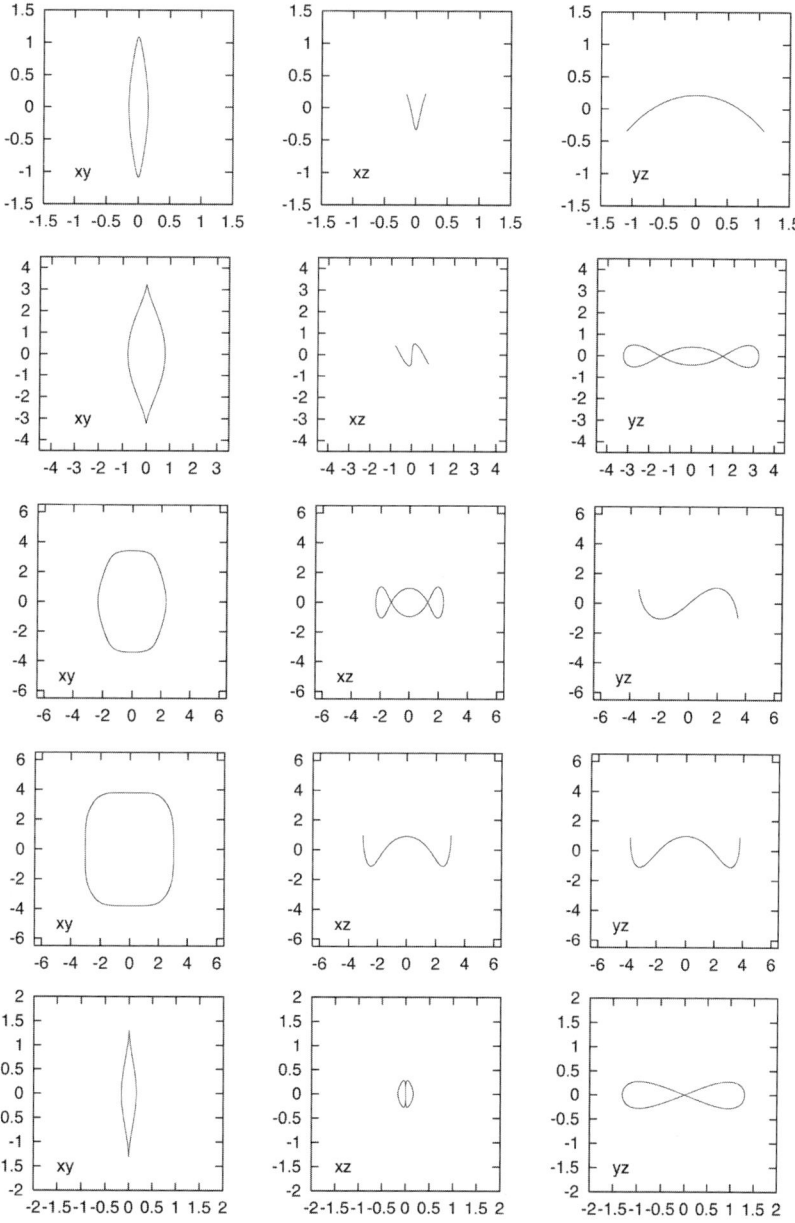

FIGURE 2. Three orthogonal views of characteristic three-dimensional orbits in barred galaxies. The bar is oriented along the y axis and has a semimajor axis equal to six length units. From *top* to *bottom*, these orbits are members of the families x1v1, x1v3, x1v4, x1v5, and x1v2.

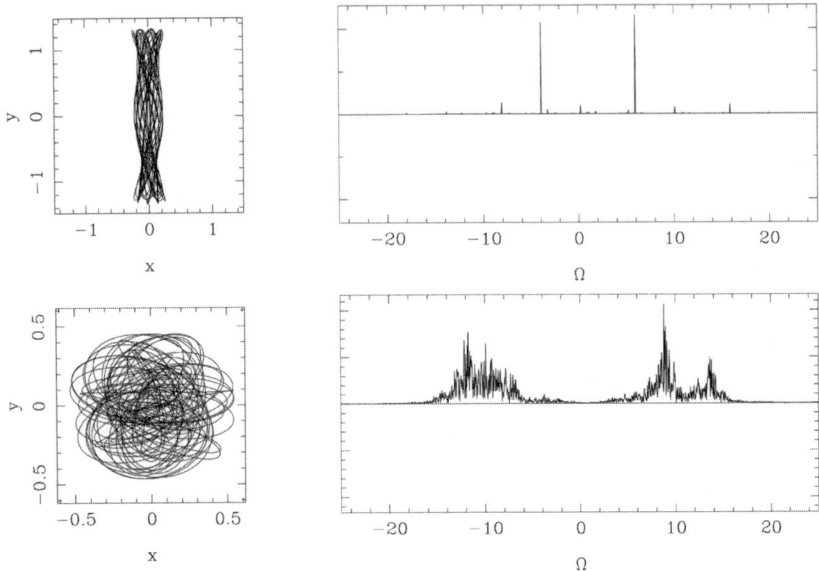

FIGURE 3. The (x,y) projection of two orbits (**left panels**) and the amplitude of the corresponding power spectra of $x + iy$ (**right panels**). The orbit described in the **upper panels** is regular, whereas that in the **lower panels** is chaotic.

one obtains a spectrum. If the orbit is regular, all the power of the spectrum lies in a few, well defined peaks, whereas if the orbit is chaotic, there is a continuum of frequencies within which the power is distributed. This is illustrated in FIGURE 3, which shows two orbits, one regular the other chaotic, and their corresponding spectra.

Kandrup et al.[27] define as the complexity $n(k)$ of an orbit, or, more precisely, of an orbital segment, the number of frequencies in its discrete Fourier spectrum that contain a fraction k of its total power. Regular orbits have small values of $n(k)$, as opposed to chaotic orbits, which have large values. The optimum value for k depends on the size of the orbital segment and the frequency of the sampling. For the cases I discuss in this paper, I use potentials from N-body simulations and consider 65,536 time steps in 40 bar rotations. For such cases, Misiriotis and Athanassoula have found $k = 0.98$ to be an optimum value. Other large values would have also been reasonable choices, since the rank–order correlation coefficient of the complexities obtained using k values between 0.90 and 0.99 is always larger than 0.9. This confirms the result found by Kandrup et al.[27] for a different potential and under different conditions. Misiriotis and Athanassoula have also found a strong correlation between the complexity found from the $x + iy$ spectrum and that calculated from the R spectrum. This is illustrated in FIGURE 4, for the disc orbits of a simulation with a strong bar. The rank–order correlation coefficient of these two quantities is, in this case, 0.99. Kandrup et al.[27] found strong correlations of the complexity with the short term Lyapunov exponents, often used to measure chaos. Thus, they concluded that the complexity $n(k)$ is a robust quantitative diagnostic of chaos. It is particularly

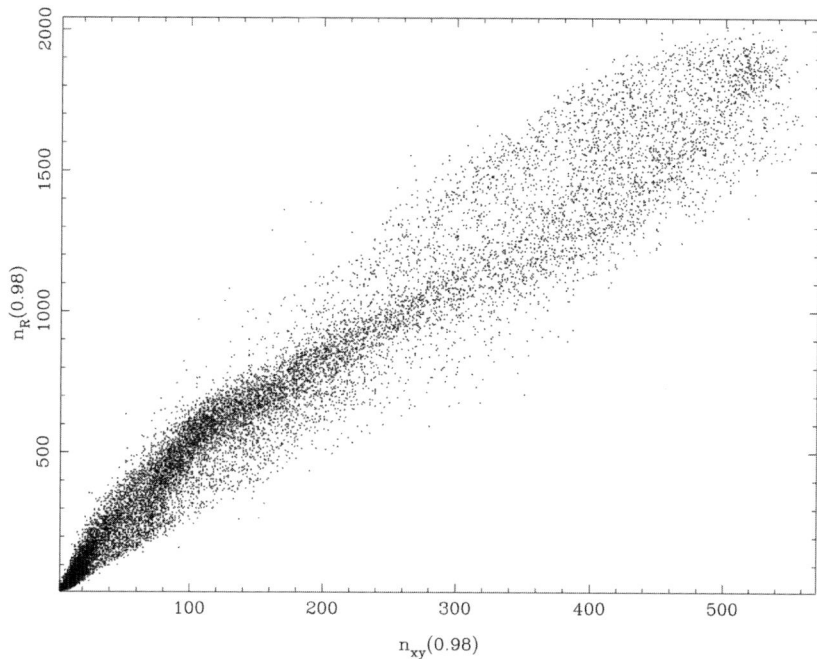

FIGURE 4. Complexity of disc orbits obtained from the R spectrum as a function of that obtained, for the same orbits, from the $x + iy$ spectrum.

straightforward to apply to N-body simulations and it executes very fast, so that it can be applied to a very large number of orbits and simulations. Thus, I adopt it for the estimates presented here. This definition shares the shortcoming of other chaos definitions, that is, there is no clear dividing line between chaotic orbits and regular, but very complex orbits.

THE EFFECT OF THE HALO

N-body simulations of the early seventies[28,29] already showed that bars form spontaneously in galactic discs. At that time the observed evidence for the existence of dark haloes around individual galaxies was hardly compelling, so the discs in these simulations are self-gravitating. Only a few years later haloes were propelled into the center of scientific discussions. Ostriker and Peebles[30] were the first to check the effect of a heavy halo on the bar instability and found it to be stabilizing. Although the number of particles in their simulations did not exceed 500, their work is very insightful. They introduced the parameter t_{op}, which is the ratio of kinetic energy of rotation to total gravitational energy, and they concluded that halo-to-disc mass ratios of 1 to 2.5 and an initial value of $t_{op} = 0.14 \pm 0.03$ are required for stability. Several later papers[31–33] confirmed the stabilizing tendency of the halo. Yet, as we see in the next section, this is an artifact, due to the fact that these simulations were either two-dimensional, or had a rigid halo, or had too few particles. Thus, the halo was not properly described and stabilized the bar. The first doubts about an entirely passive role of the halo component were voiced by Toomre.[34]

The importance of a live halo in order to model correctly the evolution of a barred galaxy was clearly demonstrated by Athanassoula.[35] Two simulations are compared in this paper. Initially they have identical disc components and their haloes have initially identical mass distributions. However, in one of the two simulations the halo is live, that is, it is composed of particles, whereas in the other it is rigid, that is, it is an imposed potential. Thus, in the former simulation the halo can absorb angular momentum, whereas in the latter it cannot. The difference in the evolution is very striking. The simulation with the live halo grows a very strong bar, which, when seen side-on, has a strong peanut shape. On the other hand, the simulation with the rigid halo has a very mild oval in the innermost regions, and hardly evolves when seen edge-on. The large difference between the results of the two simulations argues strongly that the angular momentum absorbed by the halo can be a decisive factor in the evolution of the bar component.

Athanassoula and Misiriotis[36] studied the morphology, photometry and kinematics of the bar as a function of the central concentration of the halo. They found that the bars that grow in centrally concentrated haloes are stronger, longer, and thinner than bars in less centrally concentrated halos, as can also be seen in FIGURE 5 and FIGURE 6. They called the former MH-types and the latter MD-types.[36] MH-type bars initially grow slower than MD-types, in good agreement with what was found in previous studies.[32] However, they reach higher amplitudes. Their face-on isodensity contours are rectangle-like, whereas the corresponding contours in MD-types are more ellipse-like. Their $m = 4, 6$, and even 8 Fourier components of the density are well out of the noise and their amplitude reaches a considerable fraction of the corresponding $m = 2$, contrary to MD-types in which the $m = 6$ and 8 components are negligible. The density profile along the bar major axis (face-on view) also differs in the two types of bars. In MH-types it is rather flat, with an abrupt drop at the end of the bar, whereas in MD-types it drops nearly exponentially with distance from the center (see Figure 5 in Ref. 36). Bars in MH-type models often have ansæ and/or an inner ring that is elongated, but not far from circular and that has the same major axis as the bar, as inner rings observed in barred galaxies.[37] Their side-on shape evolves first to boxy and then to peanut or X shape, in contrast to the MD-types for which the side-on outline stays boxy. The side-on velocity field of MH-types shows cylindrical rotation, whereas that of the MD-types does not. More information on these properties can be found elsewhere.[36]

It is worth noting[38] that a number of the properties of the MH-type galaxies are found in early type bars. Thus, early type bars are longer than late type bars.[39] They often have ansæ[40] and flat projected light profiles along the bar major axis, in contrast to late type bars, which have more sharply falling profiles.[39,41,42] Strong early-type bars have rectangle-like outlines[43] and $m = 6$ and 8 Fourier components of the density of considerable amplitude, contrary to late type bars in which these components are negligible.[42]

From the results summarized in this section, it becomes clear that, contrary to previous beliefs, a *live* halo can *help the bar grow*, so that bars growing in galaxies

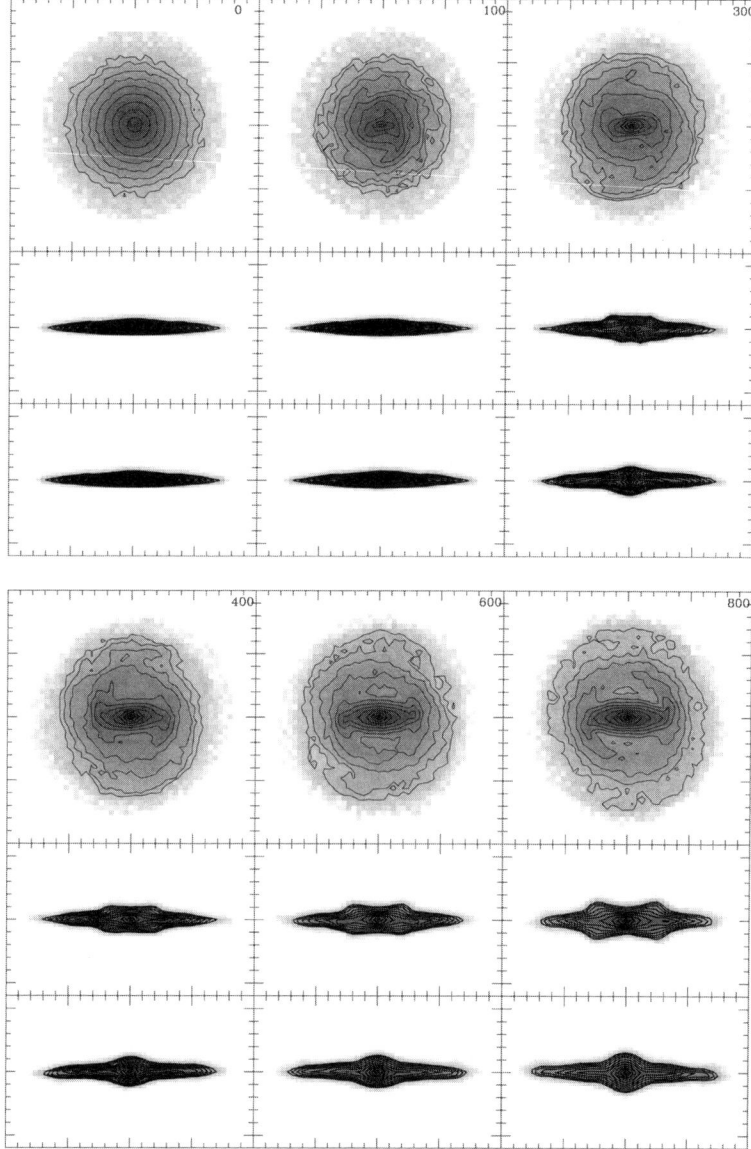

FIGURE 5. Formation of a bar in an initially axisymmetric disc. The model is of MH-type. The *upper* and *fourth rows* give the face-on views; the *second* and *fifth rows* show the side-on views, that is, the edge-on view in which the line of sight is along the bar minor axis. The *third* and *sixth rows* show the end-on views, that is, the edge-on view in which the line of sight is along the bar major axis. Time increases from *left* to *right* and from *top* to *bottom* and is given in the upper right corner of each face-on panel. Here and elsewhere in this paper, times are given in computer units as defined in Reference 36.

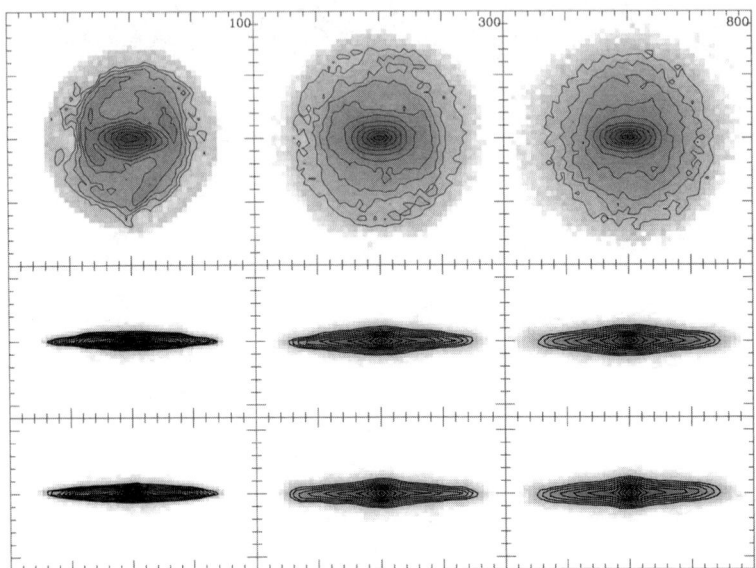

FIGURE 6. Formation of a bar in a disc dominated model (MD-type); layout as in FIGURE 5.

with responsive massive haloes can become stronger than bars growing in disc-dominated galaxies. This marks the end of a paradigm.

ANGULAR MOMENTUM EXCHANGE

The secular growth of the bar component in isolated galaxies is driven by the angular momentum redistribution within them. This was initially proposed, for galaxies with no spheroidal component, by Lynden-Bell and Kalnajs,[44] and later extended to galaxies with haloes and/or bulges by Athanassoula.[45] Angular momentum is emitted mainly by near-resonant stars in the inner disc (namely at inner Lindblad resonance, ILR), or at higher order inner resonances $(1:n)$ and, if the perturbation is growing, also by non-resonant stars in that region. It is absorbed mainly by stars at near-resonance in the outer disc—CR and outer Lindblad resonance (OLR)—and at all resonances of the halo. The latter effect of the halo can be followed analytically if the distribution function of this component depends on the energy only. A full analytical treatment of other distribution functions is not yet available, but a preliminary analysis of models with nonisotropic velocity distribution indicates that these could drive an even stronger bar growth.

The net result is a transport of the angular momentum outwards. Colder material can emit/absorb more angular momentum that hotter material. Thus, the halo is less responsive than the disc per equal amount of resonant mass. There is, however, not much material in the outer disc, where the density is very low, although the halo can be very massive. Thus, it can be that the halo absorbs more angular momentum than the outer disc, and this has proven to be the case in many N-body simulations, as is discussed in the next section. Since the bar is a negative angular momentum "perturbation,"[44,46] by losing angular momentum it becomes stronger.

The Lynden-Bell and Kalnajs formalism[44] and its extension to include spheroidal components[45] were able to predict successfully which resonances emit and which absorb angular momentum. Extending it, however, beyond such qualitative results to obtain a quantitative estimate of the bar growth is not possible since, as stressed by Weinberg,[47] it assumes that the perturbations grow slowly over a very long time and that transients can be ignored. For this reason, quantitative estimates of the bar growth have, so far, only been obtained numerically.

Tremaine and Weinberg[48] also studied the effect of the resonances, focusing on the effect of the angular momentum exchange on the bar pattern speed. They showed that the dynamical friction on the bar arises from stars that are near-resonant with the rotating bar. They derived an analogue of the Chandrasekhar formula[49] for spherical systems, valid when the angular velocity of the bar does not change too slowly. Weinberg calculated[50] that the angular momentum exchange between the bar and the halo will cause a considerable slow-down of the former within a few bar rotations. As discussed in the next section, bars in N-body simulations also present such a slowdown,[45,51–57] in good qualitative agreement with the analytical results. A quantitative comparison may not be meaningful because of the limitations underlying the theory, in particular the fact that theory has not so far treated both the change of pattern speed and the bar growth simultaneously.

RESULTS FROM N-BODY SIMULATIONS

Contrary to real galaxies, N-body simulations are well suited for studying the angular momentum exchange within a galaxy. This has been one of the main goals of References 35 and 45 and I retrace here a number of the steps made in those papers.

This work[35,45] first checked that there is a considerable amount of near-resonant material in the halo component. This can also be seen, for a strong bar simulation, in the upper panels of FIGURE 7, where I plot the mass per unit frequency ratio, M_R, as a function of the frequency ratio $(\Omega - \Omega_p)/\kappa$. Where Ω is the angular frequency of the orbit, κ is its radial frequency, and Ω_p is the bar pattern speed. It is clear that the distribution is not uniform, and that its peaks are located at the main resonances. In all simulations, the disc has a strong peak at ILR, which is made of particles trapped around this resonance and constituting the backbone of the bar. Secondary peaks can be found at other resonances—for example, inner 1:3, inner 1:4, CR, or OLR—whose existence and height varies from one simulation to another. More important, the halo component also shows similar peaks. The highest is at CR; secondary peaks can be seen at ILR and OLR. Such peaks can be seen in all the simulations I analyzed, again with varying heights.

The bottom panels of FIGURE 7 show the way the angular momentum is exchanged. For the disc component it is emitted from the region within the bar, and particularly the ILR, and absorbed at CR (and in some simulations also at OLR). However, the amount of angular momentum emitted is much more than is absorbed by the outer disc. This is understood with the help of the bottom right panel, which shows that all the halo resonances absorb a considerable amount of angular momentum, much more so than the outer disc. A more complete discussion and analysis can be found elsewhere.[35,45] Thus, simulations confirm the angular momentum exchange mechanism suggested by the analytical work and show that the halo can be an important agent in this respect.

This also explains why strong bars can not grow in simulations with rigid haloes. Indeed, as initially discussed elsewhere,[35] in such cases the halo cannot take angular momentum from the bar and, thus, it limits bar growth. The difference between the bar strength in live and in rigid haloes should be larger in cases when the role of the halo in the angular momentum exchange is more important.

The global redistribution of angular momentum was followed[45,55–57] and shows clearly that angular momentum is taken from the disc by the halo. This is in good agreement both with the theoretical predictions and with the above more detailed analysis.

Simulations also show that the bar grows as a result of the angular momentum exchange, as expected from theory. As already discussed in the previous section, they show that the bar grows particularly strong if the halo can take from it consid-

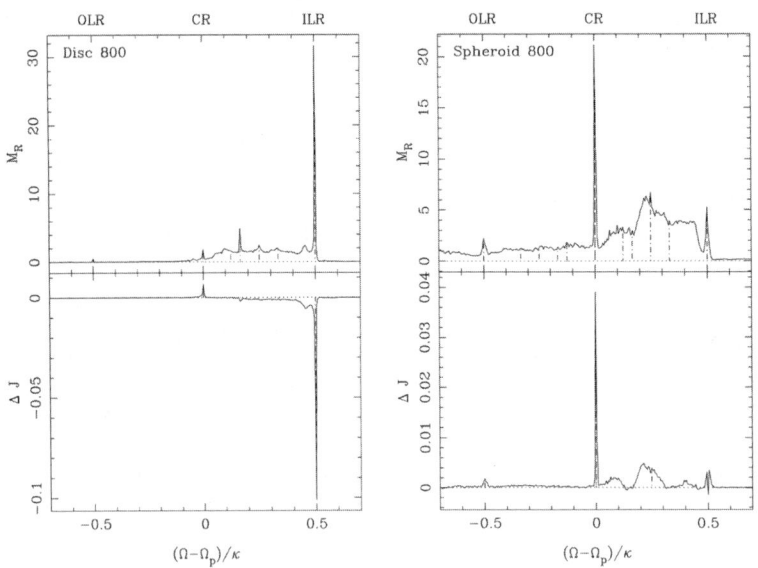

FIGURE 7. Resonances in the disc and the spheroidal component. The **upper panels** show, for the time $t = 800$, the mass per unit frequency ratio, M_R, as a function of that ratio. The frequency ratio is defined as $(\Omega - \Omega_p)/\kappa$, the ratio of the angular frequency in the frame in which the bar is at rest to the epicyclic frequency. The **lower panels** show ΔJ, the angular momentum gained or lost by particles of a given frequency ratio between times 500 and 800, as a function of that frequency ratio, calculated at $t = 800$. The **left panels** correspond to the disc component and the **right panels** to the spheroid. The *vertical dot–dashed lines* give the positions of the main resonances.

erable amounts of angular momentum. A clear correlation is found[45] between the amplitude of the bar and the angular momentum taken by the halo.

Simulations also show that the bar slows down as a result of the angular momentum exchange, as expected from theory.[45,51–57] Thus, one would expect an anticorrelation between the strength and the pattern speed of the bar, and this was indeed established[45] for a large number of simulations.

Theoretical arguments predict that the angular momentum emitted or absorbed at a given resonance depends, not only on the density of matter there, but also on how cold the near-resonant material is. This is borne out by simulations.[45] Indeed, if the disc is hot (i.e., has a high initial Q) and/or the halo is very hot, then the bar does not grow to be very strong and does not slow down much, even in cases where the halo is very dense. Examples of this are given elsewhere.[45]

The angular momentum exchange also determines the morphology of the bar.[58] Simulations where only little angular momentum has been exchanged harbour either an oval or a very short bar. Ovals are mainly found in simulations with hot discs, whereas short bars are predominantly found in simulations with hot haloes. At the other extreme, in simulations in which a large amount of angular momentum is redistributed, the bar is strong, resembling the MH models of References 35 and 36 (see also FIG. 5). As already discussed, such bars resemble the strong bars in early type barred galaxies. Thus, one can argue that a considerable amount of angular momentum has been redistributed in such galaxies between the disc and the spheroidal component. This would be partly taken by the strong bulge component these galaxies have, and partly by their halo. Extreme cases of such galaxies are some examples of bar dominated early type discs presented by Gadotti and de Souza,[59] in which the disc is not a major component any more, since a very considerable fraction of its mass is now within the bar component.

I have so far taken into account only the disc and halo (and sometimes bulge) components. Yet the complete picture of angular momentum exchange can be more complicated. Galaxies (particularly late types) have also a gaseous disc component. This may give angular momentum to the bar, and thus decrease its strength.[60] Furthermore, galaxies are not isolated universes, and thus can interact with their companions, or satellites. If the latter absorb angular momentum, then the bar can grow stronger than in the isolated disc.[60] This is in good agreement with observations that show that more bars can be found in interacting than in isolated galaxies.[61]

In isolated galaxies, angular momentum can only be redistributed between the various components, that is, there should be as much angular momentum absorbed as emitted. Thus, a very massive and responsive halo that can absorb large amounts of angular momentum is not sufficient to ensure important angular momentum redistribution. This can indeed be limited by the amount of angular momentum that the inner disc can emit, so that, if the disc has a very low surface density and/or is very hot, little angular momentum will be exchanged. In other words, it is not useful to increase the capacity of the absorbers if the emitters do not follow, and *vice versa*. Similar arguments can constrain the position of CR. If this is located in the inner parts of the disc it will privilege the absorbers to the detriment of the emitters. This occurs in simulations where the halo absorption is limited so that the contribution of the outer disc is essential. On the contrary, a CR located in the outermost parts of the disc favours emitters. This ensures a maximum exchange in cases where the halo is capable of absorbing a lot of angular momentum.

A BAR IN THE HALO COMPONENT

The halo evolves dynamically together with the disc. In simulations in which a considerable amount of angular momentum has been exchanged and which have thus formed a strong bar, the halo does not stay axisymmetric. It also forms a bar, or more precisely an oval that I call, for brevity, the halo bar to distinguish it from the bar in the disc component that I refer to as the disc bar. An example is seen in FIGURE 8, which compares the morphology of the disc bar (upper panels) to that of the halo bar (lower panels). A very clear case of such a structure is also seen in Figure 2 of Holley-Bockelmann, Weinberg, and Katz.[62]

Halo bars are triaxial, but nearer to prolate than to oblate, with their minor axis perpendicular to the disc equatorial plane. Their axial ratio in this plane (ratio of minor to major axis) is considerably larger than that of the corresponding disc bar. It increases with increasing radius, so that halo bars tend to become axisymmetric in the outer parts. Since the change in axial ratio is very gradual, it is not easy to define precisely the end of the halo bar, and thus to calculate its length, so that any

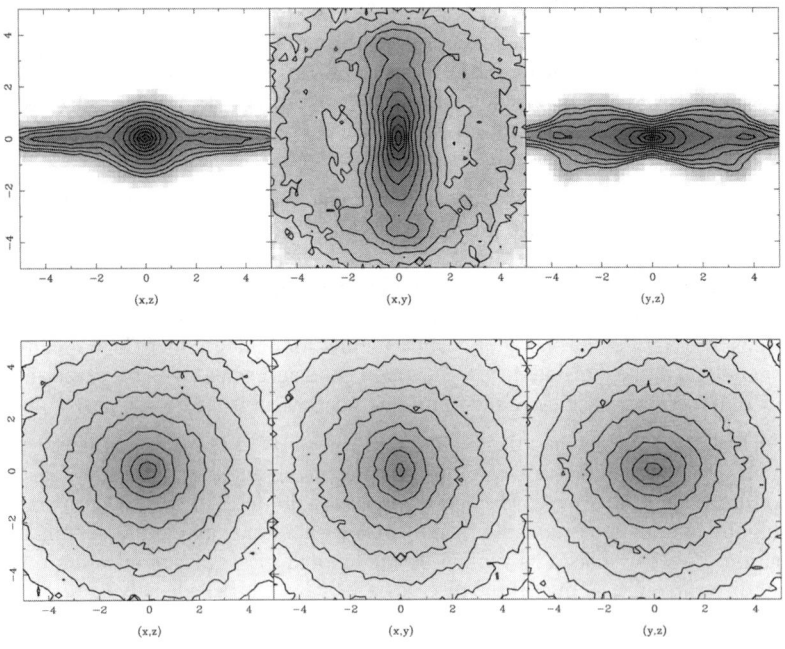

FIGURE 8. Three orthogonal views of the disc (**upper panels**) and halo components (**lower panels**). The **central panel** is a face-on view, whereas the two others are edge-on; side-on for the **right panels** and end-on for the **left panels**. Note that the halo component does not stay axisymmetric, but forms an oval in its inner parts, which I call the halo bar.

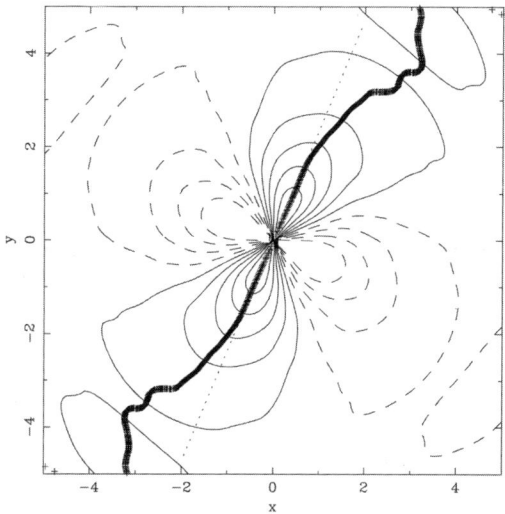

FIGURE 9. Isocontours of the $l = 2$, $m = 2$ component of the halo mass distribution on the equatorial plane. Positive isocontours are given with *solid lines* and negative ones with *dashed lines*. The *thick line* shows the phase of the halo bar and the *thin dotted line* gives the position angle of the disc bar.

measurement will have a considerable error. It is clear, however, that it is always considerably shorter than the disc bar.

The phase and amplitude of the halo bar can be best studied with the help of a decomposition into spherical harmonics. The mass distribution of the $l = 2$ and $m = 2$ component on the equatorial plane, as obtained from such a decomposition, is shown in FIGURE 9. It shows clearly that in the inner parts, where the phase of the halo bar

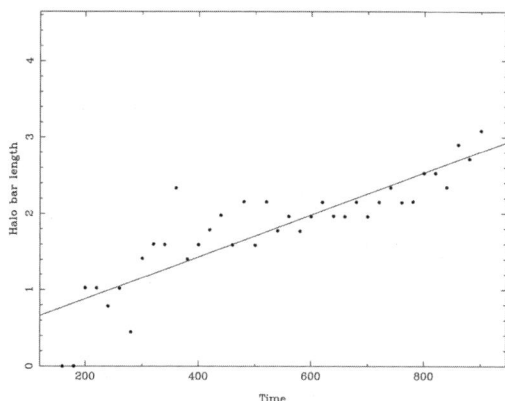

FIGURE 10. Length of the halo bar as a function of time. The solid line is a least square fit.

does not change much with radius, the halo bar has roughly the same orientation as the disc bar. This is true at all times after both bars have grown sufficiently to allow a relatively accurate measurement of their amplitude and phase. Closer examination, however, shows that the halo bar lags the disc bar slightly, by something like a couple of degrees in the innermost parts. This shift is always trailing and increases with increasing distance from the center. Thus, the $m = 2$ component of the halo continues well outside the halo bar, trailing behind the disc bar, so that, after a certain distance, the structure can be described as a trailing very open spiral.

The length of the halo bar can be defined by setting a limiting value for the phase difference between the halo and the disc bar. It is shown as a function of time for a simulation with a strong disc bar in FIGURE 10. The spread in the measurements reflects the difficulty of defining precisely the end of the halo bar. However, a least square fit shows clearly that the length of the halo bar increases with time. Comparison with similar data, but now for the disc bar, shows that the length of the halo bar increases slower than that of the disc bar.

I carried out such an analysis for a number of simulations in which the halo bar is sufficiently strong so that its properties can be measured accurately. I find that the halo bar length correlates with the disc bar length and that the regression line changes with time, as would be expected since the disc bar grows faster than the halo bar. I found also correlations between the disc and the halo bar strengths, as well as between the ellipticity of the halo bar and its flattening towards the equatorial plane. More information on the halo bar properties and their relation with the disc bar properties will be published elsewhere.

ORBITAL STRUCTURE IN THE DISC AND HALO COMPONENTS

Studies of orbital structure in three-dimensional barred potentials can give useful information on the families of periodic orbits, their stability, and the morphology of their orbits. They cannot, however, give information on how many orbits, if any, are trapped around a given periodic orbit family. This information is only available from N-body simulations. The orbital structure has been so far examined in simulations of two-dimensional or three-dimensional discs,[64–66] but not yet for their haloes. However, as discussed already, haloes also have near-resonant orbits and thus should have an interesting orbital structure. I present here preliminary results on the orbital structure of the disc and halo components. I use a method that is, in several aspects, similar to that outlined in the papers mentioned above. Namely, at a given time well after the bar has grown, the potential and forces from a given simulation are calculated on a grid, which is sometimes made bisymmetric since most of the orbital families have that symmetry. The bar pattern speed at that same time is also calculated from the simulation. In the above mentioned papers, this information is used to calculate families of periodic orbits. Unfortunately, numerical noise, which is inherent in N-body simulations, limits such studies to few orbital families and makes stability analyses very difficult, if not impossible. Instead, I take the positions and velocities of 100,000 discs and 100,000 halo particles, drawn at random from the corresponding populations at the time at which the potential and pattern speed are calculated. I use these as initial conditions and follow the corresponding orbits for 40 bar rotations.

Examining these orbits leads to a number of conclusions, some of which I briefly discuss below. A more complete analysis, comparing several simulations, will appear elsewhere.

Of course, these orbits are not the same as those of the same particles throughout the simulation. The simulation orbits start off in an axially symmetric potential and evolve as the bar first forms and then evolves. During the formation phase the potential changes drastically with time, so that any results obtained in the manner described above should be considered with great caution. The formation phase, however, is followed by a phase of calm secular evolution, in which the potential changes slowly with time. It is during this phase that the results obtained with the above method can be useful. The evolution of the potential brings about a change in the orbital structure. By studying this structure at a sequence of times, it is possible to get insight on its evolution. It should, of course, always be remembered that this is only an approximation, albeit the only one known to date that can give useful results. This approximation is also inherently present in any work on orbital structure, since such works have used only non-evolving potentials, whether these are taken from simulations, from observations, or from analytical models.

Morphology of Halo Near-Resonant Orbits

Typical halo orbits trapped around the periodic orbits of the x_1 tree are shown in FIGURE 11. They have been chosen as representative of types of orbits frequently encountered in the simulations presented here. The plots are made in a frame of reference in which the bar is at rest and the bar major axis is along the y axis.

Orbit A (top row) must be trapped around a stable periodic orbit having two oscillations in the z direction and two radial oscillations for each rotation, that is, an orbit of the x1v1 family, presumably resembling that shown in the upper panels of FIGURE 2. Orbit B (second row) is presumably trapped around an orbit of the x1v1 family *and* around its symmetric with respect to the disc equatorial plane. This orbit is located in the innermost part of the halo.

Orbit C (third row) has a considerably smaller extent in the z direction. It is also more extended in the (x, y) plane, so that it has a considerably smaller aspect ratio seen side-on—that is, in the (y, z) projection—than orbits A and B. Its shape is also different. Orbits A and B have a vertical extent that increases with increasing distance from the center, as is the case for the periodic orbits of the x1v1 family. Thus, the highest point in z is near the maximum y value, that is, toward the tips of the (x, y) loop. On the other hand, the orbit in the third row seen side-on appears relatively flat for large y values, that is, for large y values the z-extent does not increase with distance from the center, but stays roughly constant.

Orbit C can have two possible origins. The first possibility is that it is trapped around an x1v2 periodic orbit (see, e.g., the last row of panels in FIG. 2). This family has members with a similar orbital shape and extent to orbit C. It was, however, found to be unstable in the models studied to date[21,26] and thus cannot trap orbits around it. Changing the values of the parameters of the Skokos *et al.* models,[21,26] it is possible to find cases that have an x1v2 family with a stable section, but this is of a very short extent in all the cases examined. Of course the potential in the simulations may not be well described by one of the models discussed by Skokos and collaborators and thus could have considerable differences in its orbital stability properties.

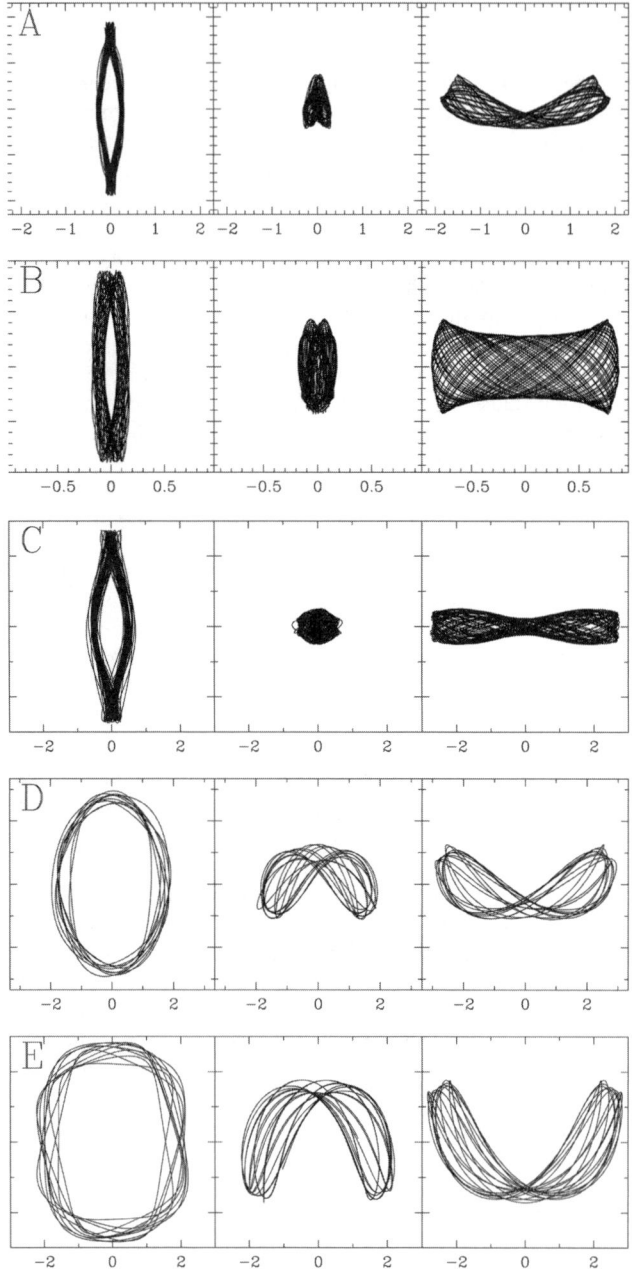

FIGURE 11. Six examples of typical halo near-ILR orbits. The (x, y), (x, z), and (y, z) projections are shown in the *left*, *middle*, and *right panels*, respectively.

The second alternative is that orbits such as C are trapped around periodic orbits of the x1v4 family. An example is given in the third row of panels in FIGURE 2. An orbit trapped around two periodic orbits of this family that are symmetric with respect to the equatorial plane will look like orbit C and have the right shape and extent. It will be possible to distinguish between the two alternatives only after a complete study of the orbital structure has been made.

Orbits such as D, although rather frequent, are even more difficult to classify. Seen face-on, such orbits show a simple oval, with no cusps or loops at its tips. The edge-on views, however, reveal that the maximum z displacement is not at the tips of the oval, but considerably displaced. Again there are two possible alternatives as to the origin of such orbits.

The first alternative is that this is due to a local twist of the isodensities, which affects the edge-on, but not the face-on view. Such a displacement is seen, for example, in the x1v4 periodic orbits (third row of panels in FIG. 2), even in cases where the isodensities have no twist.

The second alternative is that orbit D is trapped around a member of a higher multiplicity family bifurcating from the x_1, most probably one of multiplicity two. Such periodic orbits close after two rotations and four radial oscillations and thus can not be distinguished from those of the x_1 by their frequency ratio. They were found also in the models of Skokos *et al.*, but were not discussed at length since those papers concentrated on periodic orbits of multiplicity 1. Orbit E, in the lowest row of panels of FIGURE 11, provides an even stronger argument about its link with a periodic orbit of multiplicity 2. Such orbits are rather frequent among orbits having a 1:2 frequency ratio, and this argues that families with multiplicity greater than one should be considered in future orbital studies,

The orbits in FIGURE 12 are "banana-like". The orbit in the first row orbits around the L_4, or L_5, Lagrangian point and must be trapped around a long period banana orbit.[67] The example in the second row orbits around the L_1, or L_2, Lagrangian point. Although these points are known to be unstable,[68] Skokos *et al.* found a family whose members orbit around L_1, or L_2, and has considerable stable parts, They called it l_1. The morphology of the orbits of this family resembles that of the short period orbits around L_4, or L_5, rotated by $\pi/2$. The shape and orientation of orbit l (second row in FIG. 12) shows that it must be trapped around a periodic orbit of the l_1 family.

In general, the shapes of the orbits of the halo near-resonant particles are similar to the corresponding disc orbits. Although this might seem strange at first glance, it should in fact have been expected, since the periodic orbits are characteristic of the total potential, that is, they are the same for the disc and the halo.

Fraction of Chaotic Orbits

The amount of chaos as a function of bar strength was addressed elsewhere,[9] with the help of surfaces of sections. These authors plotted the fraction of the area of the surfaces of section that is covered by chaotic orbits as a function of the bar mass and axial ratio. They found that there is a clear trend, in the sense that stronger bars have a larger part of their surfaces of section covered by chaotic orbits. This was confirmed by Teuben and Sanders,[11] for a different model. Although very indicative, these results do not give a clear estimate of the amount of chaos in barred galaxies. Indeed, they do not take into account the time spent by an orbit between two consecutive

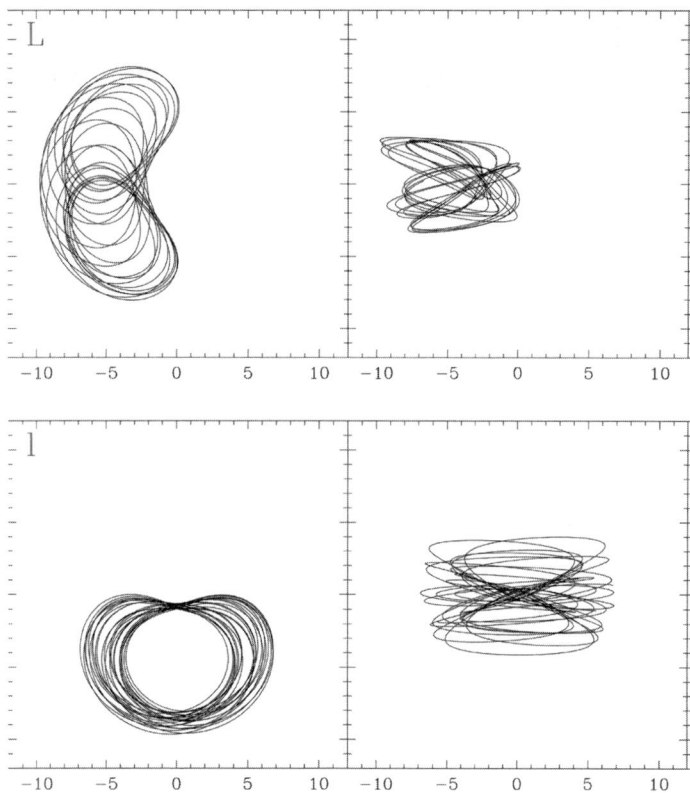

FIGURE 12. Two examples of typical halo near-CR orbits. The **left panels** give the face-on view, (x, y), and the **right panels** the end-on view, (x, z).

crossings of the surface of section[69] and, most important, they have no way of determining whether a given regular or chaotic orbit is populated or not.

This information can only be obtained from N-body simulations. Thus, I have calculated the complexity of 100,000 disc particles and 100,000 halo particles, taken at random from the corresponding population and for several times during the evolution. Initially, by construction, both the disc and the halo are axisymmetric and there is no chaos in either component. This is, however, introduced gradually as the bar first forms and then becomes stronger with time, so that towards the end of the simulation there is a considerable fraction of chaotic orbits. This is seen, for the disc orbits, in FIGURE 13. This shows, in form of a histogram, the number of disc particles as a function of their complexity, both initially and for a time toward the end of the simulations. We note that the distribution acquires a considerable tail toward complex/chaotic orbits.

The evolution of the number of disc chaotic orbits with time is further displayed in FIGURE 14. For this, I plot the number of orbits with complexity above a given

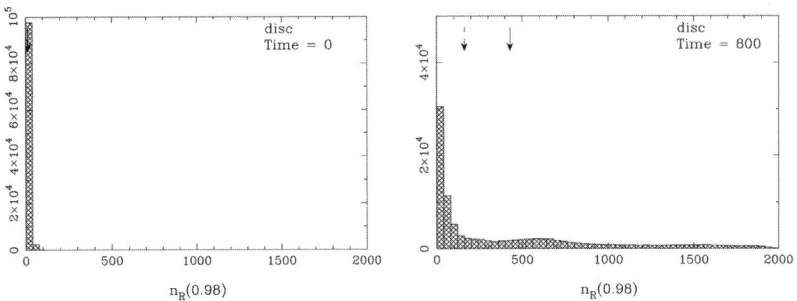

FIGURE 13. Number of disc orbits as a function of the complexity index $n_R(0:98)$. The **left panel** corresponds to the initial time and shows that all orbits are regular, that is, they have small complexity. The **right panel** corresponds to a time after a strong bar has grown and shows that many orbits have acquired very high complexity, that is, they have become chaotic. The *solid line* and *dashed-line-arrows* indicate the mean and the median values, respectively.

threshold as a function of time. Before the bar forms there are no chaotic orbits, as expected. As the bar grows, the amount of chaos increases and at the end of the simulation nearly one third of the disc orbits have a complexity greater than 100.

Where Were the Halo Near-Resonant Particles Initially?

Where do the halo particles at (near-) resonance originate from? Do all particles have equal probability of getting trapped at a resonance during the evolution, or are

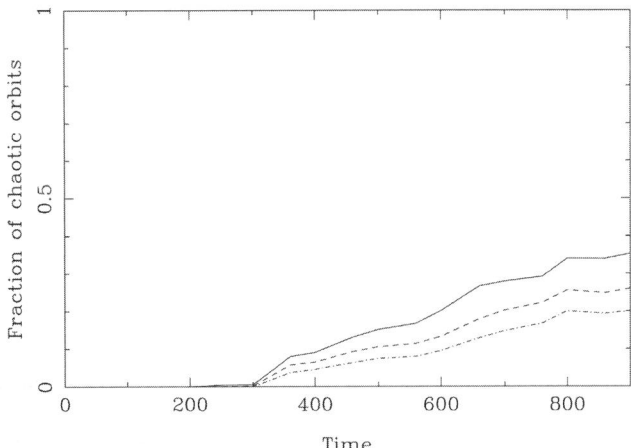

FIGURE 14. Fraction of the disc orbits that have a complexity above a given threshold as a function of time for a simulation with a strong bar. The three thresholds are 100 (*solid line*), 150 (*dashed line*), and 200 (*dot–dashed line*).

there preferred targets? To answer these questions I first found the particles that are at (near-) resonance at a time well after the bar has formed and then traced them back to the initial conditions. I could then compare the distribution of their main properties (cylindrical and spherical radius, distance from the equatorial plane, z component of the velocity and of the angular momentum, kinetic energy, etc.) with that of all the particles in the halo, both taken at $t = 0$. For CR, which is well populated and for which it is easy to make statistical tests; a Kolmogorov–Smirnov test shows beyond doubt that the two populations are not identical, that is, the particles that at later times are at CR are not initially randomly chosen from the initial halo population. The particles that later get trapped at CR do not come initially from the innermost and outermost regions. Instead, they have preferentially intermediate cylindrical and spherical radii. They have preferentially (absolutely) smaller values of v_z and considerably larger values of J_z.

THE EFFECT OF A CENTRAL MASS CONCENTRATION ON A BAR

Central mass concentrations (CMCs) can be hazardous for the growth and even for the existence of a bar. This was first discussed by Hasan and Norman[12] and Hasan, Pfenniger, and Norman,[70] who studied the orbital structure in a rigid potential with both a bar and a CMC. They showed that the CMC alters the stability of the x_1 orbits, making them largely unstable and came to the conclusion that a CMC can destroy the bar if it is sufficiently massive and/or sufficiently centrally concentrated. This work was extended with the help of N-body simulations, first by Norman, Sellwood, and Hasan[71] and later by Hozumi and Hernquist[72–74] and by Shen and Sellwood,[75] All three groups used N-body simulations with live discs and rigid haloes. In all cases the CMC was introduced gradually, in order to avoid transients.

Seeing the important role that the response of the halo can play on the evolution of barred galaxies, Athanassoula, Lambert, and Dehnen[76] revisited this problem, now using a live halo, and I summarize some of their results here. They find that the effect of the CMC depends drastically on the model. This is illustrated in FIGURE 15, which compares the effect of identical CMCs (of mass 0.05 and radius 0.01 for the middle panels, and for mass 0.1 and radius 0.01 for the lower panels) on two different barred galaxy models. The left column corresponds to an MH-type model, and the right column to an MD-type model. The difference is quite important. These CMCs decrease the strength of the bar in the MH-type models, but fully destroy it in MD-type ones. This figure also shows that the lowering of the bar strength is due to a decrease of the bar length and to more axisymmetric innermost parts. The latter can be understood since this is the vicinity of the CMC.

This difference between MH-type and MD-type models could be due to the role of the halo in the two cases. Indeed, in the MH-type haloes the inner resonances in the halo are more populated, so that the halo can absorb more angular momentum, compared to MD-type haloes.[45] This extra angular momentum is taken from the bar and will, as discussed already, tend to increase its strength and work against the CMC, whose effect is thus lessened. This is indeed what the simulations of Athanassoula *et al.* show.[76]

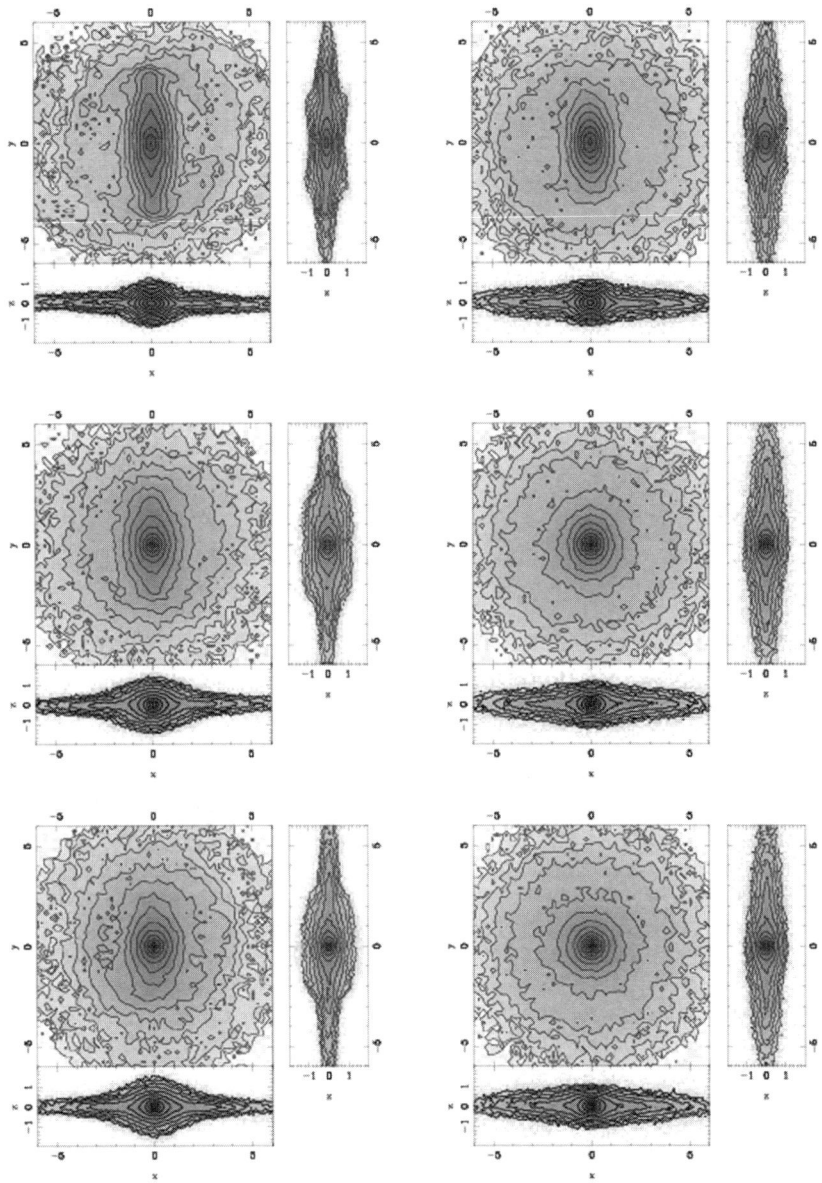

FIGURE 15. Effect of a CMC. The **left panels** correspond to MH-type models and the **right panels** to MD-type models. The **upper panel** shows the disc component at the time the CMC is introduced and the other two at $\Delta T = 300$ later. The mass of the CMC is 0.05 for the **middle panel** and 0.1 for the **lower panels**. Each *subpanel* shows one of the three orthogonal views of the disc component. The *right subpanel* gives the edge-on side-on view, the *lower left* gives the edge-on end-on view, and the *upper left* gives the face-on view. The projected density of the disc is given by grey-scale and also by isocontours (spaced logarithmically).

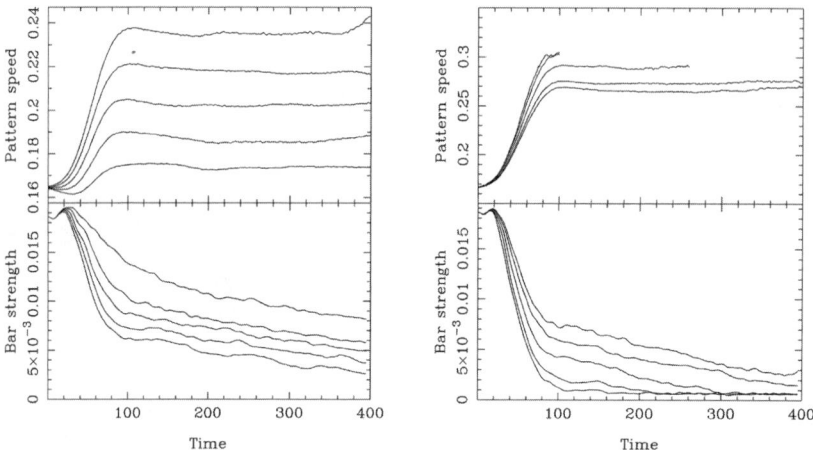

FIGURE 16. Pattern speed of the bar as a function of time (*upper panels*). For comparison, I plot in the *lower panels* a measure of the bar strength (the maximum value of the amplitude of the $m = 2$ component of the mass distribution), also as a function of time. The **left panels** refer to a series of simulations with different CMC mass (from top to bottom in the *lower panel* and from bottom to top in the *upper panel*: 0.01, 0.02, 0.03, 0.04, and 0.05; all with a CMC radius of 0.01). The **right panels** correspond to a series of simulations with different CMC radius (from top to bottom in the *upper panel* and from bottom to top in the *lower panel*: 0.01, 0.02, 0.05, 0.08, and 0.1; all with a CMC mass of 0.1). In all cases the CMC is introduced gradually in the first 100 time units.

The CMC also affects the vertical structure of the bar. FIGURE 15 shows that in MH-type models the peanut initially seen in the side-on view is converted to a boxy or elliptical shape. The radial extent, however, of this vertically protuberant part stays roughly the same. The CMC in the MD-type bar destroys the boxiness of the side-on view.

The CMC also causes an increase of the bar pattern speed. This is not always easy to measure, since, due to the CMC, the bar amplitude is severely decreased. It is, nevertheless, clear for the two sequences of simulations, compared in FIGURE 16. This effect is in agreement with the anticorrelation between the bar strength and pattern speed, found elsewhere.[45]

ACKNOWLEDGMENTS

It is a pleasure to thank many friends and colleagues for interesting and motivating discussions, in particular A. Bosma, M. Bureau, G. Contopoulos, W. Dehnen, A. Misiriotis, P. Patsis, M. Tagger, N. Voglis, G. Aronica, I. Berentzen, K. Freeman, B. Fuchs, K. Holley-Bockelmann, A. Kalnajs, A. Klypin, J. Kormendy, D. Lynden-Bell, F. Masset, I. Shlosman Ch. Skokos and M. Weinberg. I thank Jean-Charles Lambert for his invaluable help with the simulation software and the administration of the runs and W. Dehnen for making available to me his tree code and related programs. I also thank the Observatoire de Marseille, the region PACA, the INSU/CNRS and the University of Aix-Marseille I for funds to develop the computing facilities used for the calculations in this paper.

REFERENCES

1. KORMENDY, J. 1982. *In* Morphology and Dynamics of Galaxies. L. Martinet & M. Mayor, Eds. Geneva Obs. Publ., Geneva.
2. BOSMA, A. 1992. *In* Morphological and Physical Classification of Galaxies. G. Longo, M. Capaccioli & G. Busarello, Eds.: 207. Kluwer.
3. BUTA, R., D.A. CROCKER & B.G. ELMEGREEN, Eds. 1996. Barred Galaxies. Astron. Soc. Pac. Conf. Series 91.
4. ATHANASSOULA, E., A. BOSMA & R. MUJICA. 2002. Disks of galaxies: kinematics, dynamics and perturbations. Astron. Soc. Pac. Conf. Series **275**: 2002.
5. BLOCK, D., K.C. FREEMAN, I. PUERARI, *et al.* 2005. Penetrating Bars Through Masks of Cosmic Dust: The Hubble Tuning Fork Strikes a New Note. Kluwer. In press.
6. ESKRIDGE, P.B., *et al.* 2000. Astron. J. **119**: 536.
7. GROSBØL, P., E. POMPEI & P.A. PATSIS. 2002. *In* Disks of Galaxies: Kinematics, Dynamics and Perturbations. E. Athanassoula, A. Bosma & R. Mujica, Eds.: 305. Astron. Soc. Pac. Conf. Series 275.
8. BINNEY, J. 1982. Mon. Not. Roy. Astr. Soc. **201**: 1.
9. ATHANASSOULA, E., O. BIENAYMÉ, L. MARTINET & D. PFENNIGER. 1983. Astron. Astrophys. **127**: 349.
10. PFENNIGER, D. 1984. Astron. Astrophys. **141**: 171.
11. TEUBEN, P.J. & R.H. SANDERS. 1985. Mon. Not. Roy. Astr. Soc. **212**: 257.
12. HASAN, H. & C. NORMAN. 1990. Astrophys. J. **361**: 69.
13. PATSIS, P.A., E. ATHANASSOULA & A. QUILLEN. 1997. Astrophys. J. **483**: 731.
14. CONTOPOULOS, G. 2002. Order and Chaos in Dynamical Astronomy. Springer.
15. EL-ZANT, A. & I. SHLOSMAN. 2002. Astrophys. J. **577**: 626.
16. EL-ZANT, A. & I. SHLOSMAN. 2003. Astrophys. J. **595**: L41.
17. DE VAUCOULEURS, G. & K.C. FREEMAN. 1972. Vistas Astron. **1**: 163.
18. MICHALODIMITRAKIS, M. 1975. Astrophys. Space Sci. **33**: 421.
19. CONTOPOULOS, G. & T. PAPAYANNOPOULOS. 1980. Astron. Astrophys. **92**: 33.
20. FERRERS, N.M. 1877. Q. J. Pure Appl. Math. **14**: 1.
21. SKOKOS, CH., P.A. PATSIS & E. ATHANASSOULA. 2002. Mon. Not. Roy. Astr. Soc. **333**: 847.
22. ATHANASSOULA, E. 1992. Mon. Not. Roy. Astr. Soc. **259**: 328.
23. PFENNIGER, D. 1984. Astron. Astrophys. **134**: 373.
24. PATSIS, P.A., CH. SKOKOS & E. ATHANASSOULA. 2002. Mon. Not. Roy. Astr. Soc. **337**: 578.
25. PATSIS, P.A., CH. SKOKOS & E. ATHANASSOULA. 2003 Mon. Not. Roy. Astr. Soc. **342**: 69.
26. SKOKOS, CH., P.A. PATSIS & E. ATHANASSOULA. 2002. Mon. Not. Roy. Astr. Soc. **333**: 861.
27. KANDRUP, H.E., B.L. ECKSTEIN & B.O. BRADLEY. 1997. Astron. Astrophys. **320**: 65.
28. MILLER, R.H., K.H. PRENDERGAST & W.J. QUIRK. 1970. Astrophys. J. **161**: 903.
29. HOHL, F. 1971. Astrophys. J. **168**: 343.
30. OSTRIKER, J.P. & P.J.E. PEEBLES. 1973. Astrophys. J. **186**: 467.
31. EFSTATHIOU, G., G. LAKE & J. NEGROPONTE. 1982. Mon. Not. Roy. Astr. Soc. **199**: 1069.
32. ATHANASSOULA, E. & J.A. SELLWOOD. 1986. Mon. Not. Roy. Astr. Soc. **221**: 213.
33. BOTTEMA, R. 2003. Mon. Not. Roy. Astr. Soc. **344**: 358.
34. TOOMRE, A. 1977. Annu. Rev. Astron. Astrophys. **15**: 437.
35. ATHANASSOULA, E. 2002. Astrophys. J. Lett. **569**: L83.
36. ATHANASSOULA, E. & A. MISIRIOTIS. 2002. Mon. Not. Roy. Astr. Soc. **330**: 35.
37. BUTA, R.J. 1986. Astrophys. J. Suppl. **61**: 609.
38. ATHANASSOULA, E. 2003. *In* Astrophysical Supercomputing Using Particles. J. Makino & P. Hut, Eds.: 177. Astron. Soc. Pac. Conference Series, IAU Symp., 208.

39. ELMEGREEN, B.G. & D.M. ELMEGREEN. 1985. Astrophys. J. **288**: 438.
40. SANDAGE, A. 1961. The Hubble Atlas of Galaxies. Carnegie Institute, Washington.
41. OHTA, K., M. HAMABE & K. WAKAMATSU. 1990. Astrophys. J. **357**: 71.
42. OHTA, K. 1996. *In* Barred Galaxies. R. Buta, D.A. Crocker & B.G. Elmegreen, Eds.: 37. Astron. Soc. Pac. Conf. Series 91.
43. ATHANASSOULA, E., S. MORIN, H. WOZNIAK, *et al.* 1990. Mon. Not. Roy. Astr. Soc. **245**: 130.
44. LYNDEN-BELL, D. & A.J. KALNAJS. 1972. Mon. Not. Roy. Astr. Soc. **157**: 1.
45. ATHANASSOULA, E. 2003. Mon. Not. Roy. Astr. Soc. **341**: 1179.
46. KALNAJS, A.J. 1971. Astrophys. J. **166**: 275.
47. WEINBERG, M.D. 2004. Mon. Not. Roy. Astr. Soc. astro-ph/0404169.
48. TREMAINE, S. & M.D. WEINBERG. 1984. Mon. Not. Roy. Astr. Soc. **209**: 729.
49. CHANDRASEKHAR, S. 1943. Astrophys. J. **97**: 255.
50. WEINBERG, M.D. 1985. Mon. Not. Roy. Astr. Soc. **213**: 451.
51. LITTLE, B. & R.G. CARLBERG. 1991. Mon. Not. Roy. Astr. Soc. **250**: 161.
52. LITTLE, B. & R.G. CARLBERG. 1991. Mon. Not. Roy. Astr. Soc. **251**: 227.
53. HERNQUIST, L. & M. WEINBERG. 1992. Astrophys. J. **400**: 80.
54. ATHANASSOULA, E. 1996. *In* Barred Galaxies. R. Buta, D.A. Crocker & B.G. Elmegreen, Eds.: 309. Astron. Soc. Pac. Conf. Series 91.
55. DEBATTISTA, V.P. & J.A. SELLWOOD. 2000. Astrophys. J. **543**: 704.
56. O'NEILL, J.K. & J. DUBINSKI. 2003. Mon. Not. Roy. Astr. Soc. **346**: 251.
57. VALENZUELA, O. & A. KLYPIN. 2003. Mon. Not. Roy. Astr. Soc. **345**: 406.
58. ATHANASSOULA, E. 2005. Celest. Mech. Dynam. Astron. **91**: 9.
59. GADOTTI, D.A. & R.E. DE SOUZA. 2003. Astrophys. J. **583**: L75.
60. BERENTZEN, I., E. ATHANASSOULA, C.H. HELLER & K.J. FRICKE. 2004. Mon. Not. Roy. Astr. Soc. **347**: 220.
61. ELMEGREEN, D.M., B.G. ELMEGREEN & A.D. BELLIN. 1990. Astrophys. J. **364**: 415.
62. HOLLEY-BOCKELMANN, K., M.D. WEINBERG & N. KATZ. 2003. astro-ph/0306374.
63. CONTOPOULOS, G. 1980. Astron. Astrophys. **81**: 198.
64. SPARKE, L.S. & J.A. SELLWOOD. 1991. Mon. Not. Roy. Astr. Soc. **225**: 653.
65. PFENNIGER D. & D. FRIEDLI. 1991. Astron. Astrophys. **252**: 75.
66. BERENTZEN, I., C.H. HELLER, I. SHLOSMAN & K.J. FRICKE. 1998. Mon. Not. Roy. Astr. Soc. **300**: 49.
67. CONTOPOULOS, G. & P. GROSBØL. 1989. Astron. Astrophys. Rev. **1**: 261.
68. BINNEY, J. & S. TREMAINE. 1987. Galactic Dynamics. Princeton University Press.
69. BINNEY, J., O. GERHARD & P. HUT. 1985. Mon. Not. Roy. Astr. Soc. **215**: 59.
70. HASAN, H., D. PFENNIGER & C. NORMAN. 1993. Astrophys. J. **409**: 91.
71. NORMAN, C., J.A. SELLWOOD & H. HASAN. 1996. Astrophys. J. **462**: 114.
72. HOZUMI, S. & L. HERNQUIST. 1998. astro-ph 9806002.
73. HOZUMI, S. & L. HERNQUIST. 1999. *In* Galaxy Dynamics. D. Merritt, J.A. Sellwood & M. Valluri, Eds.: 259. Astron. Soc. Pac. Conf. Series 182.
74. HOZUMI, S. & L. HERNQUIST. 2005. Publ. Astronom. Soc. Jpn. Submitted.
75. SHEN, J. & J.A. SELLWOOD. 2004. Astrophys. J. **604**: 614.
76. ATHANASSOULA, E., J.C. LAMBERT, W. DEHNEN. 2005. Mon. Not. Roy. Astr. Soc. Submitted.

The Basic Dynamical Mechanism in Spiral Galaxies

DANIEL PFENNIGER AND YVES REVAZ

Geneva Observatory, University of Geneva, Sauverny, Switzerland

ABSTRACT: This paper explicates the most fundamental mechanism that rules spiral galaxies. Although spiral galaxies are complex systems for which we do not yet have a complete understanding, the dark matter being the most severe unknown, it is possible to pinpoint the few physical factors that determine their most important properties, such as bars and spiral arms. Dynamics linked to the dissipative nature of gas and its transformation into stars provides clues that spiral galaxies are driven by dissipation close to a state of *marginal stability* with respect to the dynamics in the galaxy plane. Here, we present numerical evidence suggesting that warps play a similar role but in the transverse direction. N-body simulations show that typical galactic disks are also marginally stable with respect to a bending instability, leading to typical observed warps. The frequent occurrence of warps and asymmetries in the outer galactic disks, like bars in the inner disks, give new constraints on the dark matter, but this time in the outer disks.

KEYWORDS: dynamical mechanism; spiral galaxies

INTRODUCTION

Scientific contact between Henry Kandrup and the first author began after the publication of a paper on the orbital relaxation of collisionless particles in non-integrable potentials.[1] Previous studies of individual orbits in barred galaxies had shown that these systems have typical wide resonances, in other words their phase space contains substantial regions of chaos, as well as islands of regular motion, assumed in the past to be the rule in galaxies. This means that in barred galaxies a substantial part of the stars are much more sensitive to two-body interactions or other perturbations to a smooth mass model than previously estimated with simple models, like the Chandrasekhar infinite uniform sea of stars. In a test case, we showed that chaotic orbits in a given barred galaxy model were much more sensitive to perturbations of realistic amplitude than the classical two-body relaxation time. Thus, the effective relaxation was understood to depend also on the regular or chaotic nature of motion in the unperturbed smooth system.

At the same time Gurzadyan and Savvidy[2] showed that N-body systems may be globally exponentially unstable on dynamical time scale, with no strong evidence that this global instability decreases with N. Later, we interacted with Henry Kandrup at several other conferences in Gainesville and elsewhere, especially on topics

Address for correspondence: Daniel Pfenniger, Geneva Observatory, University of Geneva, CH-1290 Sauverny, Switzerland.
daniel.pfenniger@obs.unige.ch

related to chaos, relaxation, and the discreteness effect in N-body systems studied by Henry and coauthors in a series of papers.[3-6] Henry was much interested by the interplay of order and chaos in complex systems. The situation about relaxation was then in a confused state, so we organized the Geneva Workshop, Ergodic Concepts in Stellar Dynamics,[7] where Kandrup, Gurzadyan, Miller and others could debate on the various views about relaxation. As usual in conferences, no definite solutions to the most debated problems were found at that time, but in the following years it became progressively clear that chaos was *the* factor that not only fixes the effective relaxation time of stellar systems, but, more generally, the factor that introduces in physical systems the time-scale separating deterministic and probabilistic theories. For instance the ergodic hypothesis in statistical mechanics is a good hypothesis provided the phase space of the system is mostly chaotic over the considered time-scale. The classical deterministic description of the solar system emphasized by Laplace is effective below a time-scale of a couple of million years, beyond which chaos becomes dominant for at least some of its degrees of freedom.

What follows, in this paper, would have interested Henry. The proposition discussed below is that spiral galaxies can be understood (i.e., their full complexity can be reduced to a few essential rules) as metastable self-organized systems at the interface between order and chaos. The gravitational instabilities are driven by gas dissipation and damped by the heating resulting from the gravitational dynamics reaction.

BASIC PHYSICAL INGREDIENTS IN SPIRALS

Since the work of Jeans in the early twentieth century, knowledge about galaxies has considerably improved, well beyond the pure collisionless stellar system description envisioned then. The necessity to refine the relevant physics, to identify the most fundamental factors characterizing galaxies, especially the spirals, is still there. We need, in particular, to enlarge the Hamiltonian system paradigm to dissipative dynamics and eventually to a paradigm for self-organized criticality. However, collisionless physics and relaxation in partially chaotic Hamiltonian systems, remain basic concepts upon which a refined view about galaxies can be built.

A simple tool to rank the importance of the various physical ingredients at play in galaxies (and other physical systems) is a more complete form of the virial theorem than usually discussed:

$$\frac{1}{2}\ddot{I} = \underbrace{2E_{\text{kin}}}_{>0} + \underbrace{E_{\text{grav}}}_{<0} + \underbrace{3P_{\text{int}}V}_{>0} - \underbrace{3P_{\text{ext}}V}_{<0} + \underbrace{E_{\text{mag}}}_{>0} + \ldots \approx 0. \qquad (1)$$

This allows us to rank the main energies (bulk kinetic, gravitational, internal and external pressures, magnetic, etc.) determining the system equilibrium. The equilibrium is measured by the acceleration of its moment of inertia I in the volume V. Clearly the *magnitudes of the interacting energies* (mainly bulk kinetic against gravitational energy in galaxies) rank the importance of each factor. Taylor expansion of Equation (1) in time,

$$\frac{1}{2}\{\ddot{I}(t + \Delta t) - \ddot{I}(t)\} \approx \{2\dot{E}_{\text{kin}} + \dot{E}_{\text{grav}} + 3(\dot{P}_{\text{int}} - \dot{P}_{\text{ext}})V + \ldots\}\Delta t \approx 0, \quad (2)$$

shows that an evolution along a sequence of quasisteady equilibria is determined to first order by the *magnitude of the interacting powers* (mainly in galaxies gas cooling against mechanical heating power).

The main point emphasized here is that the same rules found to so well explain, to first order, the horizontal properties of spiral galaxies, that is, gravitational physics supplemented by energy dissipation, are also able to explain the ubiquitous warps. By comparing numerical simulations of thin self-gravitating disks to the observed properties of spirals, in particular to their frequent warps, a coherent dynamical and evolutive picture of spiral galaxies emerges, with the suggestive hint that two types of dark matter are involved: (1) classical extended dark halos that are much thicker than the disks, and (2) a dark component coeval with HI disks similar to that proposed by Pfenniger and Combes.[8]

Nowadays, the need for at least two types of dark matter is actually well motivated. From the big-bang predicted baryogenesis most of the baryons remain to be found; on the other hand, also from cosmology, non-baryonic dark matter is required to obtain a coherent description of large scale structure formation and the Universe cosmological parameters. Since baryons are known to be strongly dissipative and sometimes collisional, contrary to the expected non-baryonic dark matter, we have no grounds to expect that in galaxies their respective spatial distributions should coincide.

THE ROLE OF BARS AND SPIRALS

The bar instability was used in the 1970s to support the idea that an extended dark matter halo must exist to prevent bar formation.[9] Indeed, early N-body simulations showed that bars result spontaneously from a dynamical plane instability in a collisionless disk with an initial flat core. However, because theoreticians wrongly assumed that bars were exceptional, they imagined a hot and massive collisionless component coexisting in the optical disks as a solution to prevent the quick formation of bars, despite the awareness by skilled observers, such as de Vaucouleurs, that bars are frequent. Subsequent higher resolution and infrared observations revealed that bars are in fact even more frequent and are found in a majority of spirals.

The reverse problem was, therefore, discussed many times: how bars and hot dark halos can coexist, since dark halos were then no longer viewed as hypothetical. It was also understood over the years that bar-less disks can exist when the central density profile is centrally concentrated.

Spiral density waves are a more general, but less robust, version of the bar phenomenon. The main reason for spiral formation is now well understood as resulting from the non-linear growth of a spontaneous gravitational instability in the disk plane with the same origin as for bars: a kinematically too cold disk is gravitationally unstable and the non-linear result is typically a bar in the initial flat core and spiral arms in the outer differentially rotating region. Without invoking more than Newtonian physics it was also found that the typical double exponential disks profiles are also a natural asymptotic state for a collisionless disk passing through a bar instability.[10]

The non-linear structures resulting from gravitational instability are never strictly stationary, but evolve secularly (over several rotational periods). In pure collisionless disks they tend to destroy the spiral arms and later bars, so obviously something must regenerate them in real galaxies.

The theory of bars and spirals is presently incomplete because the full self-consistent problem is highly nonlinear. Thus, no analytical theory is able yet to *predict* the full development of strongly rotating collisionless self-gravitating disks. Only brute force N-body simulations are able to do this.

Since the Newtonian physics involved in these N-body experiments is very well understood, and because the numerical codes can to a large extent be trusted because various versions implementing different techniques have been developed and checked over several decades by many groups, we can use these N-body techniques to predict and explain the behaviors of galaxies. The results of such N-body simulations should be taken as seriously as analytical developments in celestial mechanics. An example of the success of N-body techniques is the prediction that bars may evolve into peanut-shaped structures. This was first empirically found in N-body experiments,[11,12] understood theoretically,[10] and later confirmed by observations.[13]

Coupled with this well understood underlying physics, numerous independent studies of the mass to light ratio in the optical parts of spirals have determined that a substantial fraction of the gravitating mass there is well explained by the detected baryons.[14] This ensures that we have at least a basic physical understanding of the inner parts of galaxies.

THE ROLE OF GAS

Since disk galaxies as star producing systems must contain also a lot of gas, its effects must be considered in the long run. First, dust polluted gas is very efficient in loosing its thermal energy by infrared radiation, so, to a first approximation, galaxies must be seen as rotating self-gravitating objects; to a second approximation, they are also energy dissipating structures. Gravitationally bound rotating structures slowly losing energy tend to rapidly converge toward thin disks, because then angular momentum is a quantity much harder to dissipate than energy.

Consequently, the frequent occurrence of spiral arms (after all disk galaxies are called spirals) follows directly from a constant competition between the effects of cooling, driving disks toward the gravitational Safronov–Toomre instability threshold. Any additional cooling leads to strong reaction from the disk by dynamical heating. As long as gas cooling continues to be efficient, the natural long term state of spiral galaxies is, therefore, to stay close to the marginal stability threshold.

Numerous studies show that galactic disks, including the Milky Way, have disks close to a marginal stability state, with a Safronov–Toomre parameter, Q, close to unity. Because of this marginal state, spiral galaxies do react strongly to other perturbations, such as galaxy interactions, by amplifying the spiral arms.

Many studies have tried to determine whether spirals and bars result either from galaxy interactions or from a proper disk instability. The more fundamental cause of spirals and bars is actually the internal marginal stability state that causes galactic disks to be very reactive to various perturbations. Even small satellite interactions trigger large responses from a disk in the form of grand design spirals. The name of "spirals" for disk galaxies is, in the end, an excellent way to characterize their close to marginal stability state, showing both that dissipation acts and dynamics reacts.

A corollary of such a marginally stable state is that a steady state is unlikely. Instead, evolution is to be expected as long as the marginal stability state is maintained by the competing factors, gas cooling against dynamical heating.

As a byproduct the large scale dynamical instability of galactic disks leads to local interstellar gas compression, shocks, and turbulence, cascading down to smaller scale gas instabilities.[15–17] At the bottom of the cascade the most visible effect of the gas "turbulence" is star formation,[18] which implies gas consumption. The long term effect of large scale instabilities is to transform progressively the dissipative component into a collisionless stellar component. By consuming gas the cooling agent becomes rarer, and by forming stars the dynamical heating becomes more effective in counterbalancing gas cooling. In addition, the mechanical energy output produced especially by massive stars provides a second important source of heating competing gas cooling.

CONSTRAINTS ON DARK MATTER FORMS

The slow transformation of matter from gas rich, but also dark matter rich, disks to gas poor, star rich, and dark matter poor structures already indicates that the above picture is broadly consistent. Gas poor disk galaxies (S0s and Sas) have typically less prominent and open spiral arms in a more symmetric disk, whereas gas rich spirals (Sds and Scs) have large open spirals in irregular disks. The fact that along the spiral sequence the visible gas represents always a minor fraction of the mass indicates that some of the dark mass must be gas in order to be able to subsequently form all the stars that are seen in S0s and Sas.[19]

However, the fact that S0s and Sas still contain a fraction of dark mass while showing very little star formation also indicates that some of the dark mass is in a form that cannot easily form stars. About 40% of the total mass within the HI disk radius might be in a dark collisionless form. Therefore, the above considerations show that we can have a consistent dynamical picture of disk galaxies, including the gas and star formation aspects, provided that *two* forms of dark matter exist: one, close to the visible gaseous form to explain the properties of the spiral sequences as an evolutive sequence of dissipative gravitating disks, and one non-gaseous form to explain the remaining "indestructible" dark mass in the evolved part of the sequence, the S0s and Sas.

THE ROLE OF WARPS

All these considerations have been made by considering the plane dynamics of spiral galaxies, except for the bulge growth via vertical instabilities in the inner stellar disk. However, what about the dynamical effects transverse to the disks in the outer regions? Namely, a notorious puzzle in spiral galaxies is the ubiquitous warp phenomenon that has eluded a clear explanation until now. For instance, warps are unlikely to result from resonant normal modes, because the soft edges of galaxy disks damp discrete modes.[20] Normal modes generated by massive inclined dark halo[21–23] are ruled out by dynamical friction that damps the warp in a few dynamical times.[24] Only particular triaxial halos can produce a torque that leads to a warp with a straight line of node and negligible back reaction.[25] Interactions are efficient in warping disks,[26,27] however, they cannot be invoked in isolated warped galaxies.

Warps are especially obvious in the HI outer disks, but to a lesser amplitude the stellar disks are also warped. Statistics of warps in HI[28–30] and in the optical band[31–33] reveal that more than half the spiral galaxies are warped and asymmetric. Warps are also linked to large scale disk horizontal asymmetries, both signatures showing that the outer spiral disks are not as well virialized as the inner optical disks.

To answer the question about the general cause of warps, we have first tried to properly understand the dynamics of ideal isolated and purely self-gravitating disks of collisionless particles. As a second step, we will introduce energy dissipation, since disks form for the single reason that the energy dissipation rate is much larger than the angular momentum transport rate. Therefore, energy dissipation must be taken as the second most important factor in understanding galaxies, after the pure gravitational dynamics of collisionless matter.

Consequently, we have undertaken first to study in detail massive self-gravitating disks with various degrees of flattening by means of N-body simulations.[34] The simulated disks are made of a stellar bulge, an exponential disk component, and a collisionless heavy disk component proportional to the HI disk, including a density depression in the optical disk and a flaring thickness almost proportional to radius. The Milky Way is the template galaxy for guiding the choice of the various mass ratios and scale lengths. The mass components and profiles are also such that an almost flat rotation curve is obtained for any thickness of the heavy disk component. By solving the Jeans equations separately for each mass component, we can start simulations with an almost equilibrium model, but with various values for the velocity dispersion ratios $\sigma_r/\sigma_z(R)$, while retaining an initial Safronov–Toomre parameter Q well above unity on almost the full radial range (see FIGURE 1).

The main result is that, conforming to predictions made long ago by Toomre[35] and Araki[36] for self-gravitating sheets, too flat disks are unstable with respect to bending instabilities. The instability in thin sheets is only related to the velocity dispersion ratio between the vertical and radial velocity dispersions σ_z and σ_R. When $\sigma_z/\sigma_R < 0.293$ the sheet bends spontaneously with growth rates of order of Gyr. In thin disks this translates first to S-shaped warp growing modes for slightly unstable disks, and second, for strongly unstable disks, to U-shaped warp modes persisting for Gyrs (see FIGURE 2).

If, like bars and spiral arms, warps result from internal disk instabilities, not only do we obtain a unified picture of galaxies, but also several new clues about the dark matter nature and its distribution. For the same fundamental cause, the marginal stability of self-gravitating disks subject to an energy dissipation, disks spontaneously produce bars, spirals, and warps that counteract dissipation by mechanical heating.

To be in such a warped state, self-gravitating disks must first obviously have a dissipative component. Dusty gas can be identified as the primary cause of energy losses. Second, the mass distribution must be sufficiently self-gravitating and thin in

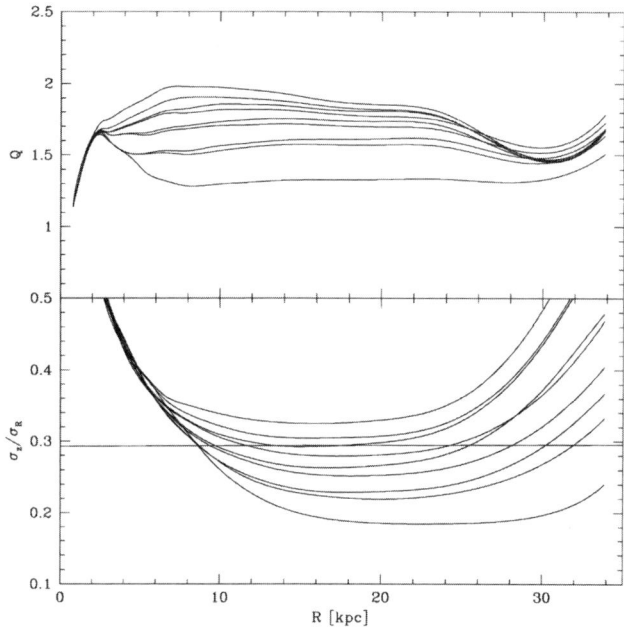

FIGURE 1. The radial stability parameter Q (**top**) and the ratio σ_z/σ_R (**bottom**) as a function of the galactic radius for various models. In both graphics, the curves correspond, from bottom to top, in the interval $R = 10$–20, to the models of increasing thickness.

order to reach the Araki stability threshold. This provides an interesting constraint for the gravitating mass.

The Araki criterion immediately tells us that a disk made of classical smooth gas would never be transversally unstable because the gas pressure would be isotropic. However, a classical gaseous disk dissipating its heat correspondingly decreases its pressure and inevitably, after some time, reaches the Safronov–Toomre radial instability threshold, at which point the subsequent nonlinear evolution depends on the detailed microscopic physics of the gas. In the galaxy disk case we know that the interstellar medium is widely non-homogeneous, which means that the isotropic

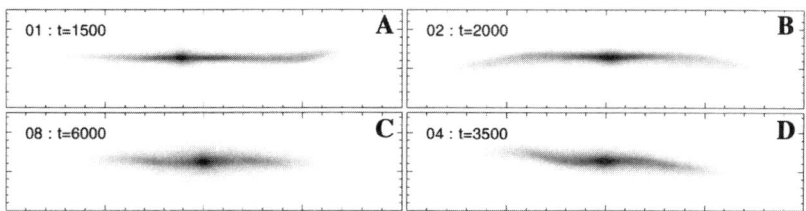

FIGURE 2. Edge-on projections of various heavy disk models having spontaneously developed a warp. The box dimensions are $100 \times 16 \, \text{kpc}^2$.

pressure assumption is not necessarily valid. We also know the most visible result of the gas instability, star formation. Once stars form they evolve as collisionless matter and disks with such fluids are well known to evolve toward anisotropic dispersions with σ_z/σ_R ratios of the order of 0.5.[10,37]

Therefore, the Araki criterion can only be met with a combination of collisionless matter property so as to be able to have rather large velocity dispersion anisotropy and a dissipative component that decreases over time faster for the velocity dispersion in z than in R. The dynamical heating produced radially by spirals and bars also increases the velocity anisotropy if the radial heating is inefficiently transferred transversally to the plane. The radial dynamics of disks indeed heats, effectively through bars and spiral arms, essentially the radial kinematics, maintaining it above $Q \sim 1$.

THE TWO TYPES OF DARK MATTER

The ubiquitous existence of warps in disk galaxies is a strong hint that they are sufficiently massive and thin to be transversally unstable to warp modes, preferentially S-shaped modes. U-shaped modes are also possible if a disk is driven sufficiently deep below the Araki threshold. If this is the case, then we must have a substantial mass component, almost as thin as HI-disks, that behaves as collisionless matter for several rotational periods. In order to regenerate warps, dissipation is essential, without it a too anisotropic disk heats dynamically until a stable thicker state is reached.

If disks react to bending instabilities by kinematic heating transverse to the disk, then one must expect that warped disks are maintained close to the marginal state, balancing gas dissipation with dynamical heating.

Then the question is if such warps may constrain the traditional thick and hot dark halos made of collisionless matter. By adding a corresponding potential to the initial N-body models, we have calculated up to what halo mass with given flattening a marginal state to bending would be retained.[34] It turns out that the effective disk thickness provides a strong constraint on the dark halo mass, but almost no constraint on its flattening. In all cases studied the exact halo flattening is very little constrained by the marginal stability state above a density flattening around 0.3–0.5, which is in fact the range usually considered in cosmological simulations. In contrast, the relative mass of a hot dark halo within a radius comparable to the HI disk radius is directly related to the precise massive disk thickness: the thicker the marginally unstable disk is, the less mass that can be contained in hot spheroidal thick halo, assuming that the warp results from a bending instability. For model parameters fitted to the Milky Way, the halo is at most as heavy as the disk.

Since we can estimate the HI disk thickness, we can give a constraint of the dark halo mass if the disk dark matter has a thickness similar to the HI. The Milky Way has a known HI thickness and a known warp, therefore, if this warp results from a disk marginal stability state, then the dark halo mass within a radius similar to the HI disk radius (about 30 kpc) is constrained to be below 0.4. This value is similar to the dark matter fraction found in evolved Sas and S0s.

CONCLUSIONS

We propose a unified picture of galactic disks where to first order they act as dissipative systems maintained for a time larger than a dynamical time in a critical state, both in the horizontal and vertical directions. Energy dissipation is constantly compensated by dynamical instabilities and stellar activity. The ubiquitous bars, spirals, and warps are a natural consequence for self-organized critical systems, as well as the fractal, turbulent interstellar gas. Warps from a bending instability demand heavier than seen outer disks by a factor 3–5, and correspondingly lighter dark halos, in agreement with studies on the degree of self-gravity of galactic disks based on radial constraints. The heavy disk dark matter must be made mostly of little collisional matter.

ACKNOWLEDGMENTS

This work was supported by the Swiss National Science Foundation. The authors declare that they have no competing financial interests.

REFERENCES

1. PFENNIGER, D. 1986. Astron. Astrophys. **165:** 74.
2. GURZADYAN, V.G. & G.K. SAVVIDY. 1986. Astron. Astrophys. **160:** 203.
3. KANDRUP, H.E. & H.J. SMITH. 1991. Astrophys. J. **374:** 255.
4. KANDRUP, H.E. & H.J. SMITH. 1992. Astrophys. J. **386:** 635.
5. KANDRUP, H.E., H.J. SMITH & D.E. WILLMES. 1992. Astrophys. J. **399:** 627.
6. KANDRUP, H.E. & M.E. MAHON. 1993. Ann. N.Y. Acad. Sci. **706:** 81.
7. GURZADYAN, V.G. & D. PFENNIGER. 1994. Ergodic Concepts in Stellar Dynamics. Lecture Notes in Physics 430, Springer-Verlag, Berlin.
8. PFENNIGER, D. & F. COMBES. 1994. Astron. Astrophys. **285:** 91.
9. OSTRIKER, J.P. & P.J.E. PEEBLES. 1973. Astrophys. J. **186:** 467.
10. PFENNIGER, D. & D. FRIEDLI. 1991. Astron. Astrophys. **252:** 7.
11. COMBES, F. & R.H. SANDERS. 1981. Astron. Astrophys. **96:** 164.
12. COMBES, F., F. DEBBASCH, D. FRIEDLI & D. PFENNIGER. 1990. Astron. Astrophys. **233:** 82.
13. BUREAU, M. & K.C. FREEMAN. 1999. Astro. J. **118:** 126.
14. SANCISI, R. 2004. *In* Dark Matter in Galaxies. S.D. Ryder, *et al.*, Eds.: 233. IAU Symposium 220.
15. FLECK, R.C. 1981. Astrophys. J. **246:** L151.
16. FLECK, R.C. 1983. Astrophys. J. **272:** L45.
17. ELMEGREEN, B. 2004. *In* Star Formation in the Interstellar Medium: In Honor of David Hollenbach, Chris McKee, and Frank Shu. D. Johnstone, *et al.*, Eds.: 117. ASP Conf. Proc., 323.
18. KLESSEN, R.S. 2004. The Relation between Interstellar Turbulence and Star Formation. Habilitation Thesis, Potsdam University.
19. PFENNIGER, D., F. COMBES & L. MARTINET. 1994. Astron. Astrophys. **285:** 79.
20. HUNTER, C. & A. TOOMRE. 1969. Astrophys. J. **155:** 747.
21. DEKEL, A. & I. SHLOSMAN. 1983. *In* Internal Kinematics and Dynamics of Galaxies. E. Athanassoula, Ed.: 177. IAU 100.
22. SPARKE, L.S. 1984. Mon. Not. Roy. Astro. Soc. **211:** 911.
23. SPARKE, L. & S. CASERTANO. 1988. Mon. Not. Roy. Astro. Soc. **234:** 873.
24. DUBINSKI, J. & K. KUIJKEN. 1995. Astrophys. J. **442:** 492.
25. PETROU, M. 1980. Mon. Not. Roy. Astro. Soc. **191:** 767.

26. HERNQUIST, L. 1991. *In* Warped Disks and Inclined Rings Around Galaxies. S. Casertano, P. Sackett & F.H. Briggs, Eds. Cambridge University Press.
27. HUANG, S. & R.G. CARLBERG. 1997. Astrophys. J. **480:** 503.
28. BOSMA, A. 1991. *In* Warped Disks and Inclined Rings Around Galaxies. S. Casertano, P. Sackett & F.H. Briggs, Eds.: 181. Cambridge University Press.
29. RICHTER, O.G. & R. SANCISI. 1994. Astron. Astrophys. **290:** L9–L12.
30. GARCIA-RUIZ, I., K. KUIJKEN & K. DUBINSKI. 1998. *In* Galactic Halos: A UC Santa Cruz Workshop. D. Zaritsky. Ed.: 385. ASP Conference Series, 136.
31. RESHETNIKOV, V. & F. COMBES. 1998. Astron. Astrophys. **337:** 9.
32. SANCHEZ-SAAVEDRA, M.L., E. BATTANER & E. FLORIDO. 1990. Mon. Not. Roy. Astro. Soc. **246:** 458.
33. SANCHEZ-SAAVEDRA, M.L., E. BATTANER, A. GUIJARRO, *et al.* 2003. Astron. Astrophys. **399:** 457.
34. REVAZ, Y. & D. PFENNIGER. 2004. Astron. Astrophys. **425:** 67.
35. TOOMRE, A. 1966. Geophys. Fluid Dyn. **46:** 111.
36. ARAKI, S. 1985. Ph.D. Thesis, Massachusetts Institute of Technology.
37. HUBER, D. & D. PFENNIGER. 2001. Astron. Astrophys. **374:** 465.

The Two Pattern Speeds of NGC 3359

VEERA BOONYASAIT,[a] P.A. PATSIS,[b] AND S.T. GOTTESMAN[a]

[a]*Department of Astronomy, University of Florida, Gainesville, Florida, USA*

[b]*Research Center for Astronomy, Academy of Athens, Athens, Greece*

> ABSTRACT: Using both observations and theoretical techniques, we show that the barred spiral galaxy NGC 3359 contains two pattern speeds. The faster pattern speed for the bar is obtained from isophotal analysis and stellar orbit theory. To explain the spiral arms and the observed velocity field of the disk, a slower pattern speed is required. Nonlinear resonance coupling is the supporting theory for the existence of two pattern speeds for the galaxy. The best match of our models with the observed data indicates a pattern speed for the bar of $39.17 \text{ km·sec}^{-1}\text{·kpc}^{-1}$ and a value between 10 and $16 \text{ km·sec}^{-1}\text{·kpc}^{-1}$ for the spiral.
>
> KEYWORDS: galactic dynamics; barred galaxies; two pattern speeds; NGC 3359

INTRODUCTION

The most important parameter for the morphology of a barred or spiral galaxy is its pattern speed (Ω_p). Models for specific barred and unbarred galaxies that differ only by their pattern speed show totally different morphologies.[1–3] Unfortunately, pattern speed is also the most difficult property of the bar, or the spiral, to ascertain from observations. Currently, three popular methods have been used to estimate the value of Ω_p.[4] The most direct method is to correlate observable features with a specific resonance and to extrapolate the pattern speed from the results. The application of this method is limited to only a few cases and is subject to observation biases and constraints (such as projection effects, asymmetries, and limited resolution). The second method, applied to bars, was introduced by Tremaine and Weinberg[5] and uses the continuity equation as a basis to determine Ω_p from long-slit spectra information. This method needs extensive observation time and is not well suited for late-type galaxies that are rich in gas. The third and most common method is a combination of observation and numerical techniques. Typically, a set of simulations is made with various Ω_p values. The models are then compared with the observed morphology and kinematics, to obtain the closest match, and hence, the best estimate of the pattern speed.[2,6–11]

Nevertheless, as initially pointed out by Sellwood and Sparke,[12] in a barred spiral galaxy the bar and spiral pattern speeds need not be synchronous. The models[12] had the same coherency between the bar and spiral arms as for single pattern speed. This smooth transition can be explained by the non-linear mode coupling between the two structures. The term was introduced by Tagger *et al.*,[13] who also advocated the

Address for correspondence: P.A. Patsis, Research Center for Astronomy, Academy of Athens, Soranou Efesiou 4, GR115-27, Athens, Greece.
ppatsis@cc.uoa.gr

possible existence of galaxies with two or more pattern speeds. The basic concept is that the resonances of the inner (i.e., the bar) and the outer (the spiral arm) modes overlap in some way (e.g., corotation of the bar could coincide with the inner Lindblad resonance of the spiral). Observational evidence for different pattern speeds has also been presented.[14,15] Moore and Gottesman[14] refer to NGC 1398 and place the outer Lindblad resonance (OLR) of the bar at the inner Lindblad resonance (ILR) of the spiral pattern. Rozas and Sempere[15] study the same galaxy we study in this paper (NGC 3359). They assume a bar corotation radius at the ILR of the spiral pattern. More examples for two pattern speeds systems can be found in the literature on "bar within bar" cases, but we do not refer to these here.

In the present study we combined observed data (photometry and kinematics) for the barred spiral galaxy NGC 3359 and we estimated the pattern speed of the bar and the spiral component. We follow the method of comparing models with observed data. Our models are based on the orbital theory (for the bar) and on SPH gaseous response models (for the spiral).

THE OBSERVED DATA

The numerical results were been compared with observed data at several wavelengths. We used information from surface photometry in U, R, I, K, and Hα, as well as Hα and 21-cm data cubes.

Observations of Atomic Neutral Hydrogen

The 21-cm data used were obtained from multiple measurements of the neutral hydrogen emission from the galaxy. All observations were performed at the Westerbrock Synthesis Radio Telescope (WSRT) in the Netherlands. Data reduction and construction of the data cubes (two spatial and one spectral, i.e., velocity, axes) were made by the observer, Dr. A.H. Broeils, who kindly gave the permission to use the HI data for this work. Two separate cubes were created having respectively 15" and 30" angular resolutions.

The HI distribution within the galaxy is quite symmetric on the global scale, although there is slightly more gas in the Southern edge of the main disk. The HI gas is clumpy in nature; some of the features are smaller than the full resolution beam size of $0.8 \times 0.8 \, \text{kpc}^2$. The densest regions observed are within the optical spiral arms that form a pseudoring. The stellar bar region contains little atomic hydrogen gas. The outer arms are composed of neutral atomic hydrogen only. The last vestige of integrated HI density resides in the Western arm, some 525" from the center. The total extent of the HI content is more than twice R_{25}. The total HI mass of the galaxy, based on the projected 30" surface density map, is 5.6×10^9 solar masses.

The isovelocity contours are similar to the typical "spider" pattern of a rotating disk that has been projected onto the sky plane (see FIGURE 1). By overplotting the velocity field on an image of the galaxy it becomes clear that there are significant non-circular motions that manifest themselves as kinks in the velocity field contour lines (see FIGURE 2). The largest deviations from circularity are near the bar and spiral arms. At the neighborhood of the arms, this velocity streaming effect is caused by the response of the gas flow as it travels through the density waves that make the arms.[16,17] The effect appears to be especially strong on the Eastern side of the galaxy where the velocity displacement is more pronounced than on the other side. The undulating velocity contours of the approaching (Southern) half is bent by a greater degree than the receding half in general. The fragmented Western arm just beyond the disk seems to have little effect on the material moving through it since the isovelocities remain fairly smooth in passage. However, the symmetric velocity contour line is skewed. The closed contour lines on the velocity map represent the maximum value of the observed rotational velocity. In the neighborhood of the galactic nucleus, the isovelocity contours bend toward the East, along the bar major axis. The contours within the bar region aligning themselves into an S-configuration, as seen in many other barred systems. This is expected for the gas flowing in elliptical orbits around the bar.[18] The contours are turned toward the bar in such a way that they appear to be pinched together at two diametrically opposing positions located Northwest and Southwest of the bar. Because this area contains mainly elliptical motions, it is excluded in the processes used to derive the global HI kinematical parameters of the galaxy.

Global Kinematical Properties and Rotation Curve.

Analysis of the kinematical data gave the kinematical and physical properties of the galaxy that are summarized in TABLE 1.

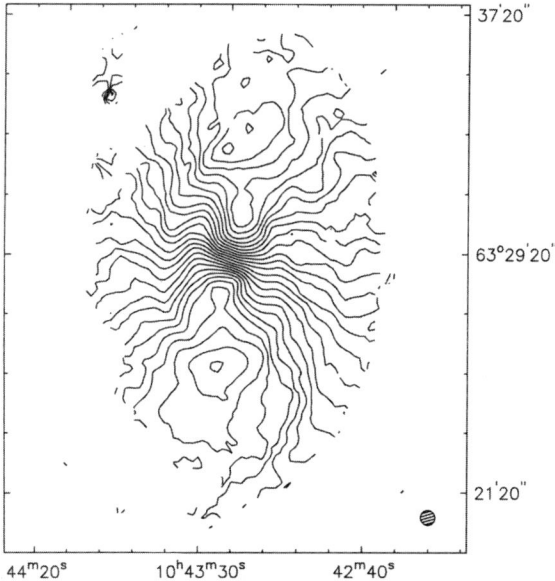

FIGURE 1. The velocity field of NGC 3359. In this and in all following images and diagrams related to the galaxy, North is up and East is to the left. The contour levels are from 850 to 1,150 km·sec^{-1} in steps of 10 km·sec^{-1}. The 15″ beam is plotted at the bottom right corner of the diagram.

TABLE 1. Kinematic and physical properties of NGC 3359, based on the 21-cm data

Parameter	Value
Kinematic center (B1950.0)	
right ascension	$10^h43^m20^s.14$
declination	$63°29'15''.8$
Systemic velocity (km·sec^{-1})	1005.2 ± 0.1
Mean inclination, i, (degrees)	52 ± 2
Mean position angle, ϕ (degrees)	-9.8 ± 0.8
Maximum rotation velocity (km·sec^{-1})	158.11 ± 0.67
HI scale length (kpc)	3.0 ± 0.1
HI disk radius (kpc)	24.0
Maximum surface density of HI ($M_\odot$ pc^{-2})	22.2 ± 0.8
Total atomic hydrogen mass ($M_\odot$)	$5.6 \pm 0.01 \times 10^9$
Total mass ($M_\odot$)	9.9×10^{10}

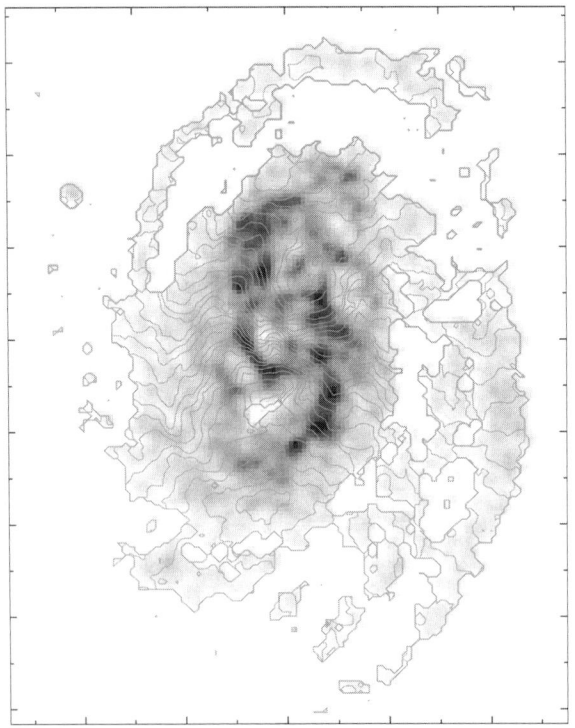

FIGURE 2. Isovelocity contours overplotted on the grayscale image of the moment zero map. Observe the characteristic kinks in the velocity field contour lines near the bar and the spiral arms.

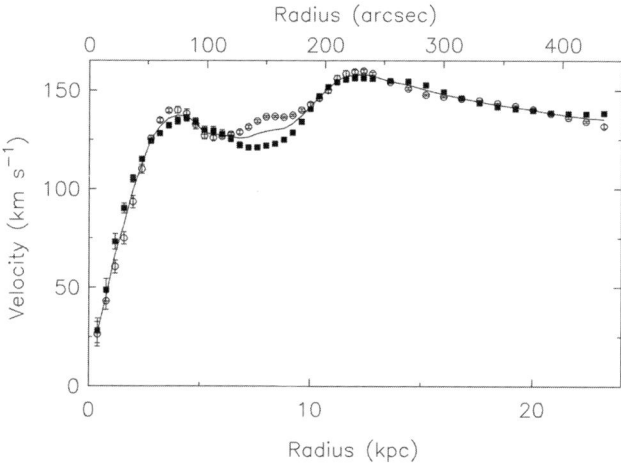

FIGURE 3. The rotation curve of NGC 3359 for both sides (*solid line*) as well as the blue- and redshifted halves of the galaxy (*open circles* and *filled squares*, respectively). The 30 arcsec data was used to construct the final 10 kpc of the diagram.

In our study we retained the distance to the galaxy, 11 Mpc, as adopted in Reference 6 so as to facilitate the comparison of our results. As a matter of fact, the distance of a galaxy does not essentially affect its dynamical modeling.[1] For the adopted distance the linear scale is 53.33 pc·arcsec^{-1}. The estimated position and inclination angles are $-9.8 \pm 0.8°$ and $52°$, respectively. The rotation curve of the galaxy was determined by using the previously derived kinematic properties and both the 15″ and 30″ velocity fields. This is shown in FIGURE 3, together with the circular velocities for the approaching and receding sides.

Overall, both sides of the galaxy appear to be rotating with similar speed as a function of radius. The exceptions are the bar region and in the area about 6.5–10 kpc. Owing to this symmetry, it can be inferred that the kinematic center must be close to or possibly coincide with the NED optical value that has been used throughout the analysis. The most striking aspects of the rotation curve are its two peaks. The second peak of the curve is the true maximum rotational velocity of the galaxy (FIG. 3). It occurs just slightly before the edge of the main HI disk at about 12.8 kpc (240″) and has a value $V_{max} = 158$ km·sec^{-1}. After this point, the rotation curve begins a long and slow decline for about 8.5 kpc (nearly two-thirds of the gas disk) until the last point is reached. The rate of change is -2.0 ± 0.1 km·sec^{-1}·kpc^{-1}. This is above the Keplerian drop-off in velocity at large radii that is expected for a system that has its total mass confined within the observed maximum radius. As in many rotation curves of other disk galaxies, the slow decline is typically conjectured to be caused by the existence of dark matter.

Observed Data from the Optical and the Near Infrared

The optical images we used were obtained from the Instituto de Astrofisica de Canarias (IAC), generously provided by Dr. John Beckman. They are U-, R-, and

I-band images, plus observations of the galactic HII regions centered around the Hα Balmer emission-line. The Hα observations include a broadband image and the Fabry–Perot interferometer data cube. Observations of the galaxy at 2.2 µm wavelength using a K-filter were taken by the Florida multi-object imaging near-IR Grism observational spectrometer (FLAMINGOS) at the 2.1-meter telescope located at KPNO in imaging mode. Observation time for the galaxy was generously provided by Drs. R. Elston and E. Lada.

The major feature of the galaxy, the bar, is found to be quite red in color despite having many bright giant HII regions located along its major axis (although none exist in the nucleus). The dust in the bar region contributes to observed colors The K-band light profile of the bar major axis was found to be flat and in agreement with published results.[19] The spiral arms are the bluest component of the galaxy due to the fact that (1) the majority of the galactic HII regions reside within them and (2) there is less dust obscuration. The color index at large radii (i.e., beyond about 150″) is reddish, suggesting that the predominant occupants of the outer disk are older stars.

Fabry–Perot observations of the Hα emission from ionized hydrogen provide an additional and vital source for studying the gas kinematics within NGC 3359. The isovels of the velocity field show strong non-circular effects caused by spiral density waves and bar streaming motions around the arms and central regions of the galaxy respectively. The deviations from circular motion are as large as 45 km·sec^{-1} (in the plane of the sky). The derived systemic velocity (1,008.6 km·sec^{-1}) and inclination (54°) of the Hα disk are in agreement with the 21-cm data.

The residual velocity field shows non-circular effects indicative of streaming motions around the bar as well as the arms. Within the bar, velocity jumps can be seen. However, the highest velocity jump that was determined by taking slices across the width of the bar was 40 km·sec^{-1}. This indicates that the shocks are moderately strong. The interstellar grains are mostly concentrated near the ends of the bar in patches, although there is substantial obscuration of light close to the galactic center. The bar does not have the typical offset curved dust lanes seen in many other late-type galaxies.

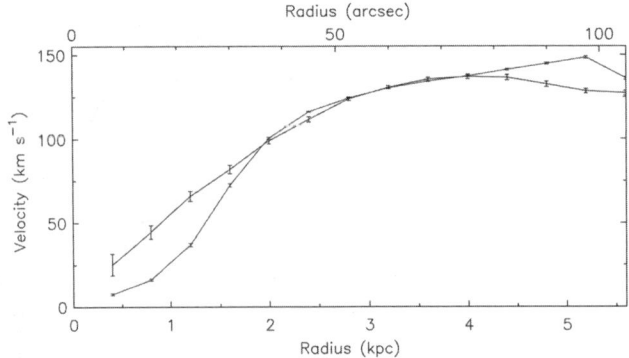

FIGURE 4. Comparison diagram of the 21-cm and Hα rotation curves. In the central region of the galaxy the HI gas (—) is moving faster than the HII (♦).

A comparison study between the HI and Hα kinematics has shown that both species of hydrogen have similar motions throughout most of the galaxy. The largest differences between the two gases are at the bar region. For the first 30″ of the area, the HI gas is moving faster than the HII (see FIGURE 4). Based on this we conclude that the ionized gas is affected to a greater degree by the local velocity fields of the HII regions that appear to have not only the usual circular velocity component but also radial and possibly vertical motions.

MODELS

The underlying potential of the galaxy is the single most important datum for the theoretical study of the stellar and gas kinematics in NGC 3359. Information about the stellar content and its mass distribution plays the central role in determining the potential because the latter governs the gravitational field that the stars dominate. The procedure that is employed here to obtain the galactic potential follows the previous work.[20] Their Fourier transform method creates a model-independent potential by convolving the surface mass density of a deprojected K image with a function that depends on the vertical scale height.

The potential in the plane of the galaxy can be expressed in the form of a Fourier series, so that

$$\Phi(r, \phi) = \Phi_0(r) + \sum_{m>0} \Phi_{mc}(r)\cos(m\phi) + \Phi_{ms}(r)\sin(m\phi). \tag{1}$$

To find the components Φ_0, Φ_{mc}, and Φ_{ms} of this equation, the potential was partitioned into concentric annuli with the width of each ring equal to one beam width (point-spread function, FWHM). A polynomial fit (of the form $\sum_n \alpha_n r^n$ was made to the components. The Fourier coefficients are also shown graphically in FIGURE 5. As expected, the axisymmetric part of the potential dominates most of the galactic gravitational field and the bar is a weak perturber of the total potential. Its maximum strength is less than 4% of the Φ_0 component. The Φ_{2c} term peaks at 32″ and crosses the Φ_{2s} term at about 65″. The latter position is close to the (deprojected) corotation radius (r_{RC}) of 63″ as stated elsewhere[21] and the value that was adopted for this work. At 65″ (3.36 kpc), $r_{RC} = 1.3 R_{bar}$, that is, within the 1.2 ± 0.2 range already proposed.[22]

A few comments should be made about the derived potential. First, it was assumed that the mass-to-light ratio (M/L) was constant throughout the disk of the galaxy. This assumption does not appear to be threatened because the color differences between the infrared images show no dramatic population change throughout the disk. Also, several potentials were created using different values of the vertical scale height h and M/L to derive theoretical circular velocities that were then compared with the observed HI and Hα rotation curves. The best estimated values for the (imperfect) match between the rotation curves are 700 pc and 1.66 for h and M/L, respectively. This potential has been used to describe the dynamics of the disk component. However, it lacks the inclusion of an explicit bulge component (however small the bulge may be) and is not accurate close to the center of the galaxy.

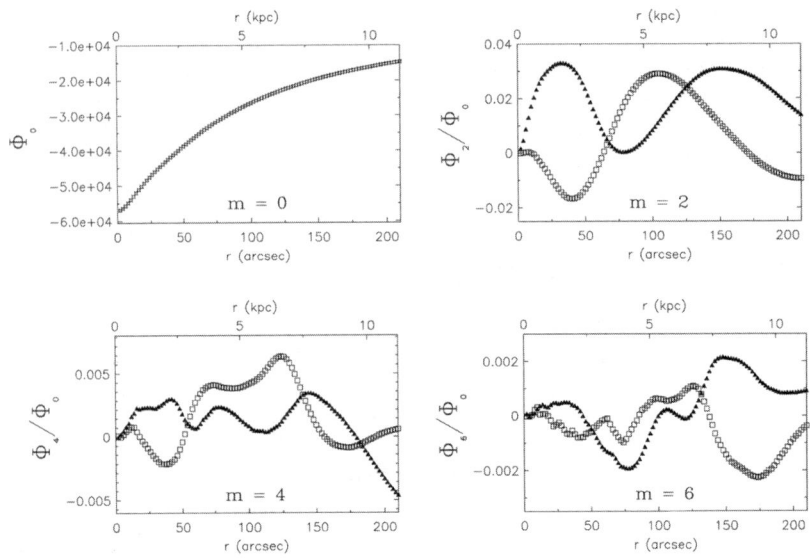

FIGURE 5. Graphical representation of the Fourier components of the potential. *Filled triangles* and *open squares* represent the cosine and sine terms of the components, respectively.

Orbital Analysis

The equations of motion were derived from the Hamiltonian,

$$H \equiv \frac{1}{2}(\dot{x}^2 + \dot{y}^2) + \Phi(x, y) - \frac{1}{2}\Omega_p^2(x^2 + y^2) = E_J, \qquad (2)$$

where (x, y) are the coordinates in a cartesian frame of reference corotating with the bar with angular velocity Ω_p, $\Phi(x,y)$ is the potential in cartesian coordinates, E_J is the numerical value of the Jacobian integral, and dots denote time derivatives; E_J is given in (km·sec^{-1})^2. We used a fourth order Runge–Kutta integration scheme with a variable step. We find the periodic orbits by using an iterative Newton method in two dimensions.

The families whose members are symmetric with respect to the bar minor axis are best followed on an (E_J, x) diagram, known as the "characteristic diagram". The characteristic of a family of periodic orbits is a curve giving the initial position along the bar minor axis, x, as a function of the Jacobi constant E_J. Orbits that are symmetric with respect to the minor axis are uniquely defined on such a diagram. For asymmetric orbits one also needs the corresponding initial velocity, $\dot{x}$, and the characteristic diagram becomes three-dimensional $(E_J, x, \dot{x})$. In our potential we have almost only asymmetric orbits due to its form. In most cases, however, we continue discussing orbits in terms of their position on the (E_J, x) diagram for reasons of continuity with previous work.

Changes in the stability of a family of periodic orbits as one of the parameters of the model varies (usually E_J) are followed by means of the characteristic diagram and the variation of Hénon's stability index.[23] According to Hénon's method, after having found a periodic orbit with initial conditions, for example, $(x_0, \dot{x}_0)$, a non-periodic orbit in its close neighborhood is integrated. Then, one considers the initial $(\delta x_0, \delta \dot{x}_0)$ and final $(\delta x_1, \delta \dot{x}_1)$ deviations of the non-periodic orbit from the initial conditions of the periodic orbit. This is done at two successive upward intersections of the non-periodic orbit with the axis $y = 0$. In this way, a $g: (\delta x_0, \delta \dot{x}_0) \rightarrow (\delta x_1, \delta \dot{x}_1)$ transformation is established, the Jacobian of which can be written as

$$J = \begin{pmatrix} a & b \\ c & d \end{pmatrix},$$

and the corresponding characteristic equation is $\lambda^2 - (a+d)\lambda + 1 = 0$. We used a deviation $\Delta = 10^{-7}$ from the periodic orbit to calculate the a, b, c, and d values. The Hénon stability index α characterizes the stability of the periodic orbits and is defined as $\alpha = (a+d)/2$. The stability condition is the condition for having two complex conjugate roots of unit modulus in the characteristic equation. An orbit is stable if $|\alpha| < 1$. The evolution of the stability of a family of periodic orbits as a function of the energy in the rotating frame, is given in a (E_J, α) diagram. Of special interest in this diagram is the case when α becomes equal to one, either by being tangent to or by intersecting the $\alpha = 1$ line, usually called the $\alpha = 1$ axis.[24] At these points, a new family is bifurcated and introduced into the system. It has the same periodicity as the parent family and inherits its stability. Thus, after a $S \rightarrow U$ transition, that is, when a stable family becomes unstable, a new *stable* family is introduced into the system, and after a $U \rightarrow S$ transition a new *unstable* family. Nevertheless, beyond their bifurcation point, the new bifurcated families may change their stability. These changes of stability of the main, as well as of the bifurcated, families are described in this paper by means of the characteristic and stability diagrams.

Periodic Orbits and their Stability

Periodic orbits provide the backbone structure to a disk galaxy. If they are stable, then they will trap non-periodic orbits around them. However, if they are unstable, stochasticity is introduced to the nearby neighborhood.[25] In the axisymmetric case one can always find the direct orbits that are circular and that make up the "central" (usually called x_1)[25] family. In the full bar potential the x_1 orbits are elongated along the bar and provide support when the perturbation is added to the total potential. For example, it was found[26] that x_1 supported the bar up the 4:1 resonance—or the end of the bar in the case of NGC 4314. The location and stability of periodic orbits in a given potential Φ can be found by using the characteristic and stability diagrams. We used these diagrams to investigate the evolution of the central family of orbits within NGC 3359. The model we studied is rotating with $\Omega_p = 39.17 \text{km} \cdot \text{sec}^{-1} \cdot \text{kpc}^{-1}$.

In an axisymmetric potential Φ_0, the x_1 family consists of direct, stable orbits that rotate circularly. FIGURE 6 shows the stability diagram of the x_1 family within the axisymmetric potential of the galaxy. The Hénon index values fluctuate between the lines of stability limit $|\alpha| = 1$. These lines denote the transition to instability and

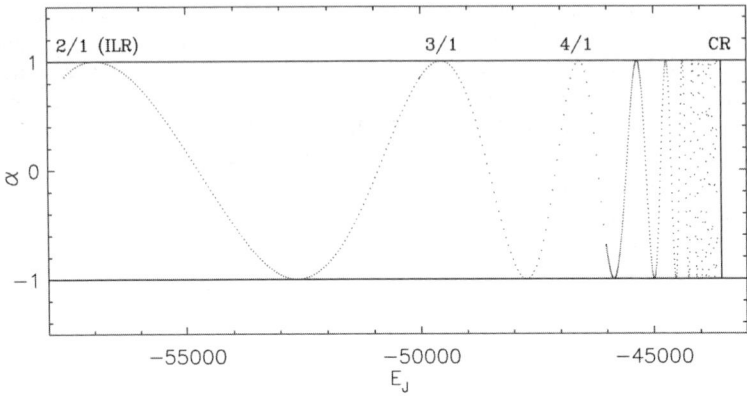

FIGURE 6. The stability diagram of the axisymmetric potential case. The area $-1 < \alpha < 1$ represents the zone of stability

occur near resonances. It can be seen that x_1 is stable everywhere. As marked in the stability diagram, the $m = 2, 3$, and 4 resonances occur at energies whose radial distances correspond to 0.11, 0.98, and 1.57 kpc, respectively. The spacings between the higher order resonances become smaller as one approaches corotation. The separation between the last marked point (the 4:1 resonance) and corotation is 1.8 kpc. Note that the quoted distances apply only to the axisymmetric case since the values will change when the nonaxisymmetric Fourier components are added. The behavior shown in the diagram is typical for axisymmetric potentials.[24]

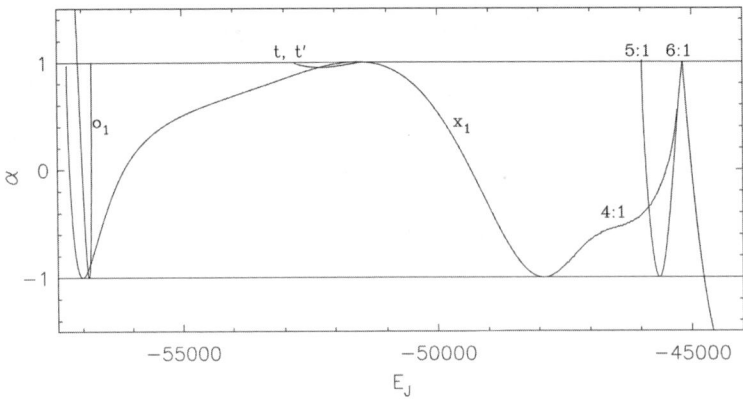

FIGURE 7. The stability diagram of x_1 and other stable families using the full potential in Equation **(1)**. The $n:1$ labels refer to the corresponding resonance regions: t, t' are 3:1- and o_1 1:1-resonance related families respectively.

Non-Axisymmetric Case

Although the Φ_2, Φ_4, and Φ_6 terms in FIGURE 5 indicate that they are rather weak as compared to the axisymmetric component of the potential, their influence on stellar orbits is significant, since they represent the bar and spiral structures.

The stability diagram of the current model (see FIGURE 7) indicates that the 3:1 resonance has shifted toward lower energy, and the stability curve intersects the $\alpha = 1$ axis at two points. The intersection of the stability curve with the $\alpha = 1$ axis is important, because at the intersections new families are bifurcated from their parent and are introduced in the system. For the case at hand, an "inverse bifurcation"[25] is present (i.e., the new branch extends toward lower energies). However, the influence of the 3:1 resonance on the dynamics of the system is only local, and hence, an in-depth study of this phenomenon is not be explored here.

The leveling off of the curve at the 4:1 resonance is not atypical.[24,27] The original x_1 stability curve also effectively ends at this region. The next curve starts at a higher stability value and ends at the 6:1 resonance, much as is shown in Figure 8 of Reference 24.

FIGURE 8 gives the characteristics of the same families included in FIGURE 7. Moving from the lowest energy in the diagram, the x_1 characteristic branches up the x axis monotonically with increasing E_J until the 4:1 gap is encountered. At $(E_J, x) = (-46,540, 1.28)$, the tip of the branch is reached and the curve begins to deviate downward with increasing energy. This gap is a typical behavior at the 4:1 resonance. The 4:1 gap displayed by the x_1 characteristic curve for NGC 3359 is of type 2.[28] Beyond the gap, a new characteristic curve for the 5:1 and 6:1 family appears. This characteristic peaks at $(E_J, x) = (-45,180, 1.4)$ and creates another (6:1) gap. In general, the central family has gaps at all even resonances.[28] A detailed study of the orbits in this potential can be found elsewhere.[29]

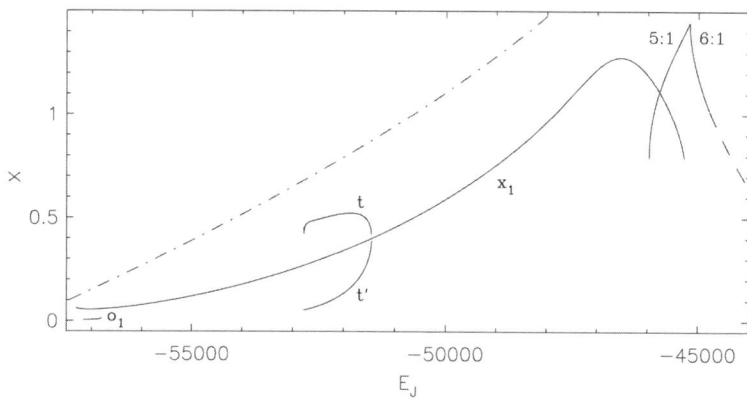

FIGURE 8. The characteristic diagram of the orbits shown in FIGURE 7. *Dashed lines* denote unstable regions of the characteristic. The *dash–dot–dash line* represents the curve of zero velocity (CZV). Note that since the orbits of the depicted families do not start perpendicular to the x-axis, this characteristic is not complete. The $n:1$ labels refer to the corresponding $n:1$-resonance regions.

The (deprojected) radius of the bar is about 2.3 kpc. The estimated Jacobi constant at this distance is approximately −44,300. All the stable periodic orbits that build up the bar structure have "energies" less than this E_j value. The x_1 makes up the predominant *central* family of such orbits. There are other families of stable orbits that are not part of x_1 but could also partially support the bar. These orbits are formed only near the resonances of the galaxy and as such, affect the dynamics of the system locally. The last set of orbits that have the longest elongation along the bar length can be found between the 4:1 and 6:1 gaps. The outer structure of the bar is defined by this family (within $E_J = -46,000$ and $-45,200$); the rectangle-like shaped orbits (see FIGURE 9) mimic the boxy isophotes near the ends of the real bar.

Gaseous Responses

Previous numerical experiments have shown that the general response of gas to a gravitational potential can be predicted if the families of periodic orbits in the underlying potential are known. However, gas exerts pressure and is naturally dissipative and viscous. These characteristics combine to alter the pattern of gas flow near resonances. Around these locations, stars can cross orbits and/or develop loops since they are considered to be collisionless. However, gas is confined to travel on noncrossing orbits.[30] Consequently, there is a gradual shift in the flowlines that causes orbital crowding and shocks. This effect has been attributed to the creation of dust lanes seen in numerical simulations.[30–35] Such differences between the stellar and gaseous components of the galaxy, therefore, merit different modeling techniques.

A modified version of the SPH code[36] used elsewhere[34] was employed to create the hydrodynamic models that are discussed below. The radius of the initial two-dimensional disk of gas particles was between 10 and 12 kpc. The gas was assumed

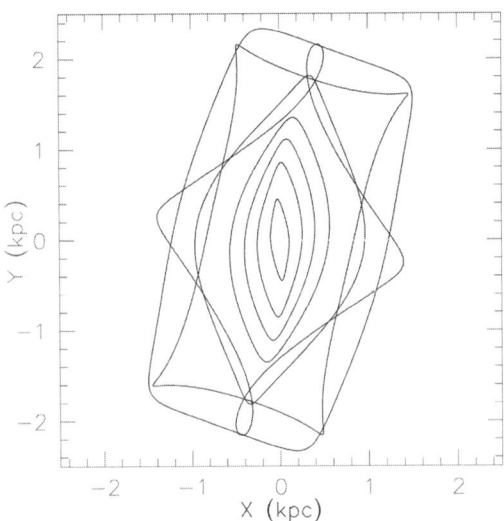

FIGURE 9. Representatives of the stable periodic orbits supporting the observed bar morphology.

to be isothermal and have a sound speed of 10km·sec^{-1}. Self-gravity of the gas was not taken into account. The particles were initially subjected to only the axisymmetric part of the potential, derived above. There were typically 15,000 to 20,000 particles at the beginning of each run. The non-axisymmetric (i.e., barred and spiral) perturbation was introduced gradually and linearly so that the gas could adjust to the forcing terms within two to three pattern rotations. The particles were initially set on circular orbits as determined by the rotation curve obtained from the axisymmetric part of the input potential. For the SPH artificial viscosity we used the values (α, β) = $(1, 2)$.[34] After each model was created, snapshots at various (time) stages of the simulations were directly compared with the HI and HII gas morphologies and the HI velocity field (the patchy nature of the HII velocity field prevented its usage). Our models differ in their pattern speed, as this is the most important parameter for determining the gaseous morphology in response models for galactic disks.[2]

The first obvious choice of pattern speed for the simulations was Ω_p = 39.17 km·sec^{-1}·kpc^{-1}, the value used to show the stellar orbits that provided bar support. As can be observed in FIGURE 10, a comparison of the morphologies of Model A and the actual galaxy show that the gas response from the simulation is markedly different nearly everywhere. The contours of the model show that the spirals that extend from the short gas bar overlap a small region of the actual arms. However, it is obvious that their origins (i.e., where they attach to the bar) are well short of the actual beginning of the observed spiral arms. The outer arms appear to originate close to the corotation radius and overlap parts of the HI regions that reside to the East and West of the pseudo HI gas-ring. These arms are long-lived features that appear to the end of the simulations ($t = 1.4$). Their morphology underlines the fact that the current pattern speed is clearly insufficient to reproduce the gas distribution of NGC 3359, although it has been shown to produce the stellar orbits that describe the structure of the bar. Thus, different pattern speeds were used to produce gas responses that have

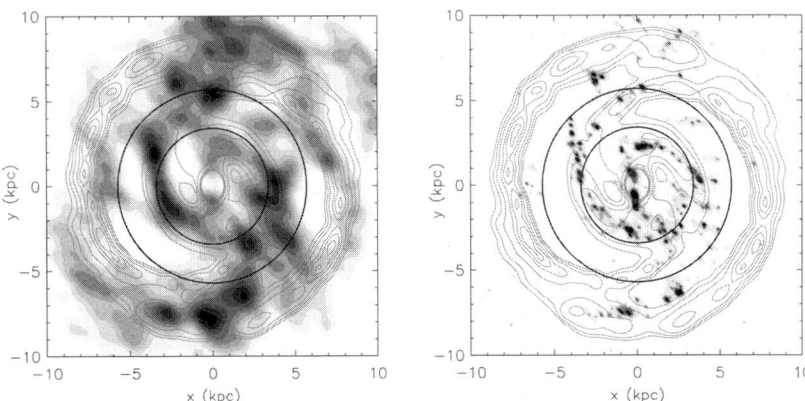

FIGURE 10. Gas response of Model A (*contour lines*) after smoothing the original particle density image with a 15-arcsec Gaussian beam. The *inner* and *outer circles* indicate the positions of the CR and OLR, respectively. The contour lines are overplotted on greyscale images of the cold atomic (**left**) and ionized (**right**) hydrogen density distribution.

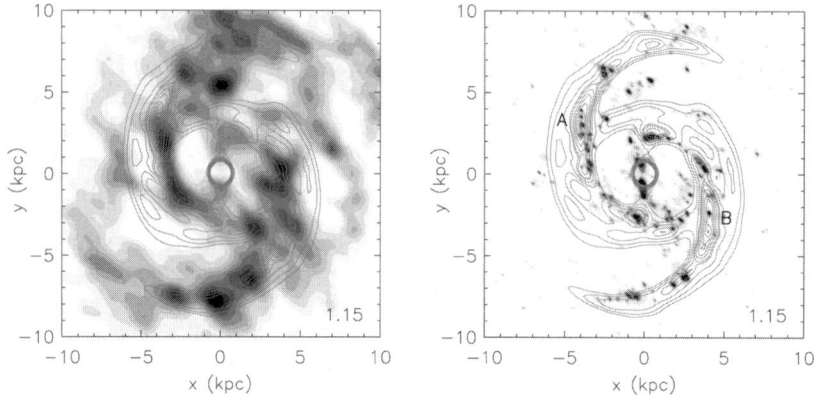

FIGURE 11. Typical gas response of Model B (*contour lines*). Explanation of A and B markers is given in the text. The underlying grayscale images are as in FIGURE 10.

much better correspondence with the major features of the galaxy outside the bar region.

We ran a large number of models and concluded that pattern speeds between 16 and 10 km·sec^{-1}·kpc^{-1} gave a much better match to the observed gas distributions. We could not point to a specific value in that range; however, we have clearly seen that density maxima on the disks of our models and the HI and HII gas morphologies differ, again significantly, for lower pattern speeds. The response of Model B, with $\Omega_p = 15.52$ km·sec^{-1}·kpc^{-1}, is shown in FIGURE 11. Contours of the model gas response are plotted on the grayscale image of the HI surface density map. It can be seen that the observed HI depression in the bar region is duplicated in the model also. The model arms attach to the ends of the bar and spiral outward with pitch angles that are in good agreement with observations. Furthermore, the middle of the arms, along with the ends of the bar, contain the greatest concentration of particles throughout the simulation. After viewing Model B in a movie sequence of snapshots, it was noticed that most of the density maxima are confined within and up to the 4:1 ultraharmonic (UHR) resonance located at $r = 5.88$ kpc. In the North, the model arm winds slightly faster than the observed arm. It is possible that a small warp in the disk is the cause of this minor discrepancy.

There is also good correspondence between the morphology of the model and the observed distribution of ionized hydrogen gas in NGC 3359 (FIG. 11). The HII gas arms are fitted especially well by the model arms. Several of the density maxima in the model overlap the star formation (SF) zones, particularly at the ends of the bar and the large HII regions located in the middle of the Eastern arm (region A). The maxima indicate shocked regions, where gas is compressed and leads to the formation of new stars. However, there is no observable SF region that corresponds to the density maximum of the Western arm (region B). Both A and B are persistent areas of density maxima where the gas particles are constantly shocked.

Model C with angular frequency 13.53 km·sec^{-1}·kpc^{-1}, as expected, does not differ greatly from the former. However, we give it here because it was the case for

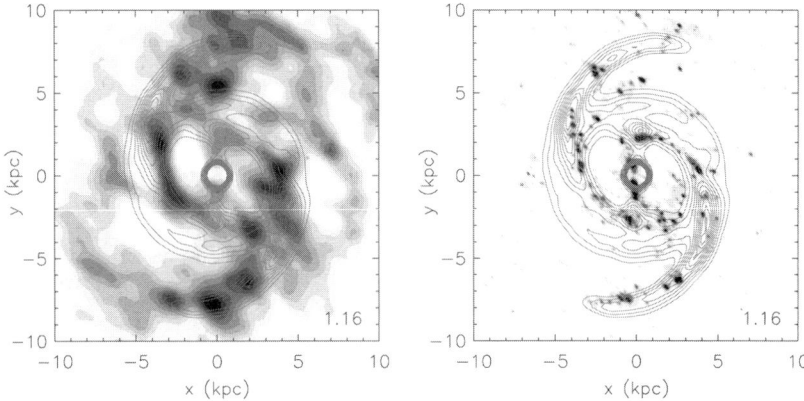

FIGURE 12. Typical gas response of Model C (*contour lines*). The underlying grayscale images are as in FIGURE 10.

which we had the best reproduction of the point where the Eastern arm breaks/bifurcates into separate segments (arm spurs) in the Northern half of the galaxy. The separation is noticeable in both data sets, although it is much easier to see in the Hα intensity map (see FIGURE 12) near $(x, y) = (-3, 3.5)$ kpc. Unfortunately, the contour line outlining the break does not trace out the bifurcated arm (spur) to large distances. On the other hand, the spiral arms are marginally more open, although they still coincide with the observed gas arms quite well.

Comparison of the Kinematics

Representative diagrams of the flow pattern of gas in the slow models are shown as velocity vectors in FIGURE 13. The orientation of the gas orbits in the center region is aligned with the x_1 family and, therefore, the flow is parallel to the bar. The vectors crowd around the ends of the bar where the gas particles are shocked and flow inward as angular momentum is lost. Such incidences also occur around the inner parts of the spiral arms. Consequently, there is a good analogy between the crowding of the flowlines and the positions of the HII regions (FIG. 13 A). This naturally explains the compression of gas that instigates the formation of new stars. Other areas of star formation that lie in low density regions may have been formed by mechanisms other than gas compression due to crowding of the orbits (e.g., spontaneous star formation). Furthermore, the kinematic information of the models has been compared with the observed data. The model velocity fields have been projected to the same sky orientation as NGC 3359 and convolved to 15 arcsec resolution to match the detail of the higher resolution data from the 21-cm observations. The RMS errors (about $5\,\mathrm{km\cdot sec^{-1}}$) associated with velocity values given here are due to the convolution process that smoothed down the original velocity values. In FIGURE 14, the velocity fields of Model A and the 21-cm data have been drawn on the grayscale image of the HI density map. Particular attention should be paid to the bar region (marked by the circle in the diagram) since the pattern speed of this model is the same as that of the

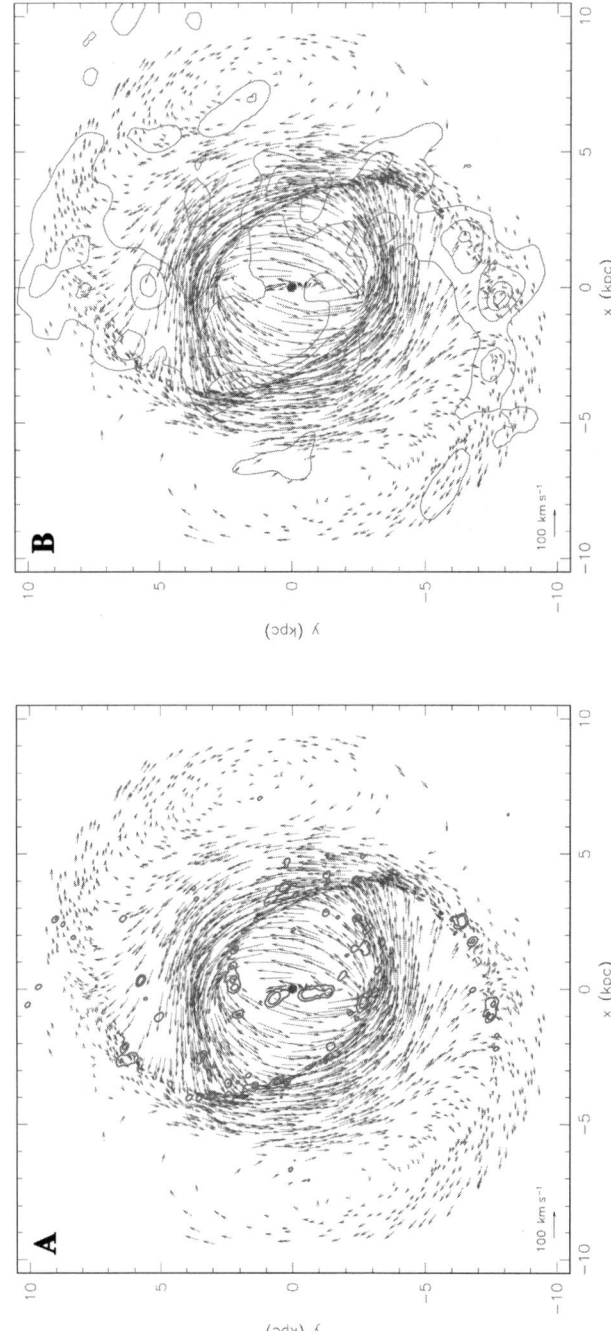

FIGURE 13. Representative example of the gas flow in the slow models. **(A)** Hα contours (at levels of 2,000 and 6,500 arbitrary units) and **(B)** HI contours (at levels of 10.9, 15.1, 18.4, and 20.5 $M_{solar} \cdot pc^{-2}$) on the vector velocity field of the model.

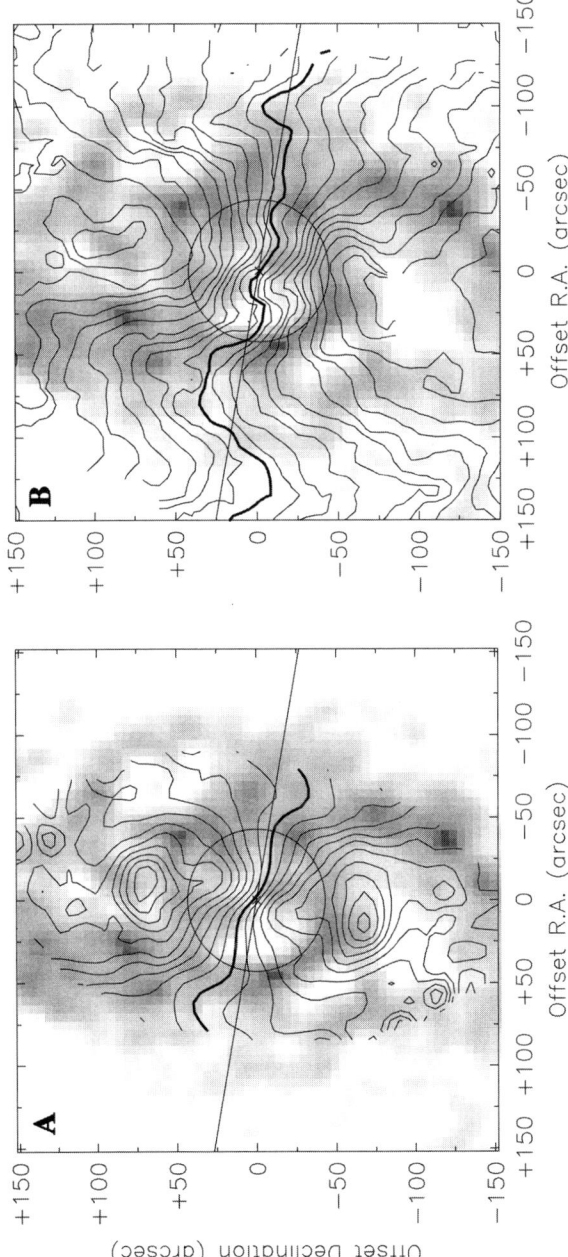

FIGURE 14. Velocity field comparison between (**A**) the $\Omega_p = 39.17 \, \text{km·sec}^{-1} \cdot \text{kpc}^{-1}$ model convolved to the 15 arcsec resolution of the HI data (**B**). The *circle* represents the bar region.

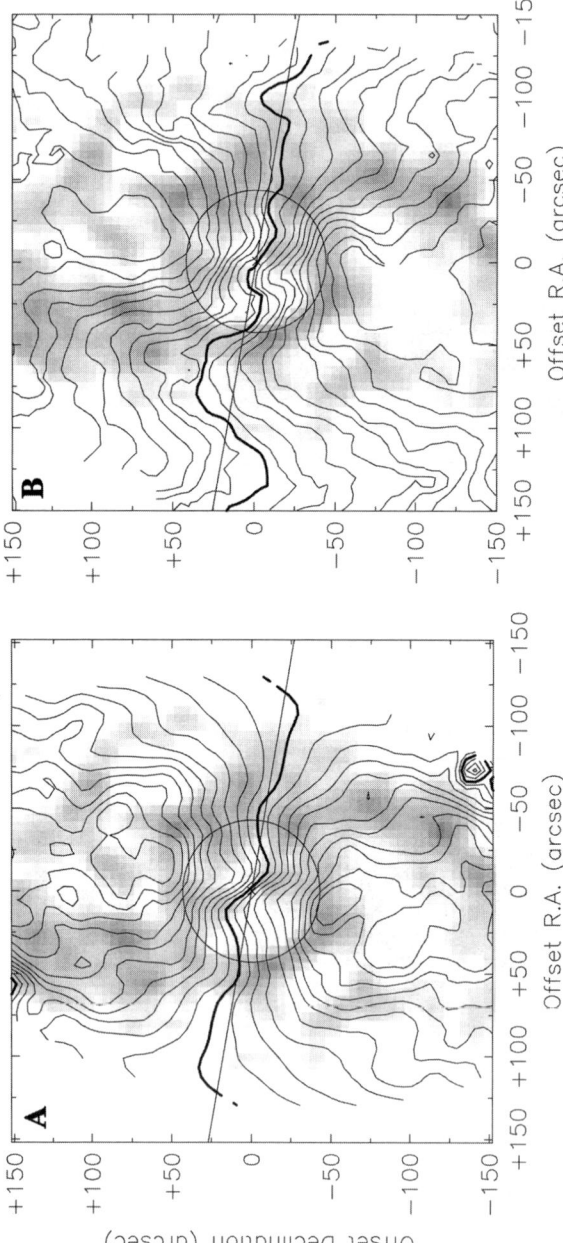

FIGURE 15. Velocity field comparison between (**A**) the $\Omega_p = 15.52 \, \text{km} \cdot \text{sec}^{-1} \cdot \text{kpc}^{-1}$ model convolved to the 15 arcsec resolution of the HI data (**B**). The *circle* represents the bar region.

bar. Within the bar dominated region, the behavior of the model isovels (particularly those West of the galactic center) is similar to the real contours. Immediately NW and SE of the nucleus and within the first 20″, the kinks in the contour lines made by the twisted elliptical gas orbits in FIGURE 14A match those of FIGURE 14B. However, the non-circular effects arising from streaming motions associated with the bar are less severe for the model NE and SW of the nucleus. The flowline of the HI gas is more disturbed than the simulated gas in its orbit around the bar. In fact, the isovels in all of the models show that the gas in the Western half of the bar region is disturbed considerably more than the opposite side. Thus, the zero velocity contour is better fitted to the Western side, owing to the fact that it, along with the real systemic velocity contour, does not intersect the minor axis line.

The corresponding diagrams for Model B are given in FIGURE 15 and are representative of the behavior of the *slow* models. Overall, the Eastern side yields a better fit. This result is not surprising because the gas response (surface density) of the model matches best with the Eastern arm of the galaxy. The bending of the isovels, caused by streaming motions along the spiral arms, are nicely duplicated in the Southeastern side. The contours in the NE region have similar kinks but are pulled up toward the major axis, unlike the observed data that bend back toward the minor axis at large radii. This dissimilarity owes its origin to the lack of sample points in the outer disk of the model. Subsequently, the convolution process used to smooth the image to 15 arcsec resolution is forced to convolve blank areas with nearby points. The model and observed contours in the West are matched primarily in the Northern half, away from the minor axis. The Southwestern isovels are similar to their diametric counterparts in that the contours develop kinks as they traverse through the spiral arm, but bend back toward the major axis shortly after returning to a more circular flow pattern. The zero (or systemic) velocity contour line of the model has a form similar to the actual line but appears to be an extended version of it.

SUMMARY AND CONCLUSIONS

The most important conclusion of our study is that NGC 3359 exhibits two pattern speeds. The morphology of Model A clearly does not show proper gas response outside the bar region. The arms are tightly wrapped and completely fail to match the open spirals of the real galaxy. Kinematically, the rotation curve is credible only within the area of the bar. These results were derived from using the bar pattern speed of $39.17 \text{km} \cdot \text{sec}^{-1} \cdot \text{kpc}^{-1}$. To model the gaseous density and velocity field of the spiral component well, an acceptable value somewhere between $\Omega_p = 10.00$ and about $16.00 \text{km} \cdot \text{sec}^{-1} \cdot \text{kpc}^{-1}$ should be used. A value from this range will suffice to properly explain the spiral pattern speed that is connected to the bar pattern speed via non-linear coupling. An exact prediction about which resonances of the two structures participate in the coupling is withheld here because the resonances can be close to one another, not to mention that the models show similar responses and so it is difficult to decide among several similar hypotheses. It is, however, encouraging to see that the end-of-the-bar close to the 4:1 resonance and spiral ILR coupling scenario is a viable option. This is obtained by combining Model A for the bar with Model C ($13.53 \text{km} \cdot \text{sec}^{-1} \cdot \text{kpc}^{-1}$) for the spiral. Within the bar region, the gas can be

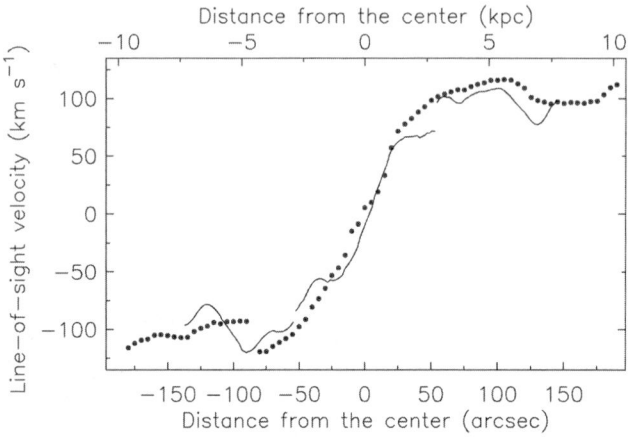

FIGURE 16. Hybrid combination of the rotation curves of Model A and Model C.

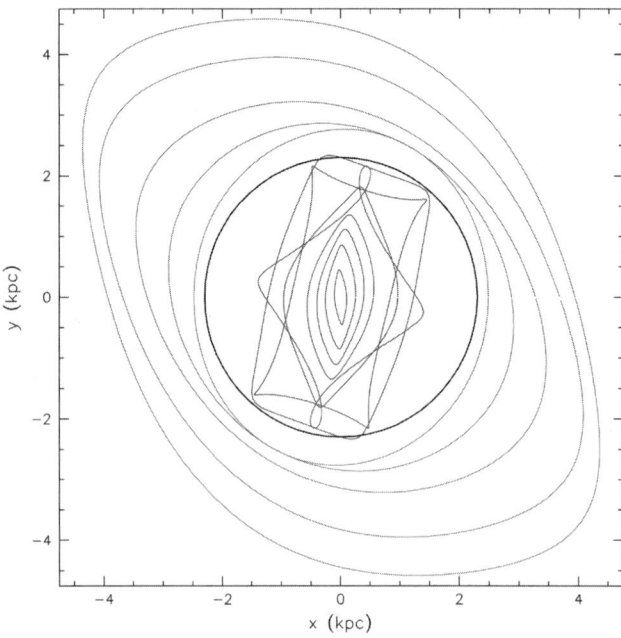

FIGURE 17. Combination of the x_1 orbits for Models A and C. This is the best theoretical representation for the (inner 5 kpc) morphology of the galaxy. The *black circle* denotes the extent of the stellar bar. Orbits inside of this circle belong to Model A and orbits inside belong to Model C.

driven by the forces produced by the structure itself, whereas just beyond the ends of the bar, the spiral potential begins to dominate the outer disk. In FIGURE 16, a hybrid combination of the two rotation curves of Model A and C is shown and supports the combination of the present resonance coupling scenario. Finally, the combined plot of the orbits for Model A (within the bar region) and C (outside the bar) is shown in FIGURE 17, This figure is a good representation of the structure of the central family of orbits within NGC 3359. For a more detailed discussion, see Boonyasait[29] and <http://www.astro.ufl.edu/~boony/research.html>.

ACKNOWLEDGMENTS

This research was supported in part by the Research Committee of the Academy of Athens.

REFERENCES

1. PATSIS, P.A., G. CONTOPOULOS & P.GROSBØL. 1991. Astron. Astrophys. **243:** 373.
2. PATSIS, P.A., P. GROSBØL & N. HIOTELIS. 1997. Astron. Astrophys. **323:** 762.
3. SKOKOS, CH., P.A. PATSIS & E. ATHANASSOULA. 2002. Mon. Not. Roy. Astr. Soc. **333:** 861.
4. KNAPEN, J.H. 1999. *In* The Evolution of Galaxies in Cosmological Timescales. J.E. Beckman & T.J. Mahoney, Eds. ASP Conf. Ser. **187:** 72.
5. TREMAINE, S. & M.D. WEINBERG. 1984. Astrophys. J. **282:** L5.
6. BALL, R. 1986. Astrophys. J. **395:** 418.
7. ENGLAND, M.N. 1989, Astrophys. J. **344:** 669.
8. LAINE, S., I. SHLOSMAN & C.H. HELLER. 1998. Mon. Not. Roy. Astr. Soc. **297:** 1052.
9. ENGLAND, M.N., J.H. HUNTER, JR. & G. CONTOPOULOS. 2000. Astrophys. J. **540:** 154.
10. KRANZ, T., A. SLYZ & H.-W. RIX. 2001. Astrophys. J. **562:** 164.
11. KRANZ, T., A. SLYZ & H.-W. RIX. 2003. Astrophys. J. **586:** 143.
12. SELLWOOD, J.A. & L.S. SPARKE. 1988. Mon. Not. Roy. Astr. Soc. **231:** 25.
13. TAGGER, M., J.F. SYGNET, E. ATHANASSOULA & R. PELLAT. 1987. Astrophys. J. **318:** L43.
14. MOORE, E.M. & S.T. GOTTESMAN. 1995. Astrophys. J. **447:** 159.
15. ROZAS, M. & M.J. SEMPERE. 2002. *In* Cosmic Evolution and Galaxy Formation: Structure, Interactions, and Feedback. J. Franco, L. Terlevich, O. Lopez-Cruz & I. Artexaga, Eds.: 117. ASP Conf. Ser. 215.
16. GOTTESMAN, S.T. & L. WELIACHEW. 1975. Astrophys. J. **195:** 23.
17. VISSER, H.C.D. 1980. Astron. Astrophys. **88:** 149.
18. VAN ALBADA, T.S. & R.H. SANDERS. 1982. Mon. Not. Roy. Astr. Soc. **201:** 303.
19. ELMEGREEN, D.M., B.G. ELMEGREEN, F.R. CHROMEY, *et al.* 1996. Astrophys. J. **111:** 1880.
20. QUILLEN, A.C., J.A. FROGEL & R.A. GONZALEZ. 1994. Astrophys. J. **437:** 162.
21. AGUERRI, J.A.L., J.E. BECKMAN & M. PRIETO. 1998. Astrophys. J. **116:** 2136.
22. ATHANASSOULA, E. 1992. Mon. Not. Roy. Astr. Soc. **259:** 328.
23. HÉNON, M. 1965. Ann. Astrophys. **28:** 499.
24. CONTOPOULOS, G. & P. GROSBØL. 1986. Astron. Astrophys. **155:** 11.
25. CONTOPOULOS, G. 2002. Order and Chaos in Dynamical Astronomy. Springer, Berlin.
26. PATSIS, P.A., E. ATHANASSOULA & A.C. QUILLEN. 1997. Astrophys. J. **483:** 731.
27. PATSIS, P.A. & P. GROSBØL. 1996. Astron. Astrophys. **315:** 371.
28. CONTOPOULOS, G. & P. GROSØL. 1989. Astron. Astrophys. Rev. **1:** 261.
29. BOONYASAIT, V. 2003. Ph.D. Thesis, University of Florida.
30. FRIEDLI, D. & W. BENZ. 1993. Astron. Astrophys. **268:** 65.
31. ATHANASSOULA, E. 1992. Mon. Not. Roy. Astr. Soc. **259:** 345.

32. ENGLMAIER, P. & O. GERHARD. 1997. Mon. Not. Roy. Astr. Soc. **287:** 57.
33. HUNTER, J.H. JR., M.N. ENGLAND, S.T. GOTTESMAN, *et al.* 1988. Astrophys. J. **324:** 721.
34. PATSIS, P.A. & E. ATHANASSOULA. 2000. Astron. Astrophys. **358:** 45.
35. SANDERS, R.H. & J.M. HUNTLEY. 1976. Astrophys. J. **209:** 53.
36. GINGOLD R.A. & J.J. MONAGHAN. 1977. Mon. Not. Roy. Astr. Soc. **181:** 375.

Evolution of Binary Supermassive Black Holes via Chain Regularization

ANDRAS SZELL,[a] DAVID MERRITT,[a] AND SEPPO MIKKOLA[b]

[a]*Department of Physics, Rochester Institute of Technology, Rochester, New York, USA*

[b]*Tuorla Observatory, Piikkiö, Finland*

> ABSTRACT: A chain regularization method is combined with special purpose computer hardware to study the evolution of massive black hole binaries at the centers of galaxies. Preliminary results with up to $N = 0.26 \times 10^6$ particles are presented. The decay rate of the binary is shown to decrease with increasing N, as expected on the basis of theoretical arguments. The eccentricity of the binary remains small.
>
> KEYWORDS: evolution; black hole; chain regularization

Coalescence of binary supermassive black holes is potentially the strongest source of gravitational waves in the universe.[1] The coalescence rate is limited by the efficiency with which massive binaries can interact with stars and gas in a galaxy and reach the relativistic regime at separations of about 10^{-3} pc. Exchange of energy between a binary black hole and stars should also leave observable traces in the stellar distribution, perhaps allowing us to infer something about the merger history of galaxies from their nuclear structure.[2] Henry Kandrup worked on this problem shortly before his death. In Supermassive Black Hole Binaries as Galactic Blenders,[3] Kandrup *et al.* investigated the effects of a massive binary on the stellar orbits near the center of spherical and nearly spherical galaxies. They showed that the periodically-varying potential due to the binary, coupled with the fixed potential from the galaxy, was effective at inducing chaos in the stellar orbits, leading to diffusion in both energy and configuration space and to ejection of stars from the nucleus. This study was a complement to previous studies based on scattering experiments,[4,5] in which the potential of the galaxy was ignored.

Another approach to the binary black hole problem is by means of direct N-body techniques.[6–8] This approach is computationally challenging because of the need to handle close interactions between the star and black hole particles with high precision. In addition, large particle numbers are required to avoid the effects of spurious relaxation.[9,10] Here, we present preliminary results of N-body integrations of the binary black hole problem, in which close interactions between the black holes and stars are handled via the Mikkola–Aarseth chain regularization algorithm.[11,12] Recently Aarseth[13] described an application of a time transformed leapfrog scheme[14] to this problem. We prefer the chain algorithm since it has proved itself in numerous applications, including one very similar to the current problem.[15] We

---

Address for correspondence: Andras Szell, Department of Physics, Rochester Institute of Technology, Rochester, NY 14623, USA. Voice: 585-475-2436.
axssps@rit.edu

incorporate the chain algorithm into a general-purpose N-body code by including the effects of nearby stars as perturbers to the chain. We present preliminary results of binary black hole evolution computed via this algorithm on a special-purpose GRAPE-6 computer with particle numbers up to 0.26×10^6.

Our basic N-body algorithm is an adaptation of the NBODY1 code of Aarseth[16] to the GRAPE-6 special purpose hardware. The code uses a fourth-order Hermite integration scheme with individual, adaptive, block time steps.[17] For the majority of the particles, the forces and force derivatives were calculated by mean of a direct-summation scheme using the GRAPE-6.

Close encounters between the massive particles ("black holes"), or between black holes and stars, create prohibitively small time steps in such a scheme. To avoid this situation, we regularized the critical interactions as follows. Let $\mathbf{r}_i$, $i = 1, \ldots, N$ be the position vectors of the particles. We first identify the subset of n particles to be included in the chain; the precise criterion for inclusion is presented below, but in the late stages of evolution, the chain always includes the two black holes as its lowest members. We then search for the particle that is closest to either end of the chain and add it; this operation is repeated, recursively, until all n particles are included. Define the separation vectors $\mathbf{R}_i = \mathbf{r}_{i+1} - \mathbf{r}_i$, where $\mathbf{r}_{i+1}$ and $\mathbf{r}_i$ are the coordinates of the two particles making up the ith link of the chain. The canonical momenta $\mathbf{W}_i$ corresponding to the coordinates $\mathbf{R}_i$ are given in terms of the old momenta by the generating function

$$S = \sum_{i=1}^{n-1} \mathbf{W}_i \cdot (\mathbf{r}_{i+1} - \mathbf{r}_i). \qquad (1)$$

Next, we apply KS regularization[18] to the chain vectors, regularizing only the interactions between neighboring particles in the chain. Let $\mathbf{Q}_i$ and $\mathbf{P}_i$ be the KS transformed $\mathbf{R}_i$ and $\mathbf{W}_i$ coordinates. After applying the time transformation $\delta t = g \delta s$, $g = 1/L$, where L is the Lagrangian of the system ($L = T - U$, where T is the kinetic and U is the potential energy of the system), we obtain the regularized Hamiltonian $\Gamma = g(H(\mathbf{Q}_i, \mathbf{P}_i) - E_0)$, where E_0 is the total energy of the system. The equations of motion are then

$$\mathbf{P}'_i = -\frac{\partial \Gamma}{\partial \mathbf{Q}_i}, \qquad \mathbf{Q}'_i = \frac{\partial \Gamma}{\partial \mathbf{P}_i}, \qquad (2)$$

where primes denote differentiation with respect to the time coordinate s. Because of the use of regularized coordinates, these equations do not suffer from singularities, as long as care is taken in the construction of the chain.

Since it is impractical to include all N particles in the chain, we must consider the effects of external forces on the chain members. Let $\mathbf{F}_j$ be the perturbing acceleration acting on the jth body of mass m_j. The perturbed system can be written in Hamiltonian form by simply adding the perturbing potential,

$$\delta U = \sum_{j=1}^{n} m_j \mathbf{r}_i \cdot \mathbf{F}_j(t). \qquad (3)$$

Only one chain was defined at any given time. At the start of the N-body integrations, there was no regularization, and all particles were advanced using the variable-time-step Hermite scheme. The chain was "turned on" at the time when one of the particles (including possibly one of the black holes) achieved a time step shorter than t_{chmin} and reached a distance from one of the black holes smaller than r_{chmin}. Each star inside a radius r_{chmin} was then added to the chain; the two black holes were always included. The values of t_{chmin} and r_{chmin} were determined by carrying out test runs; we adopted $t_{\text{chmin}} \approx 10^{-5}$–$10^{-6}$ and $r_{\text{chmin}} \approx 10^{-4}$–$10^{-3}$ in standard N-body units.

The center of mass of the chain was a pseudoparticle, as seen by the N-body code, and was advanced by the Hermite scheme in the same way as an ordinary particle. However, when integrating the trajectories of stars near to the chain, it was essential to resolve the inner structure of the chain. Thus, for stars inside a critical radius $r_{\text{crit}1}$ around the chain, the forces from the individual chain members were taken into account. The value of $r_{\text{crit}1}$ was set by the size of the chain to be $r_{\text{crit}1} = \lambda R_{\text{ch}}$ with R_{ch} the spatial size of the chain and $\lambda = 100$.

In addition, the equations of motion of the chain particles must include the forces exerted by a set of external perturber stars Whether or not a given star was listed as a perturber was determined by a tidal criterion: $r < R_{\text{crit}2} = (m/m_{\text{chain}})^{1/3}\gamma^{-1/3}R_{\text{ch}}$, where m_{chain} represents the mass of the chain, m is the mass of the star, and γ_{min} was chosen to be 10^{-6}; thus, $r_{\text{crit}2} \approx 10^{2}(m/m_{\text{chain}})^{1/3}R_{\text{ch}}$.

Membership of the chain changed under the evolution of the system. Stars were captured into the chain if their orbits approached the binary closer than R_{ch}. Stars were emitted from the chain if they got further from both of the black holes than $1.5R_{\text{ch}}$. The difference between the emission and absorption distances was chosen to avoid a too-frequent variation of the chain membership. When the last particle left the chain, the chain was eliminated and the integration turned back to the Hermite scheme, until a new chain was created.

In our numerical experiments, the typical number of chain members was 5–10. The number of perturber stars was typically 500–1,000, and the number of stars inside $r_{\text{crit}1}$ was 2,000–5,000. Using a minimum tolerance of 10^{-12} for the Bulirsch–Stoer integrator of the chain allowed us to reach typical relative accuracies of 10^{-9} in the chain integration. The relative accuracy in the conservation of the total energy of the system was determined by the Hermite scheme. For all of our numerical simulations the relative error in the total energy was less than 10^{-4} during the course of the integration.

For this set of experiments, we adopted the Dehnen density law,[19]

$$\rho(r) = \frac{(3-\gamma)M}{4\pi} \frac{a}{r^{\gamma}(r+a)^{4-\gamma}}, \tag{4}$$

with $\gamma = 0.5$, to describe the initial stellar distribution. The initial positions and velocities of the N stars were generated from the steady state phase space density $f(E)$ that reproduces the density law (4). Henceforth, we adopt units such that the gravitational constant G, the total stellar mass M, and the Dehnen scale length a are equal to one. To this model we added two black hole particles, each of mass 0.005. The black holes were placed symmetrically about the center of the galaxy, offset by a distance 0.1 from the center. The tangential velocities were set to ± 0.16 yielding nearly circular initial orbits for the two black holes about the center of the galaxy.

We integrated the above model with three different particle numbers: $N = 16{,}384$, $65{,}536$, and $262{,}144$; the latter is close to the maximum number of particles that can be handled in the GRAPE-6 memory. Integrations were carried out for 170 time units. Elapsed times were 2.5, 26, and 105 hours for the three runs.

The orbits of the two black holes initially decay, and at a time of roughly 30 they form a bound pair. After this, the semimajor axis a of the binary shrinks as the two black holes interact with stars and eject them from the nucleus via the gravitational slingshot. The instantaneous decay rate is given approximately by

$$\frac{d}{dt}\left(\frac{1}{a}\right) = \frac{G\rho}{\sigma}H, \tag{5}$$

where ρ and σ are, respectively, the density and velocity dispersion of the stars, and H is a dimensionless constant of order 16.[4,5] Because the stellar density changes with time as the binary ejects stars, the decay rate of the binary is, in general, a complicated function of time. Two limiting cases are of interest.[9] When the particle number N is small, gravitational encounters are able to scatter stars into the *loss cone* around the binary at a higher rate than they are ejected. The density near the binary remains approximately constant and the decay follows $a^{-1} \sim t$. This is the *full loss cone* regime. When N is large, encounters between stars are weak, and the binary loss cone remains nearly empty. Decay of the binary is limited by the rate at which stars diffuse into the loss cone; since the diffusion time scales approximately as N, the binary decay follows $a^{-1} \sim t/N$. This is the *diffusion* regime. Real galaxies are expected to be in the diffusion regime.[9]

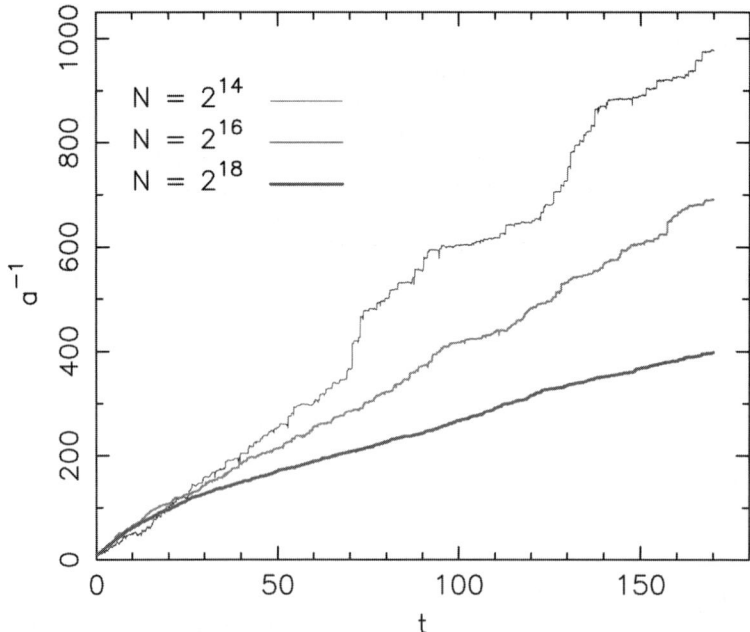

FIGURE 1. Evolution of the binary semimajor axis.

FIGURE 1 shows the time evolution of the semimajor axis of the binary in each of the three integrations. The energy of the binary evolves as an approximately linear function of the time but with a prefactor that depends on N; the approximate dependence is

$$\frac{1}{a} \approx \frac{160t}{N^{1/3}}. \qquad (6)$$

This is intermediate between the $a^{-1} \propto t$ dependence of the full loss cone limit and the $a^{-1} \propto t/N$ dependence in the diffusion limit. We conclude that, for the particle numbers considered here, replenishment of the binary loss cone is taking place, but at a lower rate than the rate at which the loss cone is being emptied. Apparently, particle numbers in excess of about 10^6 are required if N-body integrations are to be completely in the diffusive regime that is characteristic of real galaxies. Such large particle numbers can, in principle, be handled with direct-summation codes like ours if coupled with parallel hardware.[20]

FIGURE 2 shows the eccentricity evolution of the binary. The value of e at the time when the hard binary first forms, $t \approx 30$, is substantially different in the different integrations, due presumably to finite-N effects during the initial inspiral of the two black holes. Thereafter, the eccentricity fluctuates in the case of the two small-N integrations, but decreases with time in the integration with the largest N. We compared the eccentricity evolution with the predictions of scattering theory. The rate of change of e is commonly written

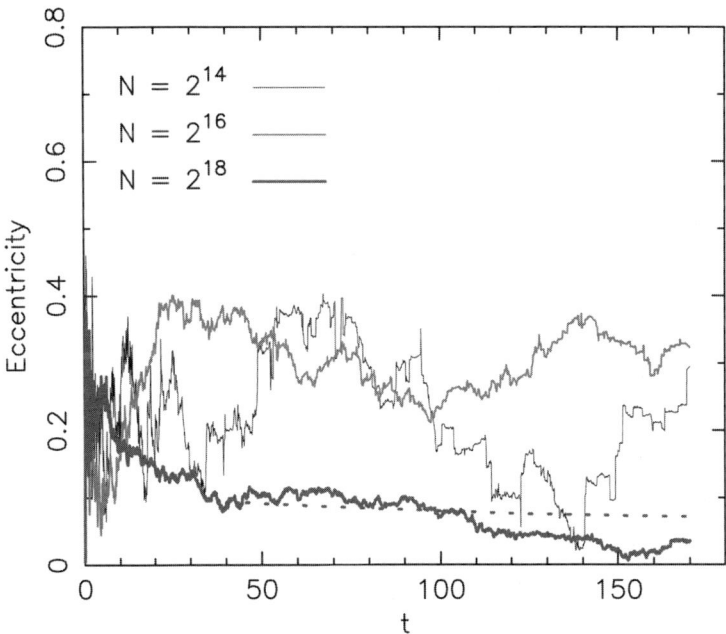

FIGURE 2. Evolution of the binary orbital eccentricity.

$$\frac{de}{dt} = K\frac{d}{dt}\ln a^{-1}, \qquad (7)$$

where $K = K(e, a)$.[4,5] The functional form of $K(e, a)$ is not well known; we adopted the expression given in Reference 5 for a hard binary. We then wrote Equation (7) as

$$e_i = e_{i-1} + K(e_{i-1}, a_{i-1})\ln\left(\frac{a_{i-1}}{a_i}\right), \qquad (8)$$

where the subscript denotes the time step. Combining Equation (8) with the N-body results for $a(t)$ (FIG. 1) we could then predict the expected evolution in e. The result for the largest-N integration is shown as a dashed line in FIGURE 2. There is reasonable agreement, but the eccentricity evolution even in the largest-N integration still exhibits substantial fluctuations. In this respect too, we are not yet in a regime where the evolution is similar to what would be expected for real galaxies. A recent N-body study[13] found a much greater degree of eccentricity evolution, although the initial orbit of the binary was highly non-circular.

As the binary decays, it ejects stars from the nucleus and lowers the nuclear density. FIGURE 3 shows initial and final density profiles for the three integrations. The net effect of a black hole binary on the stellar distribution is commonly measured in terms of the *mass deficit*, defined as the mass in stars that was removed by the binary.[2] We find mass deficits of 1.94, 1.56, and 1.17 in units of the combined black hole mass, respectively, for the integrations with $N = 0.016\times10^6$, 0.065×10^6, and

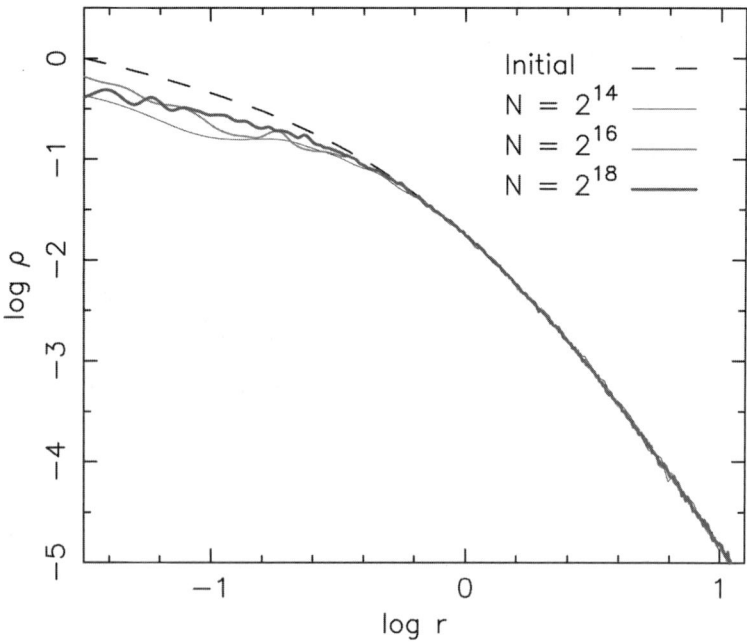

FIGURE 3. Initial (*dashed line*) and final density profiles.

0.26×10^6. These values are of the same order as the mass deficits inferred in giant elliptical galaxies.[2,21,22]

ACKNOWLEDGMENTS

This work was supported by Grants AST-0206031, AST-0420920, and AST 0437519 from the NSF, Grant NNGO4GJ48G from NASA, and Grant HST-AR-09519.01-A from STScI.

The authors declare that they have no competing financial interests.

REFERENCES

1. THORNE, K.S. & V.B. BRAGINSKII. 1976. Gravitational-wave bursts from the nuclei of distant galaxies and quasars—proposal for detection using Doppler tracking of interplanetary spacecraft. Astrophys. J. **204**: L1–L6.
2. MILOSAVIJEVIC, M., D. MERRITT, A. REST & F. VAN DEN BOSCH. 2002. Galaxy cores as relics of black hole mergers. Mon. Not. R. Astron. Soc. **331**: L51–L55.
3. KANDRUP, H.E., I.V. SIDERIS, B. TERZIC & C.L. BOHN. 2003. Supermassive black bole binaries as galactic blenders. Astrophys. J. **597**: 111–130.
4. MIKKOLA, S. & M.J. VALTONEN. 1992. Evolution of binaries in the field of light particles and the problem of two black holes. Mon. Not. R. Astron. Soc. **259**: 115–120.
5. QUINLAN, G.D. 1996. The dynamical evolution of massive black hole binaries I. Hardening in a fixed stellar background. New Astron. **1**: 35–56.
6. MAKINO, J. 1997. Merging of galaxies with central black holes. II. Evolution of the black hole binary and the structure of the core. Astrophys. J. **478**: 58–65.
7. MILOSAVLJEVIC, M. & D. MERRITT. 2001. Formation of galactic nuclei. Astrophys. J. **563**: 34–62.
8. HEMSENDORF, M., S. SIGURDSSON & R. SPURZEM. 2002. Collisional dynamics around binary black holes in galactic centers. Astrophys. J. **581**: 1256–1270.
9. MILOSAVLJEVIC, M. & D. MERRITT. 2003. Long-term evolution of massive black hole binaries. Astrophys. J. **596**: 860–878.
10. MAKINO, J. & Y. FUNATO. 2004. Evolution of massive black hole binaries. Astrophys. J. **602**: 93–102.
11. MIKKOLA, S. & S.J. AARSETH. 1990. A chain regularization method for the few-body problem. Celest. Mech. Dyn. Astron. 47: 375-390.
12. MIKKOLA, S. & S.J. AARSETH. 1993. An implementation of N-body chain regularization. Celest. Mech. Dyn. Astron. **84**: 343–354.
13. AARSETH, S.J. 2003. Black hole binary dynamics. Astrophys. Sp. Sci. **285**: 367–372.
14. MIKKOLA, S. & S.J. AARSETH. 2003. A time-trasformed leapfrog scheme. Celest. Mech. Dyn. Astron. **57**: 439–459.
15. PRETO, M., D. MERRITT & R. SPURZEM. 2004. N-body growth of a Bahcall–Wolf cusp around a black hole. Astrophys. J. **613**: 109–112.
16. AARSETH, S.J. 1999. Pub. Astron. Soc. Pac. **111**: 1333–1346.
17. AARSETH, S.J. 2003. Gravitational N-Body Simulations. Cambridge University Press, Cambridge.
18. KUSTAANHEIMO, P. & E. STIEFEL. 1965. Perturbation theory of Kepler motion based on spinor regularization. J. Reine Angew. Math. **218**: 204–219.
19. DEHNEN, W. 1993. A family of potential-density pairs for spherical galaxies and bulges. Mon. Not. R. Astron. Soc. **265**: 250–256.
20. DORBAND, E.N., M. HEMSENDORF & D. MERRITT. 2003. Systolic and hypersystolic algorithms for the gravitational N-body problem, with an application to Brownian motion. J. Comp. Phys. **185**: 484–511.
21. RAVINDRANATH, S., L.C. HO & A.V. FILIPPENKO. 2002. Nuclear cusps and cores in early-type galaxies as relics of binary black hole mergers. Astrophys. J. **566**: 801–808.
22. GRAHAM, A. 2004. Core depletion from coalescing supermassive black holes. Astrophys. J. **613**: L33–L36.

Gravitomagnetic Field and Penrose Scattering Processes

REVA KAY WILLIAMS

University of Florida, Gainesville, Florida, USA

ABSTRACT: In this paper we present theoretical model calculations involving Monte Carlo computer simulations of Compton scattering and electron–positron (e^-e^+) pair production processes in the ergosphere of a supermassive rotating black hole. Particles from an accretion disk surrounding the rotating black hole fall into the ergosphere and are scattered by particles that are confined in equatorial and nonequatorial orbits. The energy–momentum vectors are calculated for the scattered escaping particles. Particles escape with energies of about 3 GeV or greater. Importantly, these model calculations show that the Lense–Thirring effect, that is, the dragging of local inertial frames into rotation, inside the ergosphere, caused by the angular momentum of the rotating black hole, results in a gravitomagnetic force being exerted on the scattered escaping particles. Effects of this force on the Penrose scattered particles are analyzed and discussed.

KEYWORDS: gravitomagnetic field; Penrose scattering

INTRODUCTION

Results from general relativistic theoretical Monte Carlo computer simulations of Compton scattering and e^-e^+ pair production processes in the ergosphere of a supermassive (about $10^8 M_\odot$) rotating black hole reveal apparent Kerr metric equatorial reflection symmetry breaking. Particles from an accretion disk fall into the ergosphere and are scattered by particles that are confined in equatorial and nonequatorial orbits. The Penrose mechanism, in general, allows the rotational energy of a Kerr black hole to be extracted by scattered particles that escape from the ergosphere to infinity (i.e., large distances from the black hole). The results of these model calculations show that a form of the Penrose mechanism is capable of producing the observed high energy particles (up to order of GeV) emitted by quasars and other active galactic nuclei (AGN), without the necessity of the external electromagnetic field of the accretion disk. Importantly, these model calculations show that the Lense–Thirring effect,[1] that is, the dragging of local inertial frames into rotation, caused by the angular momentum of the rotating black hole, results in a gravitomagnetic (GM) force being exerted on the scattered escaping particles. Inside the ergosphere, where this dragging is severe, in appears that the GM vector field lines are frame dragged in the positive azimuthal direction, the direction of rotation of the black hole. The resulting GM force acting on the Penrose scattered particles produces

Address for correspondence: Reva Kay Williams, University of Florida, Gainesville, FL 32611, USA.
revak@vista.phys.ncat.edu

symmetric and asymmetric (or one-sided) particle emissions in the polar direction, somewhat consistent with the astrophysical jets observed in radio strong AGN. In this paper the origin of this broken symmetry is investigated, and possible explanations are provided. This paper is presented in honor of Dr. Henry Kandrup, my postdoctoral advisor and friend, who, in private communication in 1996, suggested that the asymmetry is due to the GM force field.

THE GRAVITOMAGNETIC FIELD AND BROKEN KERR METRIC EQUATORIAL REFLECTION SYMMETRY

To investigate and understand the origin of the asymmetry seen in distributions of Penrose scattered particles in the polar (Θ) direction one must look in detail at the GM force field[2]—the force suggested to be responsible for this asymmetry. The gravitational force of a rotating body of mass consists of two parts that are analogous to an electromagnetic field. The first part is the familiar gravitational force that a body of rotating or nonrotating mass M produces on a test particle of mass m

$$\left(\frac{d\mathbf{p}}{d\tau}\right)_{grav} = \frac{m}{(1-\mathbf{v}^2)^{1/2}}\mathbf{g}, \tag{1}$$

where the gravitational acceleration $\mathbf{g}$ is produced by the gradient of the *redshift factor* e^v of the Kerr metric (the inverse e^{-v} is referred to as the *blueshift factor*)[3]

$$\mathbf{g} = -\nabla \ln e^v$$

$$= -\frac{\Sigma M(r^4 - a^4) + 2Mr^2a^2\Delta\sin^2\Theta}{A\sqrt{\Delta\Sigma^3}}\hat{\mathbf{e}}_r + \frac{2Mra^2(r^2+a^2)}{A\sqrt{\Sigma^3}}\cos\Theta\sin\Theta\hat{\mathbf{e}}_\Theta \tag{2}$$

This force is analogous to the electric force field surrounding an electric charge source (i.e., like the Coulomb field between point charges) and for this reason it is sometimes referred to as the *gravitoelectric* force.

The second part of the gravitational force is less familiar. It is the additional gravitational force that a rotating mass produces on a test particle. This force, called the GM force, is produced by the gradient of $\vec{\beta}_{GM} = -\omega\hat{\mathbf{e}}_\Phi$, where ω is the frame dragging velocity of the Kerr metric. Thus,

$$\left(\frac{d\mathbf{p}}{d\tau}\right)_{GM} = \vec{\mathbf{H}}\cdot\mathbf{p}, \text{ that is, } \left\{\left(\frac{d\mathbf{p}}{d\tau}\right)_{GM}\right\}^i = H_{ij}p^j, \tag{3}$$

where

$$\vec{\mathbf{H}} \equiv e^{-v}\nabla\vec{\beta}_{GM}, \text{ that is, } H_{ij} = e^{-v}(\beta_{GM})_{j;i}, \tag{4}$$

in which the semicolon indicates the covariant derivative in three-dimensional space. The field $\vec{\mathbf{H}}$ is called the GM tensor field and $\vec{\beta}_{GM}$ is sometimes called the GM potential. Note that, like the magnetic Lorentz force, $\mathbf{f}_B = q(\mathbf{v}\times\mathbf{B})$, on a particle of charge q moving with velocity $\mathbf{v}$ in a magnetic field $\mathbf{B}$, the GM force of Equation (3) vanishes when the particle is at rest.

The justification for the term *gravitomagnetic* becomes clear when we look at the GM force at a large distance from the event horizon r_+ ($r \gg r_+$) exerted on a particle of mass m traveling with low velocity ($v \ll c$). Under these conditions, the GM force as measured by an observer at infinity is given by $\mathbf{F}_{GM} \approx m(\mathbf{v} \times \mathbf{H})$, where $\mathbf{H}$ is called the GM vector field—which contains the same information as the antisymmetric part of the GM tensor field[2]

$$H^j \equiv \varepsilon^{jkl} H_{kl} = \varepsilon^{jkl} e^{-\nu} \nabla_k (\beta_{GM})_l, \text{ that is, } \mathbf{H} \equiv e^{-\nu} \nabla \times \vec{\beta}_{GM},$$

or

$$\mathbf{H} = H^r \hat{\mathbf{e}}_\mathbf{r} + H^\Theta \hat{\mathbf{e}}_\Theta. \tag{5}$$

Thus, one sees that the GM force on a test particle of mass m is similar to the Lorentz magnetic force $\mathbf{f}_B$ on a test particle of charge q, as given above; the vectors $\mathbf{H}$ and $\mathbf{B}$ are also analogous. Furthermore, the lines of force of $\mathbf{H}$ have characteristics similar to those of a dipole magnetic field. Because of resemblances such as these (more examples are given elsewhere[2]) between the force field produced by the mass and angular momentum of the rotating black hole and a magnetic force field, the term *gravitomagnetic* is used.

To gain additional understanding of the GM force field, one can think of it as follows: just as the gravitational force causes local inertial frames to fall with acceleration $\mathbf{g}$, the GM field $\mathbf{H}$ causes local inertial frames to rotate with angular velocity $\vec{\Omega}_{GM} = -\mathbf{H}/2$. In other words, the GM field $\mathbf{H}$ can be thought of as a force field that drags local inertial frames into rotation and, as a result, produces a *Coriolis force* $m(\mathbf{v} \times \mathbf{H})$ at $r \gg r_+$. The angular rotation velocity Ω_{GM} is universally induced in all bodies and neighboring inertial frames at a given radius and is independent of the mass m of the body, just as is the case for $\mathbf{g}$.

Now consider the GM force acting on a test particle in the local nonrotating frame (LNRF), that is, the local Minkowski (or flat) spacetime, and in the local rotating frame, that is, the local Boyer–Lindquist coordinate frame (LBLF) of the observer at infinity. The LBLF is the local inertial frame that is dragged into rotation by the black hole—sometimes referred to as the Lense–Thirring effect. In the LNRF the GM vector field lines have dipole-like characteristics, as measured by a distant observer, similar to the field at $r \gg r_+$, yet distorted because of strong gravity, but with no effects of rotation, such that $\mathbf{H} = H^r \hat{\mathbf{e}}_\mathbf{r} + H^\Theta \hat{\mathbf{e}}_\Theta$. On the other hand, in the LBLF the GM vector field lines also have dipole-like characteristics, but because of strong gravity and the effects produced by the rotating black hole near the event horizon, the GM vector field acquires an azimuthal component H^Φ due to the frame dragging, such that $\mathbf{H} = H^r \hat{\mathbf{e}}_\mathbf{r} + H^\Theta \hat{\mathbf{e}}_\Theta + H^\Phi \hat{\mathbf{e}}_\Phi$.

If we define the components of the GM tensor in terms of components of the vectors $\mathbf{H}$ and $\tilde{\mathbf{H}}$ as follows:

$$\begin{aligned} H^r &\equiv H_{\Theta\Phi}, & \tilde{H}^r &\equiv -H_{\Phi\Theta}, \\ H^\Theta &\equiv H_{\Phi r}, & \tilde{H}^\Theta &\equiv -H_{r\Phi}, \\ H^\Phi &\equiv H_{r\Theta}, & \tilde{H}^\Phi &\equiv -H_{\Theta r}, \end{aligned} \tag{6}$$

where the tensor is antisymmetric whenever $H^i = \tilde{H}^i$, then the GM force exerted on a test particle in the LNRF is[4]

$$\mathbf{f}_{GM} = \overleftrightarrow{\mathbf{H}} \cdot \mathbf{p}$$

$$= (H_{r\Phi}p^{\Phi})\hat{\mathbf{e}}_{\mathbf{r}} + (H_{\Theta\Phi}p^{\Phi})\hat{\mathbf{e}}_{\Theta} + (H_{\Phi r}p^r + H_{\Phi\Theta}p^{\Theta})\hat{\mathbf{e}}_{\Phi} \quad (7)$$

$$\propto (-\tilde{H}^{\Theta}P_{\Phi})\hat{\mathbf{e}}_{\mathbf{r}} + (H^r P_{\Phi})\hat{\mathbf{e}}_{\Theta} + (H^{\Theta}P_r - \tilde{H}^r P_{\Theta})\hat{\mathbf{e}}_{\Phi},$$

and the GM force exerted on a test particle in the LBLF is given by

$$\mathbf{F}_{GM} = \overleftrightarrow{\mathbf{H}} \cdot \mathbf{p}$$

$$= (H_{r\Theta}p^{\Theta} + H_{r\Phi}p^{\Phi})\hat{\mathbf{e}}_{\mathbf{r}} + (H_{\Theta r}p^r + H_{\Theta\Phi}p^{\Phi})\hat{\mathbf{e}}_{\Theta} + (H_{\Phi r}p^r + H_{\Phi\Theta}p^{\Theta})\hat{\mathbf{e}}_{\Phi} \quad (8)$$

$$\propto (H^{\Phi}P_{\Theta} - \tilde{H}^{\Theta}P_{\Phi})\hat{\mathbf{e}}_{\mathbf{r}} + (H^r P_{\Phi} - \tilde{H}^{\Phi}P_r)\hat{\mathbf{e}}_{\Theta} + (H^{\Theta}P_r - \tilde{H}^r P_{\Theta})\hat{\mathbf{e}}_{\Phi}.$$

Note: the symbol $\propto$ indicates that we are assuming (1) the space metric tensor g^{jk} that multiplies the covariant momentum components has been defined by a local orthonormal (Lorentz) tetrad h^{μ}_{α} so that, as usual, $g^{\mu\nu} = h^{\mu}_{\alpha}h^{\nu}_{\beta}\eta^{\alpha\beta}$, where $\eta^{\alpha\beta}$ are the Minkowski metric components; and (2) the space momentum transformations from the LBLF to the global Boyer–Lindquist coordinate frame (BLF, the frame of the observer at infinity) are standard,[5] that is, are similar to transformations from the LNRF to the global BLF.[3] It is understood that these assumptions are at least valid for the purposes of our analysis. Note that the GM force in Equation **(8)** is derived here only to show how the particles might be affected by the GM field, but such effects, shown in the resulting distributions of the escaping particles as measured by a BLF observer, are intrinsically incorporated into the calculations through physical processes occurring in the Kerr metric spacetime geometry.

From a geometric analysis of the characteristics of the antisymmetric GM tensor components of $\overleftrightarrow{\mathbf{H}}$ in the BLF (for $r \gg r_+$, see **(5)**), and the components of $\overleftrightarrow{\mathbf{H}}$ in the LNRF from which the frame dragging is removed (see Eq. **(7)**), and the probable condition, in the LBLF, that the GM tensor field lines are frame dragged in the azimuthal direction, we deduce that the GM tensor in the LBLF will be somewhat similar to that in the LNRF in respect to the dipole-like characteristics; the differences, in general, depend on the frame dragging velocity ω. The tensor components of Equations **(7)** and **(8)**, when evaluated at $r \sim r_+$, reveal the following:[4]

$$H^r \leq 0 \text{ and } \tilde{H}^r \leq 0 \text{ for } \Theta \leq 90°$$
$$H^r \geq 0 \text{ and } \tilde{H}^r \geq 0 \text{ for } \Theta \geq 90°$$
$$H^{\Theta} > 0 \text{ for } \Theta \leq 90° \text{ and } \Theta > 90°$$
$$\tilde{H}^{\Theta} < 0 \text{ for } \Theta \leq 90° \text{ and } \Theta > 90° \quad (9)$$
$$H^{\Phi} < 0 \text{ for } \Theta \leq 90° \text{ and } \Theta > 90°$$
$$\tilde{H}^{\Phi} > 0 \text{ for } \Theta \leq 90° \text{ and } \Theta > 90°.$$

To describe the escaping particle polar momentum distributions of the Penrose processes, a general analysis of the GM force acting in the polar direction,[6]

$$(F_{GM})_{\Theta} \propto (H^r P_{\Phi} - \tilde{H}^{\Phi}P_r), \quad (10)$$

with

$$H^r \leq 0 \text{ for } \Theta \leq 90°$$
$$H^r \geq 0 \text{ for } \Theta \geq 90° \qquad (11)$$
$$\tilde{H}^\Phi > 0 \text{ for } \Theta \leq 90° \text{ and } \Theta > 90°$$

gives us the following conditions:

1. At $\Theta = 90°$, the force on the particle is $(F_{GM})_\Theta > 0$ for $P_r < 0$ or $(F_{GM})_\Theta < 0$ for $P_r > 0$.
2. At $\Theta > 90°$, $P_r > 0$, and $P_\Phi > 0$, the force on the particle is $(F_{GM})_\Theta > 0$ for $H^r P_\Phi > \tilde{H}^\Phi P_r$ and $(F_{GM})_\Theta < 0$ for $H^r P_\Phi < \tilde{H}^\Phi P_r$.
3. At $\Theta < 90°$, $P_r > 0$, and $P_\Phi > 0$, the force on the particle is $(F_{GM})_\Theta < 0$.
4. At $\Theta > 90°$, $P_r < 0$, and $P_\Phi > 0$, the force on the particle is $(F_{GM})_\Theta > 0$.
5. At $\Theta < 90°$, $P_r < 0$, and $P_\Phi > 0$, the force on the particle is $(F_{GM})_\Theta < 0$ for $H^r P_\Phi > \tilde{H}^\Phi |P_r|$ and $(F_{GM})_\Theta > 0$ for $H^r P_\Phi < \tilde{H}^\Phi |P_r|$.

PENROSE SCATTERING DISTRIBUTIONS: SYMMETRIC AND ASYMMETRIC POLAR JETS

We now use the five conditions in the items above to explain the escaping particle distributions in the spectra shown in FIGURES 1–3. The momentum components (P_Φ and P_r) and the polar angle (Θ) are to be compared with those of the initial and escaping particles. Notice that using Equation (11) in Equation (10), that the first term maintains equatorial reflection symmetry, whereas the second term introduces asymmetry. This second term is expected to be effective only near the event horizon (r_+), that is, inside the ergosphere. If P_r is less than zero and allowed to increase in magnitude for the various cases considered, such as, the case of the initial radially infalling photon, $(P_{ph})_r < 0$ and $(P_{ph})_\Theta = (P_{ph})_\Phi = 0$, displayed in FIGURE 1. In this case, the target particle has a relatively low energy–momentum, and Condition 1 shows that $(F_{GM})_\Theta$ exerts a dominant force on the photon, increasing in the positive $\hat{e}_\Theta$ direction, and thus, initializing the asymmetry that should eventually be seen in the high energy Penrose Compton scattered (PCS) photon distributions. This would show that there is coupling between the incident photons and the GM field. When the energy–momentum of the target electrons is allowed to increase, as is shown in FIGURE 2 (i.e., for nonequatorially confined targets), the first term in Equation (10) dominates, so that symmetry should be reestablished to some extent according to Conditions 1–5. However, the dynamics intrinsic to the orbit of the particle eventually dominate and the escaping PCS photon polar jets acquire the Kerr reflection symmetry of the target particle orbits. The initial asymmetry remains for the highest energy PCS photons according to Condition 2, yet tends toward symmetry according to Condition 5.

A similar coupling of the GM field with Penrose pair production (PPP, $\gamma\gamma \to e^-e^+$) should also exist. The Kerr metric reflection symmetry appears to exist to some extent in FIGURE 3A and B, perhaps because of a force balance in Conditions 1–3. That is, the initial asymmetry, favoring the positive $\hat{e}_\Theta$ direction, caused by the GM field acting on the radially infalling incident photons according to Condition 1

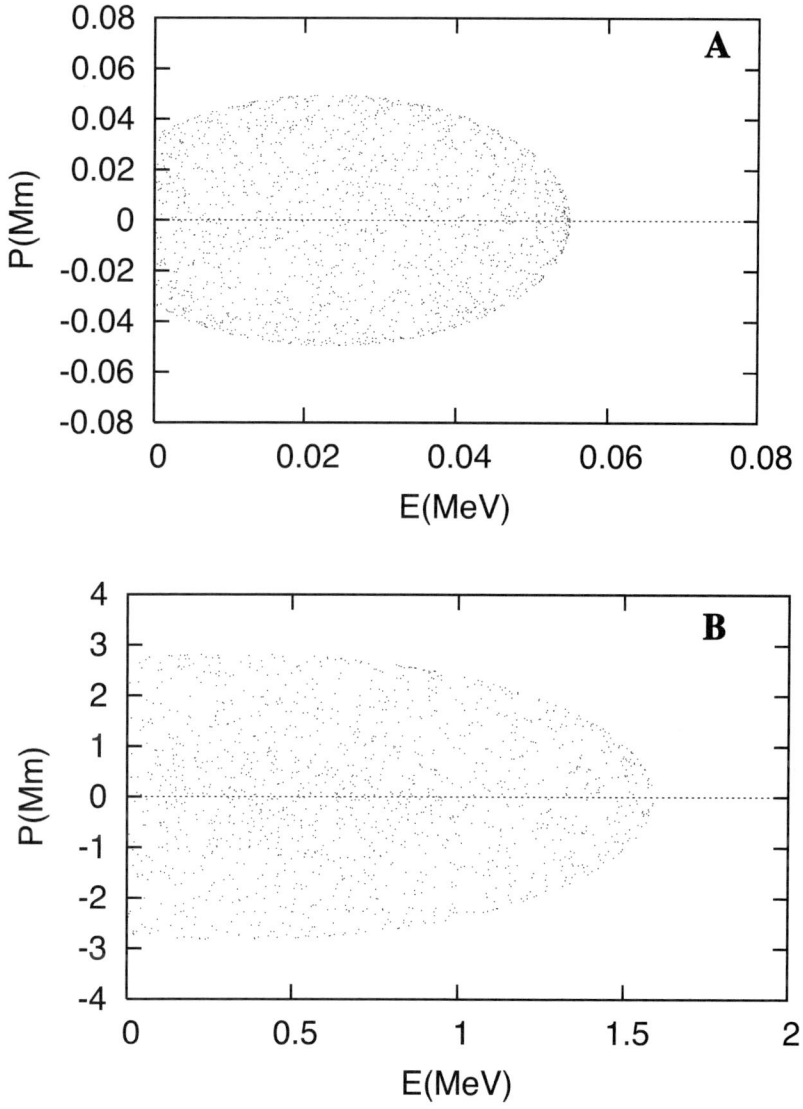

FIGURE 1. Compton scattering, scatter plots showing polar coordinate space momentum components: $(P'_{ph})_\Theta \equiv (Q'_{ph})^{1/2}$ versus E'_{ph} of the escaping PCS photons after 2,000 events (each *point* represents a scattered photon) at $r_{mb} \approx 1.089 M$. The various cases are defined by the following parameters: E_{ph}, initial photon energy; E_e, target electron orbital energy; $Q_e^{1/2}$, corresponding polar coordinate momentum $(P_e)_\Theta$ of the target electron; and N_{es}, number of photons escaping. **(A)** $E_{ph} = 0.511$ keV, $E_e \approx 0.539$ MeV, $Q_e^{1/2} = 0$, and $N_{es} = 1,706$. **(B)** $E_{ph} = 0.15$ MeV, $E_e \approx 0.539$ MeV, $Q_e^{1/2} = 0$, and $N_{es} = 1,442$.

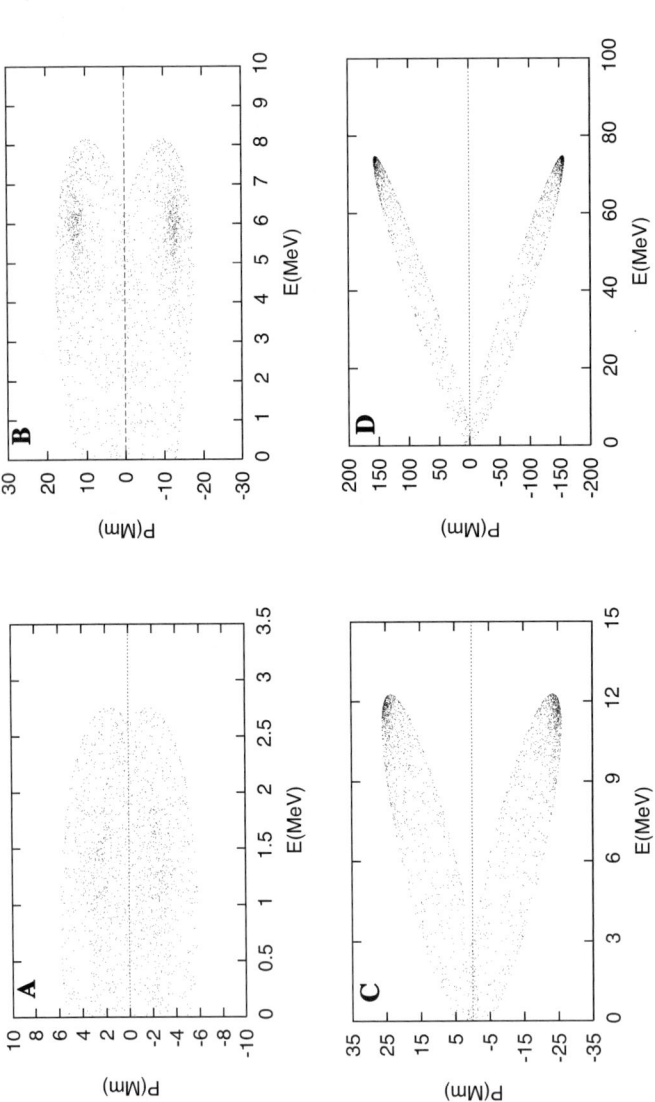

FIGURE 2. Compton scattering, scatter plots showing polar coordinate space momentum components: $(P'_{ph})_\Theta \equiv (Q'_{ph})^{1/2}$ versus E'_{ph} of the escaping PCS photons after 2,000 events (each *point* represents a scattered photon) at $r_{mb} \approx 1.089 M$. The various cases are defined by the following parameters: E_{ph}, initial photon energy; E_e, target electron orbital energy; $Q_e^{1/2}$, corresponding polar coordinate momentum $(P_e)_\Theta$ of the target electron; and N_{es}, number of photons escaping. **(A)** $E_{ph} = 0.15$ MeV, $E_e \approx 1.297$ MeV, $Q_e^{1/2} = \pm 2.479 Mm_e$, and $N_{es} = 1,628$. **(B)** $E_{ph} = 0.15$ MeV, $E_e \approx 5.927$ MeV, $Q_e^{1/2} = \pm 12.43 Mm_e$, and $N_{es} = 1,843$. **(C)** $E_{ph} = 0.03$ MeV, $E_e = 11.79$ MeV, $Q_e^{1/2} = \pm 24.79 Mm_e$, and $N_{es} = 1,971$. **(D)** $E_{ph} = 0.03$ MeV, $E_e \approx 74.29$ MeV, $Q_e^{1/2} = \pm 156.4 Mm_e$, and $N_{es} = 1,975$.

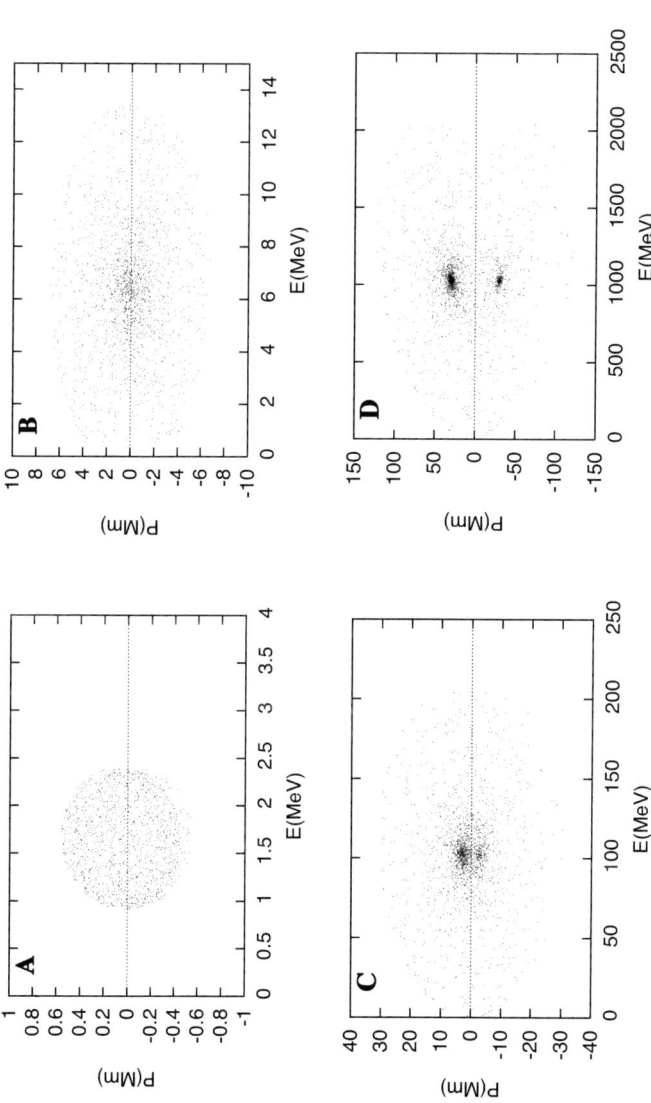

FIGURE 3. Penrose pair production ($\gamma\gamma \to e^- e^+$), scatter plots showing polar coordinate space momentum components: $P_\Theta \equiv Q^{1/2}$ versus $E_\mp$ of scattered escaping $e^- e^+$ pairs after 2,000 events (each *point* represents an electron) at r_{ph}. The various cases are defined by the following parameters: $E_{\gamma 1}$, the infalling photon energy; $E_{\gamma 2}$, the target photon orbital energy; $Q_{\gamma 2}^{1/2}$, corresponding polar coordinate momentum ($P_{\gamma 2}$) Θ of the target photon; and N_{es}, the number of $e^- e^+$ pairs escaping. (**A**) $E_{\gamma 1} = 0.0035\,\text{MeV}$, $E_{\gamma 2} \approx 3.40\,\text{MeV}$, $Q_{\gamma 2}^{1/2} = \pm 0.0987\,Mm_e$, and $N_{es} = 1{,}326$. (**B**) $E_{\gamma 1} = 0.03\,\text{MeV}$, $E_{\gamma 2} \approx 13.54\,\text{MeV}$, $Q_{\gamma 2}^{1/2} = \pm 0.393\,Mm_e$, and $N_{es} = 1{,}850$. (**C**) $E_{\gamma 1} = 0.03\,\text{MeV}$, $E_{\gamma 2} \approx 2.146\,\text{GeV}$, $Q_{\gamma 2}^{1/2} = \pm 62.28\,Mm_e$, and $N_{es} = 1{,}997$. (**D**) $E_{\gamma 1} = 0.03\,\text{MeV}$, $E_{\gamma 2} \approx 206.7\,\text{MeV}$, $Q_{\gamma 2}^{1/2} = \pm 6.0\,Mm_e$, and $N_{es} = 1{,}984$.

appears to be approximately balanced by $(F_{GM})_\Theta < 0$ according to Conditions 2 and 3. However, as the energy momentum of the target particle increases (FIG. 3C) and D), the asymmetry, favoring the positive $\hat{e}_\Theta$ direction, increases because the GM field acts on the radially infalling photons according to Condition 1 and on the high energy escaping e^-e^+ pairs according to Condition 2. In this case, $(F_{GM})_\Theta > 0$ dominates for $H^r P_\Phi > H^\Phi P_r$ (note that P_Φ increases with increasing E). In this case, the asymmetry, seen in the polar jet distributions above and below the equatorial plane, reaches a maximum ratio of about 3:1 (see FIG. 3).

In closing this section, at least in the case of PPP ($\gamma\gamma \to e^-e^+$), I have demonstrated the apparent reflection symmetry breaking in the Kerr metric, above and below the equatorial plane, and thought to be caused by the frame dragged GM force field, near the event horizon, acting on particles of the Penrose process. This general relativistic field is discussed and analyzed further in the following sections as we consider the exotic characteristics of a rotating black hole. A possible explanation is also given as to why polar coordinate asymmetry does not seem to appear in escaping distributions of PCS photons (compare FIGS. 1 and 2).

DISCUSSION

We now look at the general trajectory of freely falling observers to investigate the cause of the asymmetry in more detail. The geodesic equation of motion for a test particle in an arbitrary gravitational field and an arbitrary coordinate system is

$$m_0 \frac{dP^\mu}{d\tau} + \Gamma^\mu_{\nu\alpha} P^\nu P^\alpha = f^\mu, \tag{12}$$

where $\lambda = \tau/m_0$ and f^μ is a contravariant force vector. The term f^μ is non-negligible when the particle is not much smaller than the characteristic dimensions of the gravitational field (as in the case of the spin motion of the Moon in the gravitational field of the rotating Earth). The term is also non-negligible when an electromagnetic field is present, significant, and the moving particle has charge; but this is not the case when measuring the behavior of a moving particle (spinning or not spinning) in a strong rotating gravitational field.[2] In the latter case, the field will be felt and measurable locally in the rotating frame, that is, an observer will measure the force on the particle relative to this rotating frame and the force on the particle due to its motion in this rotating frame. However, these forces should be "contained" or accounted for in the left-hand side of Equation (12).

Equation (12), with $f^\mu = 0$, is true for the test particles in the gravitational field of the nonrotating (Schwarzschild) or the rotating (Kerr) black hole. In our case, the Kerr black hole is rapidly rotating ($a = 0.998M$), giving rise to a strong GM force field. It appears that the effects of this field on a moving test particle in the rotating frame and resulting in a Coriolis-like force (see (3) and (4)), must be included in Equation (12), with $f^\mu = 0$.[2] If the f^μ term is important, then the principle of equivalence (special relativity holds in the freely falling frame) or the principle of general covariance (a physical law equation in a gravitational field must (1) hold in the absence of gravity, i.e., in inertial frames; and (2) be generally covariant, i.e., preserve its form under a general coordinate transformation $x \to x'$) must be applied to the infinitesimal elements of which the particle is composed, although $f^\mu = 0$ or

$f^\mu \neq 0$ might give a reasonable phenomenological representation of the motion of the whole particle.[7]

In general, the GM force on a test particle as measured by an observer in an arbitrary local frame coordinate system, from Equations (3) and (4), is given by

$$\left(\frac{dp^i}{d\tau}\right)_{GM} = H_{ij}p^j. \tag{13}$$

One can see that this force, like the Coriolis force, arises from the motion of the test particle with momentum p^j. Now examine the polar coordinate component of the GM force (see (8)) on the test particle in the *local* Kerr metric spacetime (i.e., in the local frame-dragged rotating frame) in BL coordinates at an instant of time,

$$\left(\frac{dp^\Theta}{d\tau}\right)_{GM} = H_{\Theta r}p^r + H_{\Theta\Phi}p^\Phi$$
$$= H_{\Theta r}g^{\hat{r}\hat{r}}p_r + H_{\Theta\Phi}g^{\hat{\Phi}\hat{\Phi}}p_\Phi, \tag{14}$$

where

$$H_{\Theta r} \equiv -\tilde{H}^\Phi = -e^{-\nu}\Gamma^{\hat{\Phi}}_{\hat{\Theta}\hat{r}}\beta_\Phi, \tag{15}$$

$$H_{\Theta\Phi} \equiv H^r = -e^{-\nu}\left(\frac{\partial\beta_\Phi}{\partial\Theta} - \Gamma^{\hat{\Phi}}_{\hat{\Phi}\hat{\Theta}}\beta_\Phi\right), \tag{16}$$

and

$$\Gamma^{\hat{\Phi}}_{\hat{\Theta}\hat{r}} = \frac{1}{2}g^{\hat{\Phi}\hat{\Phi}}\left(\frac{\partial g_{\hat{\Phi}\hat{r}}}{\partial\Theta}\right), \tag{17}$$

$$\Gamma^{\hat{\Phi}}_{\hat{\Phi}\hat{\Theta}} = \frac{1}{2}g^{\hat{\Phi}\hat{\Phi}}\left(\frac{\partial g_{\hat{\Phi}\hat{\Phi}}}{\partial\Theta}\right) + \frac{1}{2}g^{\hat{\Phi}\hat{r}}\left(\frac{\partial g_{\hat{r}\hat{\Phi}}}{\partial\Theta}\right). \tag{18}$$

Note that the hat (^) indicates a locally measured expression. Recall from Equation (4), the GM tensor is given by $H_{ij} = e^{-\nu}(\beta_{GM})_{j;i}$, where $e^{-\nu} = \sqrt{-g^{tt}}$ is the so called blueshift factor. At the photon orbit r_{ph}, $e^{-\nu} \approx 52$ and at the marginally bound orbit r_{mb}, $e^{-\nu} \approx 32$. This means that the magnitude of the GM tensor components increases toward the event horizon. Notice that the GM force requires the metric component $g_{\hat{\Phi}\hat{r}}$ to be nonzero in order for polar asymmetry to exist. This component was set equal to zero in the conventional derivation of the global Kerr metric[8] by assuming equatorial reflection symmetry. This assumption does not appear to be valid locally inside the ergosphere, where $g_{tt} = 0$ at the so called stationary or static limit,[3,5] and where the dragging of local inertial frames becomes so severe that the time coordinate basis vector changes from timelike to spacelike (which means that its magnitude changes from real to imaginary). It is a region in which test particles can have negative energies as measured by an observer at infinity. Resulting from this Penrose analysis it appears that inside the ergosphere the local $g_{\hat{\Phi}\hat{r}}$ metric component is not equal to zero, thus allowing equatorial reflection symmetry to break down locally and thus appear upon transformation to the global BLF. This appears to be what we find here, where polar asymmetry appears to be present. (Note, it seems reasonable to expect, based on my model calculation findings, that the term $g_{\hat{\Phi}\hat{r}}$ will be proportional to $\cot\Theta$.) It is beyond the scope of this paper to derive the local frame dragged metric tensor $g_{\hat{i}\hat{j}}$, and indeed this may be neither practicable nor practical.

My model calculations[3] use equations of motion for the trajectories of test particles determined from the Lagrangian for the geodesics (see Ref. 3 and references therein). My calculations depend directly on (1) the frame dragging velocity,

$$\omega = \frac{2Mar}{(r^2+a^2)^2 - a^2(r^2 - 2Mr + a^2)\sin^2\Theta}, \quad (19)$$

(2) the metric components g_{rr}, $g_{\Theta\Theta}$, $g_{\Phi\Phi}$, and g^{tt} through the transformations between the local nonrotating frame (LNRF) and the BLF (i.e., frame of the observer at infinity), and (3) derivations of the conserved energy E and azimuthal angular momentum L for equatorially and nonequatorially confined circular orbits.[3,6] Moreover, I show elsewhere[4] specifically how ω of Equation **(19)** enters into the four-momentum components of the scattered particles. These components reveal to us the effects of the frame dragging and the GM force acting on the test particles.

Now we take a closer look at Equation **(12)**, with $f^\mu = 0$, and evaluate the equation of motion in the $\hat{e}_\Theta$ direction. We find that the only nonzero terms involving the space momentum component in the radial direction are

$$m_0 \frac{dP^\Theta}{d\tau} + \Gamma^\Theta_{rr} P^r P^r + 2\Gamma^\Theta_{r\Theta} P^r P^\Theta + 2\Gamma_{r\Phi} P^r P^\Phi + \ldots = 0, \quad (20)$$

where the last term (proportional to $\partial g_{r\Phi}/\partial\Theta$, see below) is assumed to be zero in the global Kerr metric, as stated above, which appears to be a valid assumption. The radial momentum component P_r of a particle appears to be important in equatorial reflection symmetry breaking, as seen in the emitted particles from Penrose processes **(8)**. Substitution of the Kerr metric components into Equation **(20)** yields

$$m_0 \frac{dP^\Theta}{d\tau} = \frac{2a^2\cos\Theta\sin\Theta}{r^2 - 2Mr + a^2}(P^r)^2 - \frac{2r}{r^2 + a^2\cos^2\Theta} P^r P^\Theta$$

$$+ \frac{2}{r^2 + a^2\cos^2\Theta} \frac{\partial g_{r\Phi}}{\partial\Theta} P^r P^\Phi + \ldots, \quad (21)$$

where $P^i = g^{ik}P_k$ (i.e., $P^r = g^{rr}P_r$, $P^\Phi = g^{\Phi\Phi}P_\Phi = g^{\Phi\Phi}L$, and $P^\Theta = g^{\Theta\Theta}P_\Theta$) and $\partial g_{r\Phi}/\partial\Theta \equiv 0$.

Since there are no terms in Equations **(20)** or **(21)** that depend linearly on just P_r, and since there are no terms depending linearly on just P_μ in general among the other force components on the left hand side of Equation **(12)**; that is, there are no terms resembling the GM force acting a particle (**(7)** and **(8)**), we can safely presume that the GM force acting on a particle is measured locally, thus validating the above reasoning concerning the Kerr metric in the local rotating frame (i.e., the LBLF).

In concluding this section, it appears that my calculations best reveal the "true" behavior of the particles in a Kerr metric inside the ergosphere, that is, where the local inertial frame dragging is severe. Moreover, I have checked my calculations thoroughly without finding any computational error that would cause an artificial asymmetry to appear. The polar symmetry and asymmetry found in the particle distributions for the specific examples of FIGURES 1–3 are discussed in the next section.

ADDITIONAL DISCUSSION AND CONCLUSIONS

On Penrose Compton Scattering

After careful consideration of Equation (3.39) in Reference 3 and the distribution of the PCS photons in the electron rest frame into the four geometric quadrants surrounding the Kerr black hole as chosen by the Monte Carlo method, I conclude that my original assumption of equatorial reflection symmetry[9] appears to be valid, at least in the case of Penrose Compton scattering (PCS). However, if broken equatorial reflection symmetry were present, perhaps because of a general relativistic dynamical effect, a simple assumption that allows computation of the polar coordinate momentum in only one hemisphere would not allow us to explore the asymmetry. This is my reason for calculating the polar coordinate momentum in all four quadrants; that is, if asymmetry does exist I will account for its presence. For example, we show below, in the case of PPP ($\gamma\gamma \rightarrow e^-e^+$), that the polar direction asymmetry appears to be a true effect.

From an examination of the standard transformation law for components of a tensor[5] between the global asymptotically flat spacetime (or inertial frame) of the observer at infinity and the local nonrotating frame, that is, the rest frame of the local frame-dragged inertial observer,[3] and an examination of the Lorentz transformations between the local nonrotating frame and the rest frame of the orbiting material target particle or the center of momentum (CM) frame of a massless scattered particle, we find that if rotational and/or frame dragging effects, due to the GM field, are important, then the effects mathematically would only appear when performing a four-dimensional Lorentz transformation from the electron rest frame (or the center of momentum frame) to the laboratory frame of the observer in the LNRF.

In the rest frame of the electron for PCS the particle number difference in the upper hemisphere ($-\hat{e}_\Theta$) and the lower hemisphere ($+\hat{e}_\Theta$) is 48 for 2,000 scattering events; that is, 24 more particles are scattered into the $-\hat{e}_\Theta$ direction than in the $+\hat{e}_\Theta$ direction. Since this is true for all the PCS cases irrespective of the initial energies (even for $\gamma\gamma \rightarrow e^-e^+$), then it is safe to conclude that the difference is due to a statistical error and not a true asymmetry. As related to the number of particles scattered above or below the equatorial plane for 2,000 scattering events, the statistical error is defined as $1,000 \pm 24$, based on the applied Monte Carlo method, for equatorial reflection symmetry.

When we perform a four-dimensional Lorentz transformation from the electron rest frame to the laboratory frame, in the case of equatorial target electrons, the polar coordinate momentum components do not change. However, in the case of the non-equatorially confined target electrons, the polar coordinate values between frames change and some change signs. The changes give symmetric distributions for the escaping particles above and below the equatorial plane, as measured by an observer at infinity, within a statistical error of ± 24 (the difference of the allowed particle number distribution in the rest frame of the electron above or below the equatorial plane, as defined by the Monte Carlo method).

In the case of PCS, polar asymmetry is not detected, which is believed to be due to frame dragging (compare FIGS. 1 and 2). A reason for this might be that measurable effects at this scattering radius and relatively small energies are smaller than the above identified statistical error. Nevertheless, the fact remains that according to

Equations **(8)** and **(10)**, the GM force in the polar direction produces vortical jets[6,10] proportional to the local average azimuthal momentum component of the escaping particles, an effect that increases with increasing energy. The existence of such vortical jets was predicted by Fernando de Felice in 1992[11] from studying the general geometry of the Kerr metric.

On the Penrose Pair Production

From a thorough investigation as to where the asymmetry originates (FIG. 3), I am compelled to conclude that the apparent symmetry braking in the Kerr metric is a real effect, which may be caused by frame-dragged GM force field lines in the azimuthal direction, as described above, or another "strange effect" that arises when particles at the exotic photon orbit are involved. Equation (3.121) of Reference 3, which gives the scattering angle of the e^-e^+ pairs relative to the equatorial plane, within statical errors gives symmetric distributions about the equatorial plane in the CM frame. As in the PCS case, the particle number difference in the upper hemisphere ($-\hat{e}_\Theta$) and the lower hemisphere ($+\hat{e}_\Theta$) is 48 for 2,000 scattering events. This is interpreted to mean that the particle numbers emitted into the polar directions will differ by ±24 from the true value, which we can attribute to "noise". Overall, it is found that the asymmetry in the polar direction of the e^+e^- pairs produced increases from a factor of about 1.12 to a maximum factor of about 2.6, whereas the energy of the initial orbiting target photon increases from about 3.4 MeV to about 3,893 MeV. Once the maximum factor (a 2.6:1 ratio for the jet pointed in the $+\hat{e}_\Theta$ direction to the jet pointed in the $-\hat{e}_\Theta$ direction) is reached, as the energy of the target photon continues to increase to the maximum energy allowed by the code (about 108 GeV), the ratio does not change, but the polar and azimuthal coordinate momenta increase in strength. Perhaps, this exotic behavior might in some way be related to the maximum latitudinal angle (about 0.5°) that the photon orbits above and below the equatorial plane, irrespective of how large the orbital energy of the target photon. Note, such unstable, marginally bound target photons move along *sphere-like* circular orbits.[3,4,12]

Another interesting feature we find from the model calculations, as can be seen in FIGURE 3, is that the polar asymmetry is not so apparent for the low energy PPP process (compare FIG. 3A and B), but is very apparent for the high energy PPP process (compare FIG. 3C and D). Perhaps this finding is important in an explanation for the nearly symmetric (perhaps force balanced) distribution illustrated in FIGURE 3B. For completeness, the ratios ($+\hat{e}_\Theta$ to $-\hat{e}_\Theta$) of the escaping polar-jet distributions in electron particle number, per 2,000 events, are 699/627 = 1.12, 967/883 = 1.095, 1,259/725 = 1.74, and 1,428/569 = 2.51 for FIGURE 3A–D.

The asymmetry seen in FIGURE 3 is consistent with the GM force acting according to the Conditions 1–5 stated above. For the lower energy case, since most of the e^-e^+ pairs escape with $\Theta \sim 90°$ and $|(P_{\gamma 1})_r| < |(P_\mp)_r|$, by Condition 1 $(F_{GM})_\Theta < 0$ for $(P_-)_r > 0$ will dominate to cancel out the initial asymmetry resulting from the infalling photon energy as its energy increases from $E_{\gamma 1} = 0.0035$ MeV to 10 MeV, with $E_{\gamma 2} = 3.40$ MeV held constant. Over this range of initial photon energy $E_{\gamma 1}$, asymmetry decreases until symmetry is attained and then the polar distributions appear to *flip* hemispheres. The same is true for $E_{\gamma 1} = 0.0035$–10 MeV, with $E_{\gamma 2} = 10.75$ MeV held constant, but the polar distributions do not appear to *flip*. Note, for the cases discussed above, the number of particles escaping decreases as $E_{\gamma 1}$ increases, and therefore, statistical error (or noise) can render particular findings uncertain. However, these statistical errors become negligible for $E_{\gamma 2} \gtrsim 340\,\text{MeV}$ and the trend continues; that is, the asymmetry continues decreasing slightly as $E_{\gamma 1}$ is allowed to increase for a given value of $E_{\gamma 2}$.

The most prominent feature is the increasing asymmetry, favoring the $\hat{e}_\Theta$ direction when $E_{\gamma 1} = 0.03\,\text{MeV}$ is held constant and $E_{\gamma 2}$ is allowed to increase from 3.40 to 3,893 MeV, with the $+\hat{e}_\Theta$ to $-\hat{e}_\Theta$ jet ratio increasing from about 1 to 2.6. This can be explained by the initial asymmetry of $E_{\gamma 1} = 0.03\,\text{MeV}$ and $(P_{\gamma 1})_r < 0$ according to Condition 1, and then amplified by Condition 2, with Condition 3 falling short of reestablishing symmetry. Subsequently, the maximum ratio is reached and maintained at about 3:1 as $E_{\gamma 2}$ increases to the limit defined by the code (about 108 GeV).

Additional investigation of the origin of this polar asymmetry is beyond the scope of this paper. I leave this matter now, yet, the subject remains somewhat open. If, however, the asymmetry proves to be a true effect, occurring most prominently at the photon orbit in the PPP ($\gamma\gamma \rightarrow e^-e^+$), this might lead to a unique characteristic to identify a rotating black hole source.

ACKNOWLEDGMENTS

I first thank God for His thoughts and for making this research possible. Next, I thank Dr. J.R. Ipser for his helpful comments that led to my finding specifically and analytically where the polar coordinate momenta appear to break equatorial reflection symmetry.

The authors declare that they have no competing financial interests.

REFERENCES

1. THIRRING, H. & J. LENSE. 1918. Phys. Z. **19**: 156.
2. THORNE, K.S., R.H. PRICE & D.A. MACDONALD. 1986. Black Holes: The Membrane Paradigm. Yale University Press, New Haven.
3. WILLIAMS, R.K. 1995. Phys. Rev. D **51**: 5387.
4. WILLIAMS, R.K. 2005. Preprint (astro-ph/0203421).
5. BARDEEN, J.M., W.H. PRESS & S.A. TEUKOLSKY. 1972. Astrophys. J. **178**: 347.
6. WILLIAMS, R.K. 2004. Astrophys. J. **611**: 952.
7. WEINBERG, S. 1972. Gravitation and Cosmology: Principles and Applications of the General Theory of Relativity. John Wiley & Sons, New York.
8. CHANDRASEKHAR, S. 1992. The Mathematical Theory of Black Holes. Clarendon Press, Oxford.
9. WILLIAMS, R.K. 1991. Ph.D. Thesis, Indiana University.
10. DE FELICE, F. & L. CARLOTTO. 1997. Astrophys. J. **481**: 116.
11. DE FELICE, F. & A. CURIR. 1992. Class. Quantum Grav. **9**: 1303.
12. WILKINS, D.C. 1972. Phys. Rev. D **5**(4): 814.

Self-Gravity Driven Instabilities at Accelerated Interfaces

ROBERT M. HUECKSTAEDT,[a] JAMES H. HUNTER, JR.,[b]
AND RICHARD V.E. LOVELACE[c]

[a]*Applied Physics Division, Los Alamos National Laboratory,
Los Alamos, New Mexico, USA*

[b]*Department of Astronomy, University of Florida, Gainesville, Florida, USA*

[c]*Departments of Applied Physics and Astronomy, Cornell University,
Ithaca, New York, USA*

> ABSTRACT: Nonlinear hydrodynamic flows are ubiquitous in the interstellar medium (ISM). Such flows play an important role in shaping atomic and molecular clouds and determining the initial conditions for star formation. One mechanism by which nonlinear flows arise is the onset and growth of interfacial instabilities. Any interface of discontinuous density is subject to a host of instabilities, including Rayleigh–Taylor, Kelvin–Helmholtz, and Richtmyer–Meshkov. As part of an ongoing study of structure formation in the ISM, Hunter, Whitaker, and Lovelace discovered an additional density interface instability. This instability is driven by self-gravity and termed the self-gravity interfacial instability (SGI). The SGI causes any displacement of the interface to grow on roughly a free-fall time scale, even when the perturbation wavelength is much less than the Jeans length. Numerical simulations have confirmed the expectations of linear theory, including the near scale invariance of the growth rate. Here, we build upon previous work by considering an initial condition in which the acceleration due to self-gravity is non-zero at the interface.
>
> KEYWORDS: hydrodynamics; instabilities; ISM; evolution

INTRODUCTION

Hydrodynamic instabilities play a key role in shaping the interstellar medium in ways that can lead to star formation. Atomic and molecular gas flows are affected by global instabilities (such as Jeans) and interfacial instabilities. Examples of interfacial instabilities include Kelvin–Helmholtz instability, due to velocity shear, and Rayleigh–Taylor (RT) instability, due to a lighter fluid accelerating a heavier fluid. In their investigation of Kelvin–Helmholtz and coupled instabilities, Hunter, Whitaker, and Lovelace[1,2] identified a new interfacial instability that acts upon a density interface. The driving mechanism is the self-gravity of the fluids, a force usually ignored in studies of interfacial instabilities in the ISM.

Address for correspondence: James H. Hunter, Jr., Department of Astronomy, University of Florida, Gainesville, FL 32611, USA.
hunter@astro.ufl.edu

The self-gravity interfacial instability (SGI) persists in the static limit and for wavelengths less than a Jeans length. Hunter et al.[1] used normal mode analysis to derive a linear growth rate for the SGI in compressible media. When taken in the incompressible, static limit, the growth rate in planar geometry for a perturbation amplitude proportional to $e^{-i\omega}$ is given by

$$\omega^2 = \frac{-2\pi G(\rho_2 - \rho_1)^2}{\rho_2 + \rho_1} + \frac{gk(\rho_2 - \rho_1)}{\rho_2 + \rho_1}. \tag{1}$$

In Equation (1), k is the horizontal perturbation wave number ($k > 0$), g is a constant background acceleration, and G is the gravitational constant. The mass densities of the lower and upper fluids are specified as ρ_1 and ρ_2, respectively. The second term gives the growth rate for the incompressible RT instability, and the first term is the incompressible growth rate for the SGI. Both instabilities persist in the static limit, but several important differences are evident. Self-gravity knows no preferred direction, so the SGI is destabilizing across any density interface. An interface is RT unstable only if $g(\rho_2 - \rho_1) < 0$, such that the heavy fluid sits "on top" of the light fluid. The SGI growth rate depends on the absolute densities of the fluids and their ratio, but not on the perturbation wavelength. The RT growth rate changes with perturbation wavelength and density ratio, but it does not depend on the absolute densities in the fluids. Given these dependencies, the SGI is expected to grow faster than the RT instability for a fixed value of g when the perturbation wavelength is long enough, such that $\lambda = 2\pi/k > g/G|\rho_2 - \rho_1|$. Numerical simulations[3] reveal another key difference. The RT instability is characterized at early times by dense spikes penetrating the tenuous fluid. The SGI appears to evolve in the opposite manner; tenuous spikes stream into the denser fluid.

In planar geometry, the growth rate for the SGI depends only weakly upon the Jeans criterion for the fully compressible case and not at all on the perturbation wavelength in the incompressible limit. The underlying reason that self-gravity is able to drive an instability for any wave number is that the configuration is not one of minimum energy. In this paper, we expand our study to consider initial conditions in which the self-gravity induced acceleration is non-zero at the interface. Defining γ to be the ratio of the specific heat at constant pressure to that at constant volume, we limit our examination to the case $\gamma = 2$. This is a stiffer equation of state than is realistic for the ISM, but this choice allows for relatively simple and spatially bounded expressions for the initial equilibrium distributions of density, pressure, and gravitational acceleration.

The outline for this paper is as follows. After deriving background equilibrium solutions in the next section, the numerical method is outlined. Selected simulations are then presented and, finally we summarize our results and discuss their astronomical implications.

EQUILIBRIUM CONFIGURATION

In order to isolate the gravitational instabilities, simulations are begun from a state of hydrostatic equilibrium. The geometry of the problem is shown in FIGURE 1. The interfacial values of the equilibrium densities and temperatures in the two regions are $\rho_1(0) > \rho_2(0)$ and $T_1(0) < T_2(0)$, and the interfacial pressure P_0 is the

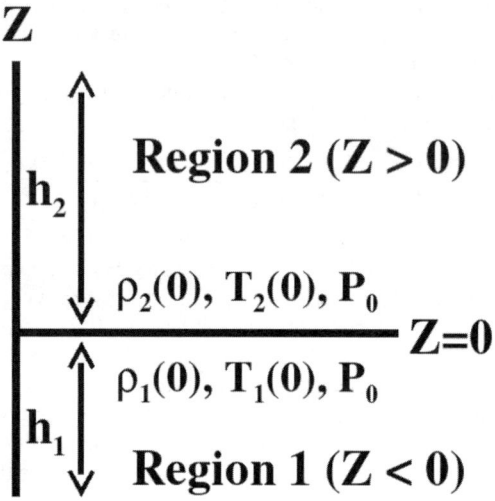

FIGURE 1. Schematic of the equilibrium configuration.

same in both media. The densities, temperatures, and pressures lapse to zero at heights h_1 below and h_2 above the interface.

Adopting a polytropic equation of state ($P = \kappa\rho^\gamma$), exact equilibrium solutions for a self-gravitating gas exist in the one-dimensional, unperturbed, planar problem when $\gamma = 2/3, 1, 2$, and ∞. In previous work,[3] we examined the case in which the self-gravitational acceleration is zero across the interface. We looked at both the analytic solution for $\gamma = 2$ and a numerically solved solution for $\gamma = 1.4$, and concluded that changing γ over this range has no effect on the behavior of the SGI. Here, we relax our boundary condition and allow the acceleration to be non-zero at the interface for gasses with $\gamma = 2$. In this case, the hydrostatic and Poisson equations reduce to

$$\frac{d\rho_n(z)}{dz} = K_n f_n(z) \tag{2}$$

and

$$\frac{df_n(z)}{dz} = -4\pi G \rho_n(z), \tag{3}$$

where $n = 1$ for region 1 and $n = 2$ for region 2, $K_n = \rho_n(0)^2/2P_0$, and $f_n(z)$ are the self-gravitational accelerations. The solutions for the densities and accelerations have the forms

$$\rho_n(z) = \rho_n(0)\cos\left(\frac{z}{l_n}\right) + \rho_n(0)\frac{f_0}{\beta}\sin\left(\frac{z}{l_n}\right) \tag{3}$$

and

$$f_n(z) = f_0 \cos\left(\frac{z}{l_n}\right) - \beta \sin\left(\frac{z}{l_n}\right). \tag{5}$$

The acceleration is continuous across the interface, $f_1(0) = f_2(0) = f_0$. The constant β is defined by

$$\beta = 2\sqrt{2\pi G P_0}. \tag{6}$$

The gravitational scale heights obey the relation

$$l_n = \frac{1}{\rho_n(0)}\sqrt{\frac{P_0}{2\pi G}}. \tag{7}$$

Defining h_1 and h_2 as the absolute values of the heights at which the density and pressure tend to zero in the corresponding regions, the surface densities are

$$\sigma_2 \equiv \int_0^{h_2} \rho_2(z)dz = \frac{\beta}{4\pi G}\sin\left(\frac{h_2}{l_2}\right) + \frac{f_0}{4\pi G}\left[1 - \cos\left(\frac{h_2}{l_2}\right)\right] \tag{8}$$

and

$$\sigma_1 \equiv \int_{-h_1}^{0} \rho_1(z)dz = \frac{\beta}{4\pi G}\sin\left(\frac{h_1}{l_1}\right) - \frac{f_0}{4\pi G}\left[1 - \cos\left(\frac{h_1}{l_1}\right)\right]. \tag{9}$$

Using Equations (5), (8), and (9), it can be verified that the Gauss theorem is satisfied,

$$f_2(h_2) - f_1(h_1) = -4\pi G(\sigma_2 + \sigma_1). \tag{10}$$

By symmetry, $f_2(h_2) = -f_1(h_1)$, which leads to the expression

$$f_2(h_2) = -2\pi G(\sigma_2 + \sigma_1). \tag{11}$$

Upon integrating Equation (3) from $z = 0$ to $z = h_2$, we find

$$f_2(h_2) - f_0 = -4\pi G \sigma_2, \tag{12}$$

or

$$f_0 = 4\pi G \sigma_2 + f_2(h_2) = 4\pi G \sigma_2 - 2\pi G(\sigma_2 + \sigma_1), \tag{13}$$

or

$$f_0 = 2\pi G(\sigma_2 - \sigma_1). \tag{14}$$

Recalling that the densities in media 1 and 2 lapse to zero at $z = -h_1$ and $z = h_2$, respectively, it follows from Equation (4) that

$$\tan\left(\frac{h_2}{l_2}\right) = -\frac{\beta}{f_0} \tag{15}$$

and

$$\tan\left(\frac{h_1}{l_1}\right) = \frac{\beta}{l_1}. \tag{16}$$

Hereafter, we define $\theta_1 = h_1/l_1$ and $\theta_2 = h_2/l_2$, both greater than zero. The quadrants in which these angles are defined depends on the sign of f_0. If f_0 is greater than zero, θ_1 is in the first quadrant and θ_2 is in the second quadrant. Defining $\psi = \sqrt{\beta^2 + f_0^2}$, we have the relations $\sin\theta_1 = \beta/\psi$, $\cos\theta_1 = f_0/\psi$, $\theta_2 = \pi - \theta_1$, $\sin\theta_2 = \sin\theta_1$, and $\cos\theta_2 = -\cos\theta_1$. Therefore,

$$\sigma_2 = \frac{\beta^2 + f_0^2 + f_0\psi}{4\pi G \psi}, \tag{17}$$

and
$$\sigma_1 = \frac{\beta^2 + f_0^2 - f_0\psi}{4\pi G\psi}. \tag{18}$$

Consequently,
$$f_0 = 2\pi G(\sigma_2 - \sigma_1) = \frac{4\pi G f_0 \psi}{4\pi G \psi} = f_0, \tag{19}$$

an identity. The same process can be applied for the case $f_0 < 0$, with similar results. Therefore, the solutions form a consistent set.

In view of these results, we adopt the following strategy. We set the molecular weight to $\mu = 2$ for both media. We specify the densities and temperatures at the interface and calculate P_0, β, l_1, and l_2. Then, we select f_0 and compute the density distributions from Equation (4). The pressure and temperature distributions follow from the equation of state. The final step is to use Equation (5) to calculate the boundary values of $f_n(z)$ for use by the gravity solver.

Due to discretization of the distributions across the grid, a truly static state is not achieved. We deem the setup to be sufficiently static if the motions induced in the unperturbed case are negligible compared to any imposed velocity perturbations. We choose the set of parameters: $\rho_1(0) = 1.0 \times 10^{-20}\,\text{g cm}^{-3}$, $T_1(0) = 20\,\text{K}$,

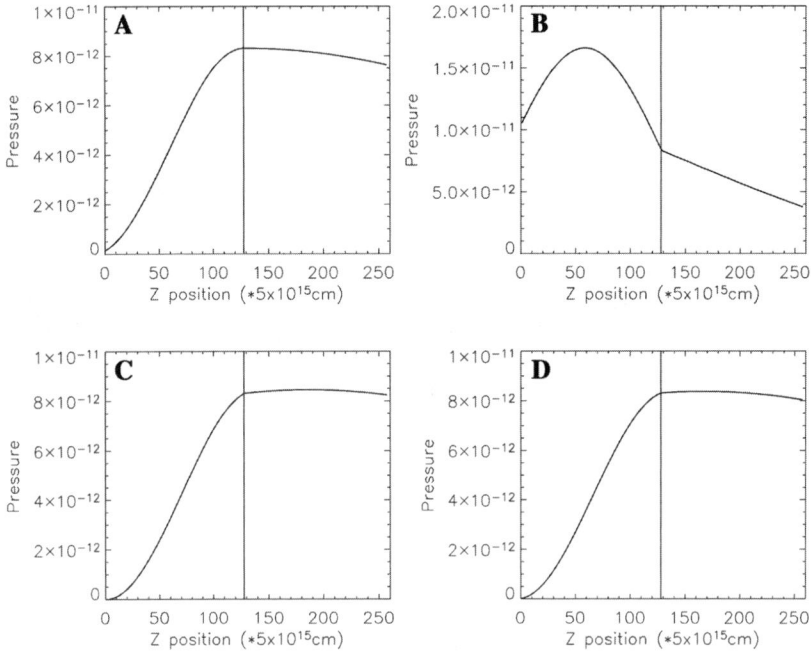

FIGURE 2. Initial pressure distributions for unperturbed backgrounds. Profiles are shown for values of f_0 (in units of cm sec^{-1}) (**A**) 0.0, (**B**) $-\beta = -3.7344 \times 10^{-9}$, (**C**) 4.8515×10^{-10}, and (**D**) 3.23437×10^{-10}.

$\rho_2(0) = 0.2 \times 10^{-20}\,\mathrm{g\,cm^{-3}}$, and $T_2(0) = 100\,\mathrm{K}$. The pressure profiles for four different choices of f_0 are shown in FIGURE 2. The adiabatic sound speed for a temperature of 20 K is $c = \sqrt{\gamma P_0 / \rho_1(0)} = 40{,}777\,\mathrm{cm\,sec^{-1}}$. If allowed to run to five e-folding times $(1.057 \times 10^{14}\,\mathrm{sec})$, the highest velocities observed throughout the grid are less than $100\,\mathrm{cm\,sec^{-1}}$. (This represents the worst case; most static models show lower velocities.) This is an order of magnitude or more lower than the initial velocity perturbation amplitude. Deviations from the static solutions are not large enough to affect the results of the perturbed simulations.

MODELS

Theoretical growth rates are compared to growth rates obtained through numerical modeling using the computational fluid dynamics library (CFDLib), which was developed at the Los Alamos National Laboratory.[4] CFDLib is a finite-volume code well suited for problems of all flow speeds. The self-gravitational potential is solved in two-dimensions using the MUDPACK multigrid code developed at the National Center for Atmospheric Research.[5,6] Models are run on a two-dimensional Cartesian mesh of size 257×257. Simulations repeated on a 513×513 grid show a difference in fine-scale structure, but not in growth rate. The normal velocity components of the gas are confined by reflective boundary conditions on all sides; whereas, the gravitational potential solver uses periodic boundary conditions along the side boundaries and specified gradient conditions along the top and bottom.

Perturbed simulations are begun with a velocity perturbation along the interface of the form

$$v(x,z) = v_0 \cos(kx) e^{-kz}, \qquad (20)$$

with v_0 set to 5% of the sound speed. We use a velocity perturbation instead of a spatial perturbation for two reasons. Although the grid resolution is sufficient to determine growth rates and get a sensible picture of nonlinear structure, it is too coarse to impose a spatial perturbation without giving rise to spurious instabilities due to the square cell structure. Also, without careful consideration, imposing a perturbation across an interface gives rise to a decaying as well as a growing mode. The effect of the decaying mode upon the velocity is easily seen and considered in determining growth rates, as summarized by Hueckstaedt and Hunter.[3]

We consider two different wavelengths, $\lambda = 6.425 \times 10^{17}\,\mathrm{cm}$ for two waves across the grid (k2) and $\lambda = 4.283 \times 10^{17}\,\mathrm{cm}$ for three waves across the grid (k3). Each choice of wavelength allows us to calculate the acceleration that would give rise to a Rayleigh–Taylor instability with the same growth rate as the SGI, $g = \sqrt{2G(\rho_1(0) - \rho_2(0))\lambda}$. Since we wish to compare our new results to both the pure SGI ($f_0 = 0$) and RT ($g > 0$, no self-gravity) cases, we choose values of f_0 to match these g values, $f_0 = 4.8515 \times 10^{-10}\,\mathrm{cm\,sec^{-1}}$ for case k2 and $f_0 = 3.2343 \times 10^{-10}\,\mathrm{cm\,sec^{-1}}$ for case k3. We also consider the case $f_0 = -\beta = -3.7344 \times 10^{-9}\,\mathrm{cm\,sec^{-1}}$. This choice results in equal weight being given to the sine and cosine terms in Equations (4) and (5).

RESULTS

SGI results for both k2 and k3 wavelengths are shown in FIGURE 3 and FIGURE 4. Density contours are plotted for various times, with white for high and black for low densities. A small variation in linear growth rate is observed among various wavelengths due to a small degree of compressibility allowed in the simulations. After the onset of nonlinearity, longer wavelength perturbations grow faster. As is characteristic of the SGI, the tenuous fingers grow faster than the dense spikes.

Various values of g are used to develop RT models with the same growth rates as the SGI cases. The results are shown in FIGURE 5 and FIGURE 6. The values of g are chosen so that the same linear growth rate results for both the k2 and the k3 cases. As for the SGI case, faster growth is seen during the nonlinear phase for the longer wavelength perturbation. As is typical for RT growth, the dense fingers out pace the tenuous bubbles at early times.

FIGURE 7 shows the evolution for the k2 case with $f_0 = -\beta$. The direction of f_0 is from the tenuous to the dense material. This produces a stabilizing effect in the RT sense. The stable RT-like term quenches the unstable SGI term since $\beta > G(\rho_2 - \rho_1)\lambda$. Decreasing the wavelength to the k3 case enhances the overall stability.

Unlike the previous case, a value of $f_0 > 0$ enhances rather than quenches perturbation growth. Any value of f_0 greater than zero adds a RT-like component to perturbation growth. FIGURE 8 shows the evolution of the k2 model with $f_0 = 4.8515 \times 10^{-10}$ cm sec^{-1}, a value chosen to match g in the RT model shown in

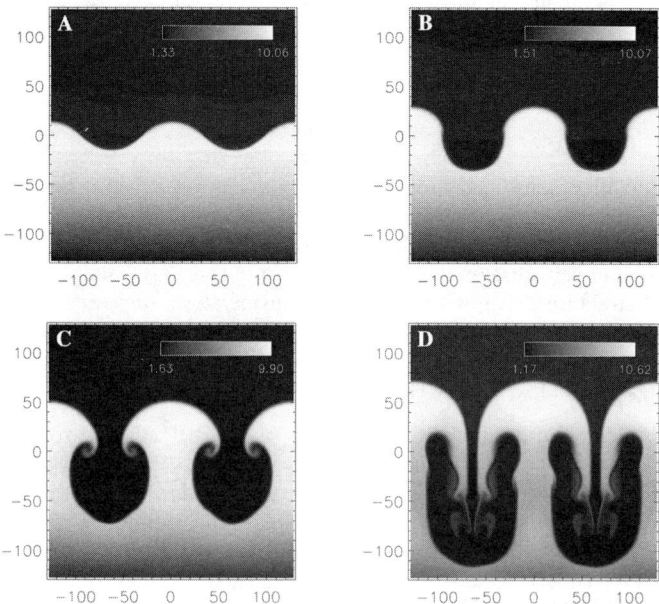

FIGURE 3. Density contours for $f_0 = 0.0$ cm sec^{-1} at (**A**) $2.0t_e$, (**B**) $3.0t_e$, (**C**) $4.0t_e$, and (**D**) $5.0t_e$. The density scales show the density values $\times 10^{-21}$. The e-folding time is $t_e = 2.1145 \times 10^{13}$ sec.

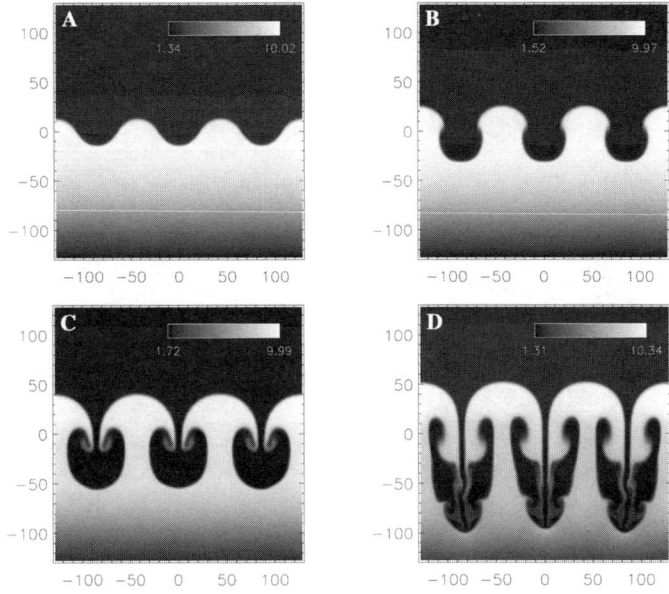

FIGURE 4. Density contours for $f_0 = 0.0\,\mathrm{cm\,sec^{-1}}$ at (**A**) $2.0t_e$, (**B**) $3.0t_e$, (**C**) $4.0t_e$, and (**D**) $5.0t_e$. The density scales show the density values $\times 10^{-21}$. The e-folding time is $t_e = 2.1145\times 10^{13}\,\mathrm{sec}$.

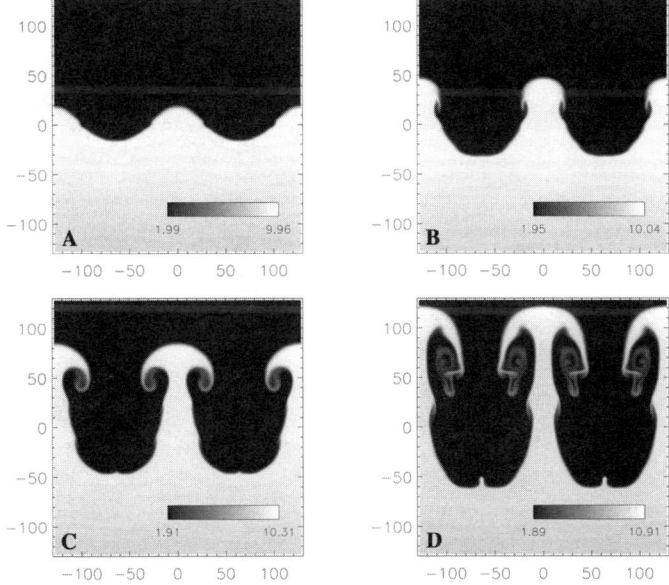

FIGURE 5. Density contours for RT with $g = 4.8515\times 10^{-10}\,\mathrm{cm\,sec^{-1}}$ at (**A**) $2.0t_e$, (**B**) $3.0t_e$, (**C**) $4.0t_e$, and (**D**) $5.0t_e$. The density scales show the density values $\times 10^{-21}$. The e-folding time is $t_e = 2.1145\times 10^{13}\,\mathrm{sec}$.

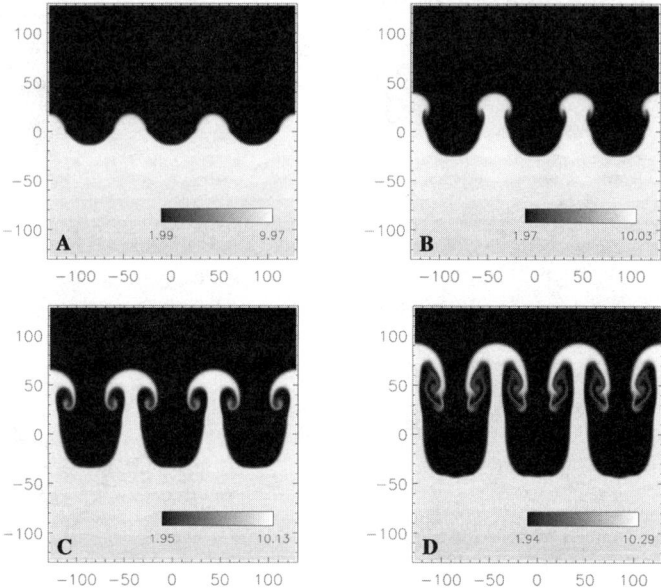

FIGURE 6. Density contours for RT with $g = 3.2343 \times 10^{-10}$ cm sec^{-1} at **(A)** $2.0 t_e$, **(B)** $3.0 t_e$, **(C)** $4.0 t_e$, and **(D)** $5.0 t_e$. The density scales show the density values $\times 10^{-21}$. The e-folding time is $t_e = 2.1145 \times 10^{13}$ sec.

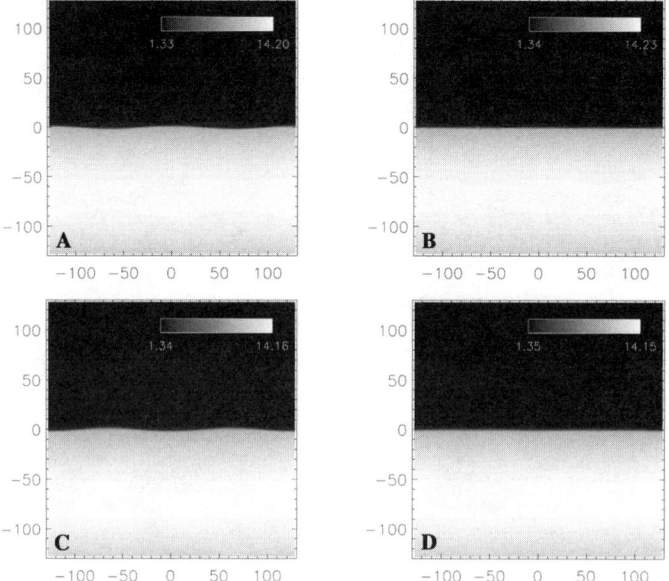

FIGURE 7. Density contours for $f_0 = -\beta = -3.7344 \times 10^{-9}$ cm sec^{-1} at **(A)** $0.5 t_e$, **(B)** $1.0 t_e$, **(C)** $1.5 t_e$, and **(D)** $2.0 t_e$. The density scales show the density values $\times 10^{-21}$. The e-folding time is $t_e = 2.1145 \times 10^{13}$ sec.

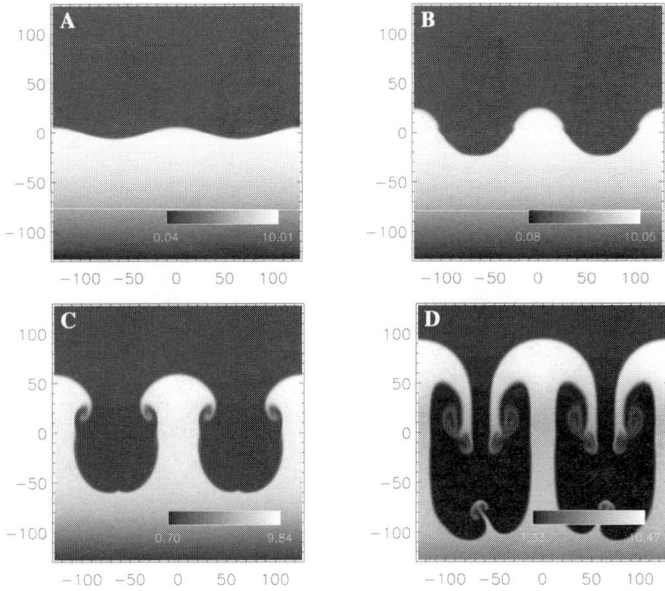

FIGURE 8. Density contours for $f_0 = 4.8515 \times 10^{-10}$ cm sec^{-1} at **(A)** $1.0 t_e$, **(B)** $2.0 t_e$, **(C)** $3.0 t_e$, and **(D)** $4.0 t_e$. The density scales show the density values $\times 10^{-21}$. The e-folding time is $t_e = 2.1145 \times 10^{13}$ sec.

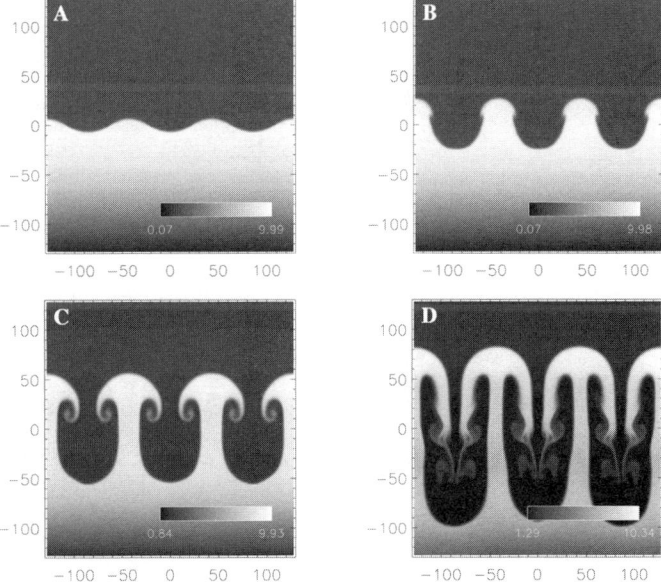

FIGURE 9. Density contours for $f_0 = 4.8515 \times 10^{-10}$ cm sec^{-1} at **(A)** $1.0 t_e$, **(B)** $2.0 t_e$, **(C)** $3.0 t_e$, and **(D)** $4.0 t_e$. The density scales show the density values $\times 10^{-21}$. The e-folding time is $t_e = 2.1145 \times 10^{13}$ sec.

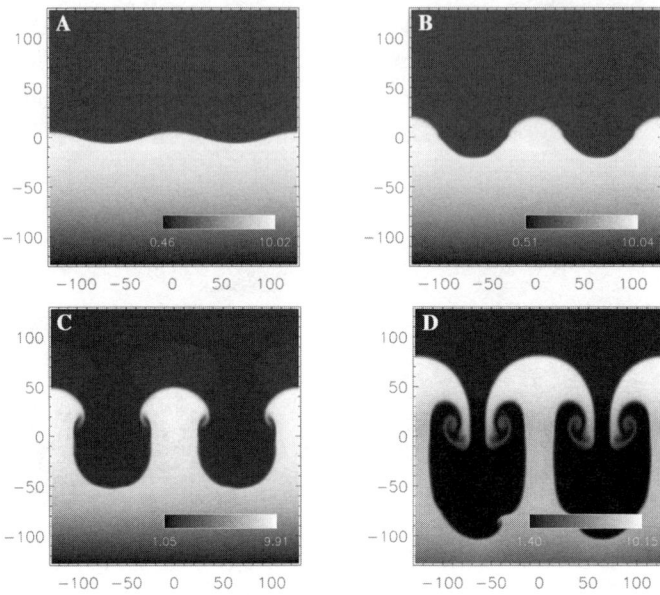

FIGURE 10. Density contours for $f_0 = 3.2343 \times 10^{-10}$ cm sec^{-1} at **(A)** $1.0 t_e$, **(B)** $2.0 t_e$, **(C)** $3.0 t_e$, and **(D)** $4.0 t_e$. The density scales show the density values $\times 10^{-21}$. The e-folding time is $t_e = 2.1145 \times 10^{13}$ sec.

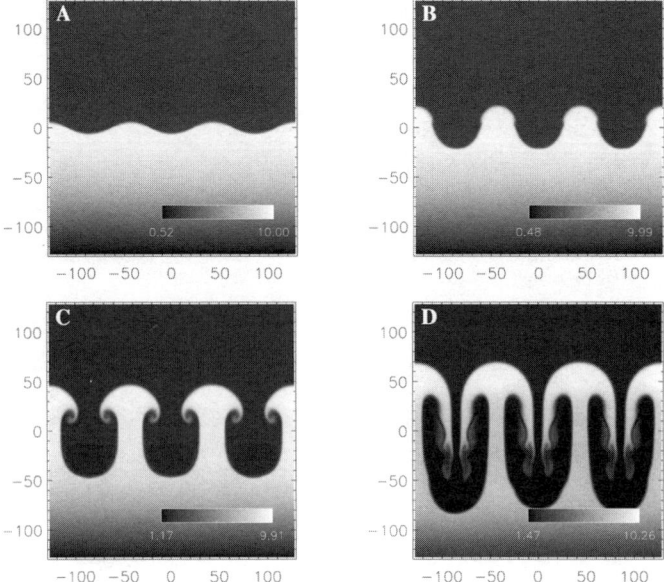

FIGURE 11. Density contours for $f_0 = 3.2343 \times 10^{-10}$ cm sec^{-1} at **(A)** $1.0 t_e$, **(B)** $2.0 t_e$, **(C)** $3.0 t_e$, and **(D)** $4.0 t_e$. The density scales show the density values $\times 10^{-21}$. The e-folding time is $t_e = 2.1145 \times 10^{13}$ sec.

FIGURE 5. The perturbation growth is faster than that for either the SGI or RT cases. The structure is different as well. Dense and tenuous fingers grow at nearly the same rate, resulting in a more symmetric shape about the interface.

Because the RT growth rate has a wavelength dependence absent for the SGI, decreasing the wavelength for the same f_0 should give greater emphasis to any RT-like component to the growth. FIGURE 9 shows the result for $f_0 = 4.8515 \times 10^{-10}$ cm sec^{-1} and a shorter wavelength. The growth rate is a bit higher than in the k2 case, consistent with the presence of a RT-like component to the growth. The near equality in growth rate for dense and tenuous fingers remains.

Similar results are seen for lower f_0. FIGURE 10 and FIGURE 11 show results for $f_0 = 3.2343 \times 10^{-10}$ cm sec^{-1}. This value corresponds to the value of g used for the RT case in FIGURE 6. As for the previous case, with $f_0 > 0$, the growth rate is larger than that for the SGI or RT individually. Again, the shorter wavelength perturbation grows faster, and the growth of dense and tenuous features proceed at the same rate.

We measure the velocity magnitude over time to determine numerical growth rates for each model. We then normalize all growth rates by dividing by the incompressible, linear SGI rate determined from Equation (1), $\omega_{SGI} = t_e^{-1} = 4.73 \times 10^{-14}$ sec^{-1}. We define two ratios for each model, Ω_S for the growth of dense spikes, and Ω_B for the growth of tenuous bubbles. This language is consistent with typical RT descriptions. The results for eight models are summarized in TABLE 1. We do not include the $f_0 = -\beta$ case since it did not result in perturbation growth. The growth rates are presented graphically in FIGURE 12. The numerical growth rates are determined using velocity data obtained between $1.0 t_e$ and $2.0 t_e$ for the SGI models (1 and 2) and data obtained between $0.6 t_e$ and $1.6 t_e$ for the others.

The SGI models are different from the others in that tenuous bubbles grow faster than dense spikes. SGI and RT models show large differences in the growth rates for spikes and bubbles. These gaps narrow for those models with $f_0 > 0$. The overall peak-to-peak growth for models in which a self-gravitating gas feels an acceleration in the RT unstable direction is considerably higher than that for the SGI or RT instability alone. Since $f_n(z)$ is slowly varying near the interface, the appearance of a RT-like term for $f_0 > 0$ is not surprising. The values of f_0 were chosen to match the g values that give equivalent growth rates for the SGI and RT instabilities, so we

TABLE 1. Model parameters

Model	f_0	g	λ	Ω_S	Ω_B	Figure
1	0.0	0.0	6.425	0.704	0.892	3
2	0.0	0.0	4.282	0.578	0.825	4
3	n/a	4.8515	6.425	0.949	0.745	5
4	n/a	3.2343	4.283	0.938	0.706	6
5	4.8515	0.0	6.425	1.201	1.158	8
6	4.8515	0.0	4.283	1.319	1.233	9
7	3.2343	0.0	6.425	1.033	1.036	10
8	3.2343	0.0	4.283	1.151	1.136	11

NOTE: units for f_0 and g, 10^{-10} cm sec^{-2}; λ, 10^{17} cm.

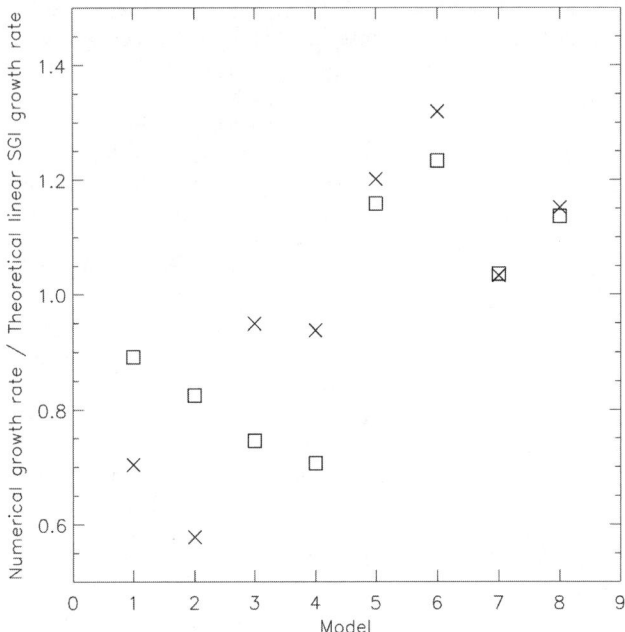

FIGURE 12. Growth rate results for eight perturbed models. Model numbers correspond to TABLE 1. The growth rates for each model as determined from velocity magnitude growth have been normalized to the theoretical, incompressible, linear SGI growth rate: ×, dense spike; □, tenuous bubble.

expect from Equation (1) that model 5 and model 8 would both result in a value of $\Omega = \sqrt{2} \approx 1.41$. Both the spike and bubble growths taken separately give a lower value of about 1.2. However, if spike and bubble are taken together, models 1–4 give an average of $\Omega_S + \Omega_B = 1.52$. By multiplying this number by $\sqrt{2}$, we arrive at the prediction $\Omega_S + \Omega_B = 2.15$ for models 5 and 8. The numerical values are 2.36 for model 5 and 2.29 for model 8, a reasonable match considering the effect on the average by the anomalously low Ω values produced for model 2.

CONCLUSIONS

We conclude that a non-zero, self-gravity induced acceleration at a density interface has the effect of quenching the SGI for values of f_0 such that $f_0(\rho_2 - \rho_1) < 0$ (the acceleration points in the direction from tenuous to dense material) and $\beta > G(\rho_2 - \rho_1)$ (the RT-like term is stronger than the SGI term). An acceleration in the direction from dense to tenuous enhances perturbation growth above that of the SGI alone. The growth rates for $f_0 > 0$ are consistent with predictions made using Equation (1) for a hybrid SGI-RT instability with $g = f_0$. Perturbation growth in these cases occurs with more symmetry between dense spikes and tenuous bubbles than is observed for pure SGI or RT. A more detailed treatment of these hybrid instabilities will appear elsewhere.

Gravitation processes play a key role in the evolution of the ISM and star formation, both as principal and coupling forces driving structure evolution. Ultimately, we seek a description of the ISM that includes all coupled phenomena. However, no such goal is achievable without a careful analysis of the underlying forces. We have taken up the study of self-gravity interfacial instabilities and their behavior when coupled to other instabilities as a step toward achieving an integrated picture of the ISM. We are confident that modern computing methods put this lofty goal within reach of the astrophysics community.

ACKNOWLEDGMENTS

The authors are honored to pay tribute to Henry Kandrup, a treasured teacher, colleague, and friend. We thank B.A. Kashiwa and N.T. Padial for help in modifying CFDLib for astrophysical use. Los Alamos National Laboratory is operated by the University of California for the U.S. Department of Energy under Contract W-7405-Eng-36.

The authors declare that they have no competing financial interests.

REFERENCES

1. HUNTER, J.H., JR., R.W. WHITAKER & R.V.E. LOVELACE. 1997. Astrophys. J. **482**: 852.
2. HUNTER, J.H., JR., R.W. WHITAKER & R.V.E. LOVELACE. 1998. Astrophys. J. **508**: 680.
3. HUECKSTAEDT, R.M. & J.H. HUNTER, JR. 2001. Mon. Not. Roy. Astr. Soc. **327**: 1097.
4. KASHIWA, B.A., N.T. PADIAL, R.M. RAUENZAHN & W.B. VANDERHEYDEN. 1994. *In* Symposium on Numerical Methods for Multiphase Flows. C.T. Crowe, Ed. ASME, New York
5. ADAMS, J. 1989. Appl. Math. Comput. **24**: 113.
6. ADAMS, J. 1991. NCAR Technical Note 357+STR.

Dynamics of Intracluster Gas and Bulk Motions in Clusters

RENATO DUPKE

Department of Astronomy, University of Michigan, Ann Arbor, Michigan, USA

ABSTRACT: In this paper I briefly discuss progress in recent lines of research on the internal dynamics of clusters of galaxies that have been made possible thanks to improvements in current X-ray spectrometers that enable them to perform detailed spatially resolved spectroscopy. In particular, I focus on the study of bulk motion in intracluster gas and the nature of features called cold fronts.

KEYWORDS: dynamics; intracluster gas; bulk motions

Clusters of galaxies are the largest gravitationally bound systems in the universe. As such, they are interesting not just for their intrinsic nature, structure, and evolution, but also for their use as tools in cosmology. For example, if they represent a fair sample of the early universe their baryonic mass fraction f_b value should be equal to Ω_b/Ω_m, where Ω_b and Ω_m are the baryon and total mass densities of the universe, respectively, normalized to the critical density.[1] The baryon fraction can be combined with the value of Ω_b as predicted from light element abundances through big bang nucleosynthesis,[2] to provide constraints on Ω_m[3,4] (see FIGURE 1). Since galaxies amount to only a small fraction of the baryonic mass in clusters, the precision with which the baryon fraction in clusters is estimated depends heavily on how well we can determine the intracluster gas mass, which is the dominant baryonic component and makes up typically 15–20% of the total cluster mass.

The intracluster gas is very hot (10^7–10^8 K), highly ionized, and emits mostly in X-rays through a combination of Bremstrahlung and line emission (see FIGURE 2). The X-ray emitting mass is calculated through a combination of X-ray surface brightness profile and spectral fittings. Typically, the surface brightness (proportional to $n^2 T^{1/2}$ for a hydrogen plasma) profile is fitted with some deprojected King-like profile proportional to $(1 + (r/r_0)^2)^{-3\beta+0.5}$, where β is the ratio of specific energies of galaxies and gas ($\mu m_p \sigma_{\rm gal}^2 /kT$). To determine the total mass, the X-ray method usually relies on the assumption of hydrostatic equilibrium (HE), which gives the mass enclosed within radius r as $M(r) \propto -rT(\Delta rT + \Delta r\rho)$, where T and ρ are the gas temperature and density, respectively, and ΔrX is the logarithmic derivative ($d\ln X/d\ln r$).

The degree of applicability of the above mentioned working assumptions is an important current issue, since it is now known that clusters are not spherically symmetric virialized systems as was previously thought, but instead they are dynamically active and have significant substructures, as shown by recent X-ray and optical

---

Address for correspondence: Renato Dupke, Department of Astronomy, University of Michigan, Ann Arbor, MI 48109, USA.
 rdupke@umich.edu

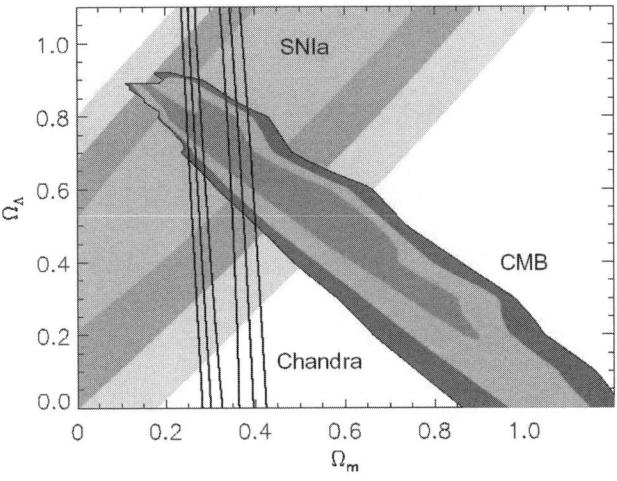

FIGURE 1. Constraints on Ω_m as derived from clusters of galaxies, overlaid on 1, 2, and 3σ contour plots from studies of the anisotropies of the CMB and from SN Ia. (Reproduced from Ref. 4 with permission.)

studies. Shocks, cold fronts, buoyant bubbles, temperature anisotropies, and velocity gradients are often found in these systems. Detailed knowledge of the dynamical phenomenology of clusters is important, not just to understand their origin and evolution, but also to use them as cosmological tools. The primary process of formation and growth of galaxy clusters is the merging of collapsed subsystems. This formation

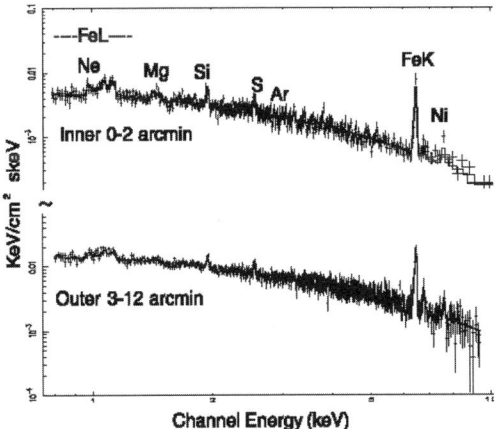

FIGURE 2. Illustration of X-ray spectra with ASCA GIS. Spectra of inner 2' (*top*) and outer (3') (*bottom*) of Abell 496. The main spectral lines and line complexes are indicated.

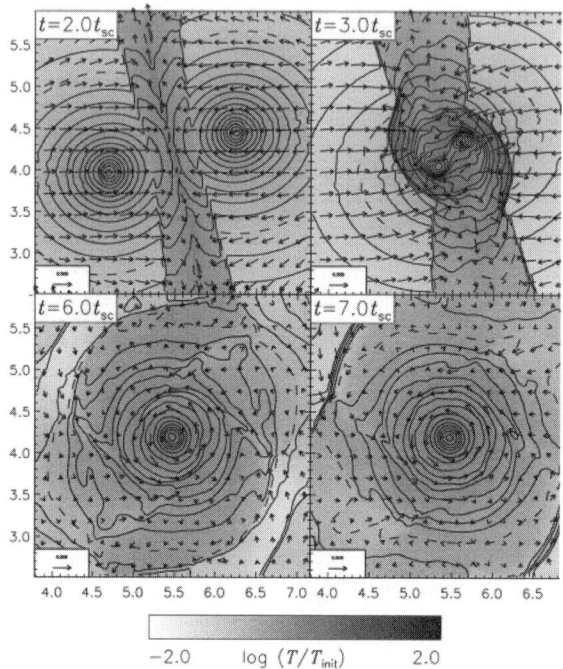

FIGURE 3. Off-center collision of clusters where initial impact parameter is about 10 core radii. Shading indicates the logarithm of the temperature, t_{sc} is the crossing time (Gyr). Velocity arrows are drawn for every eighth cell, with the fiducial arrow at the corner of each plot representing $v \sim 2{,}000$ km/sec. (Reproduced from Ref. 5 with permission.)

process is within the framework of the so-called bottom-up hierarchical scenario for the origin of clusters, where larger structures build up from small to large scales, via merger/accretion of smaller size systems. It is predicted that these mergers will be more frequent along larger filamentary structures that characterize the "cosmic web".

The most notable expected dynamical process of the ICM is clusters off-center merging (for a recent review see Sarazin[6]). This process is believed to create strong temperature, density, gas velocity, and also chemical substructures.[5,7] The link between temperature substructure and the merger stage is often made by comparison with hydrodynamic simulations. There is currently an enormous variety of cluster formation/merger simulations in the literature,[5,8–19] see also <vce.physics.lsa.umich.edu>. These simulations provide, among other quantities, temperature and surface brightness (or gas density) maps that can be compared with observations. In particular, off-center merging imparts angular momentum to the intracluster gas, which, according to recent numerical and hydro simulations, can last several Gyr[5,16,20–22] (see FIGURE 3). Measurements of bulk gas flows are crucial to track the evolutionary stage of clusters breaking the usual degeneracies that appear when comparing only at projected temperatures and densities to the simulations.

BULK MOTIONS

To date most of evidence of gas bulk flows in clusters in indirect (e.g., A2218,[23] A2142, A3667, and 1E0657-56[24–26]). We have recently shown that it is possible to measure bulk gas velocities *directly* with currently available spectrometers.[27,28] We determine gas velocities by measuring redshift differences in the emission lines in the X-ray spectra from cluster gas in different directions (regions). A line centroid can be measured with a precision $\Delta V \approx 127 \Gamma_{ev} E_{kev}^{-1} N^{-1/2}$ km·sec^{-1}, where N is the number of photons in the line and Γ_{ev} is the FWHM of the line, or if the line is narrower than the instrumental width, it is the FWHM of the instrument, and E_{kev} is the line energy. Currently, the energy resolution (FWHM) of ACIS-S3 on-board *Chandra* varies from 3% to 3.6% at 5.9 keV depending on the chip row, <http://space.mit.edu/ACIS/acismemo10.html>. This would, in principle, allow us to measure a line centroid (FeK$_\alpha$) to a precision of 500–1,000 km/sec with only a few hundred line photons. However, the ability to measure velocities with this precision is not merely a matter of collecting a sufficient number of photons. The main difficulty one faces in determining bulk velocities to know precisely the conversion between pulse-height and energy (gain), how it changes across the detectors and also over time. The gain fluctuations are typically on the same order or larger than the velocities one is trying to measure, so that taking the gain fluctuations into account properly is crucial to determine reliable velocity gradients.

The advanced satellite for cosmology and astrophysics (ASCA) was the first satellite to have the minimum conditions for velocity studies, high energy resolution, recorded gain variations across the detectors, and reasonably good gain stability. The first cluster where ICM velocity gradients were found was Perseus, where we discovered, using ASCA, evidence for large-scale gas bulk motions of more than 1,500 km/sec at the 90% confidence level, consistent with either ICM circulation or streaming motions due to a large merger event.[27] In Centaurus, where excellent data from the ASCA solid state spectrometers (SIS) were available, we found a significant (greater than 99% confidence level) velocity gradient at small scales (less than 5′),[28] see FIGURE 4A.

Since the main uncertainties of measuring velocities using X-ray spectroscopy is knowing well the intrinsic variations of the gain, it is extremely important: (1) to corroborate the measurements using other instruments, since it is unlikely the gain systematics will be the same, and (2) to develop strategies to minimize the dependence on gain. We used (1) and (2) for the Centaurus clusters by using the Chandra satellite to perform two observations of the cluster, one centered in the region of maximum velocity and the other centered in the region of minimum velocity using the same chip position,[29] thus avoiding having to deal with intrachip gain variations. Chandra off-center pointings allowed us not just to confirm but also to improve ASCA measurements. The maximum velocity difference found was $(3.4 \pm 1.0) \times 10^3$ km/sec and showed that the velocity distribution is more consistent with the presence of "eddies" than with "bulk rotation". Furthermore, the characteristic energy associated with this eddie is about 10^{61} $(\mu/0.6)(n_{gas}/0.01\,\mathrm{cm}^{-3})(V_{circ}/1200\,\mathrm{km/sec})^2(L/4')^3$ ergs! The presence of multiple eddies in clusters may provide non-thermal pressure support that can significantly affect mass and energetics estimates.

The analysis of multiple pointings with Chandra and XMM specifically tailored to minimize gain variations can benefit from a prior knowledge of the configuration

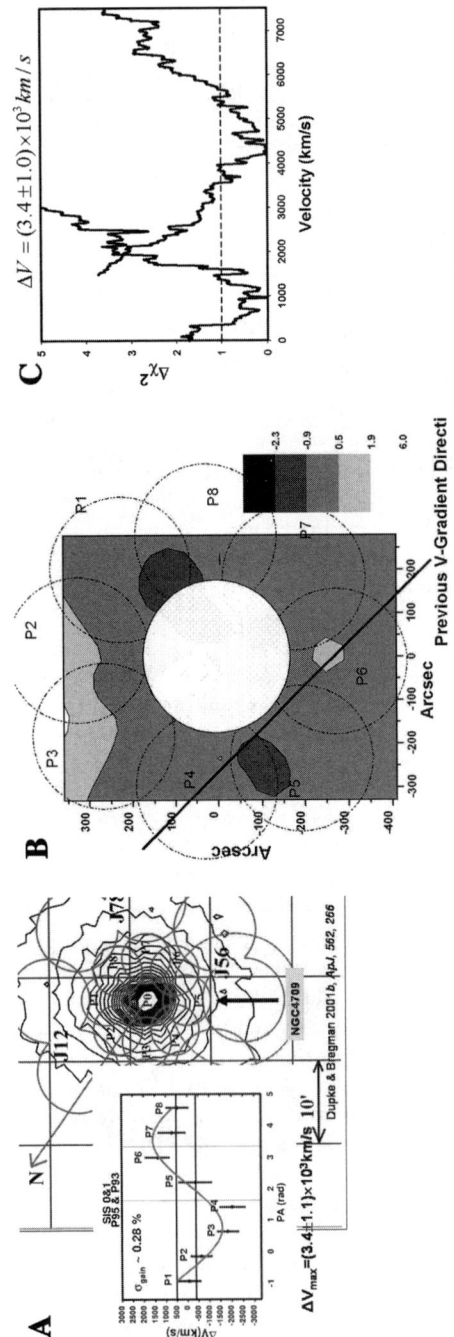

FIGURE 4. (**A**) Average azimuthal velocity distribution of Centaurus obtained with ASCA. The errors also include a 1 σ instrumental gain variation of 0.28%. The regions are circles 5′ from the cluster center and with 3′ radius. The *dark solid horizontal lines* represent the confidence limits for the central region. The region between the two *vertical lines* shows the approximate direction of the incoming group Cen 45, centered on NGC4709. The *solid line* shows the best-fit for a solid-body cosine function with a corresponding maximum rotational velocity of about 1,600 km/sec. North is indicated. Regions P1–P8 are overlaid on the X-ray image. (**B**) Chandra confidence (velocity divided by the 1σ errors) map for velocity measurements in a double pointing observation. *Dark shading* indicates negative velocities (with respect to systemic) and *lighter shading*, positive velocities. The previous ASCA pointings are overlaid and the direction of maximum velocity gradients derived with ASCA is also shown. The core of the cluster is excluded from the analysis (masked with a *white circle*) since it requires more complex spectral models, which may bias redshift measurements. North is up. (**C**) Chi-square distribution for the two regions inside P4/P5 and P6 that show velocity gradient of 3,400 km/sec inside P4 and P6. The difference is about 5σ detection.

of the velocity gradients and ASCA can provide it for clusters with very high bulk velocities. After the initial success in detecting velocity gradients in Perseus and Centaurus we carried out a systematic search in the ASCA archives for clusters with velocity substructure. We have recently finished this search[30] and we found three more clusters where we detected internal velocity gradients greater than 1,000 km/sec (at 90% confidence level). Their internal velocity distribution (normalized by the errors) are shown in FIGURE 5.

The cluster found with the highest and most reliable velocity gradient is Abell 576. Abell 576 is a richness class 1 cluster with low central gas temperatures and average metal abundances. It has an optical redshift of 0.0389. ASCA velocity analysis of this cluster found a significant velocity gradient of more than 4,000 km/sec[30] with respect to the inner Eastern regions. The high velocity gradient found in Abell 576 adds to the body of indirect evidence to the presence of high dynamical activity in this cluster. Rines et al.[31] determined the mass profile of A576 using the infall pattern in velocity space for more than 1,000 galaxies in a radius of $4h^{-1}$·Mpc from the center of the cluster. They found that the mass of the central Mpc was more than twice that found from X-ray measurements, suggesting that non-thermal pressure support may be biasing the X-ray derived mass. Their result is also in agreement with previous mass estimates of Mohr et al.,[32] who found a high velocity tail, separated by about 3,000 km/sec from the mean of the cluster. Kempner et al.[33] analyzed the Chandra observations of the core of this cluster and found sharp edges corresponding to jumps in gas density and pressure roughly in the N–S direction and suggested that the core substructures are caused by a current merger with tangential core velocities of about $750\,\text{km·sec}^{-1}$.

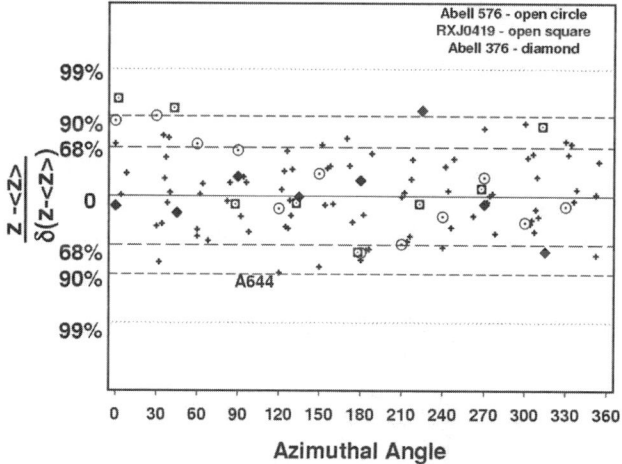

FIGURE 5. Error weighted distribution of relative velocities: we show all subregions, for all clusters in the ASCA sample. The 68% (*long dashed lines*), 90% (*short dashed lines*), and 99% (*dotted lines*) distribution boundaries are also shown. The clusters with the highest significant velocity structures are indicated by different symbols: Abell 576 (*open circles*), RX J0419+0225 (*open squares*), and Abell 376 (*diamonds*).

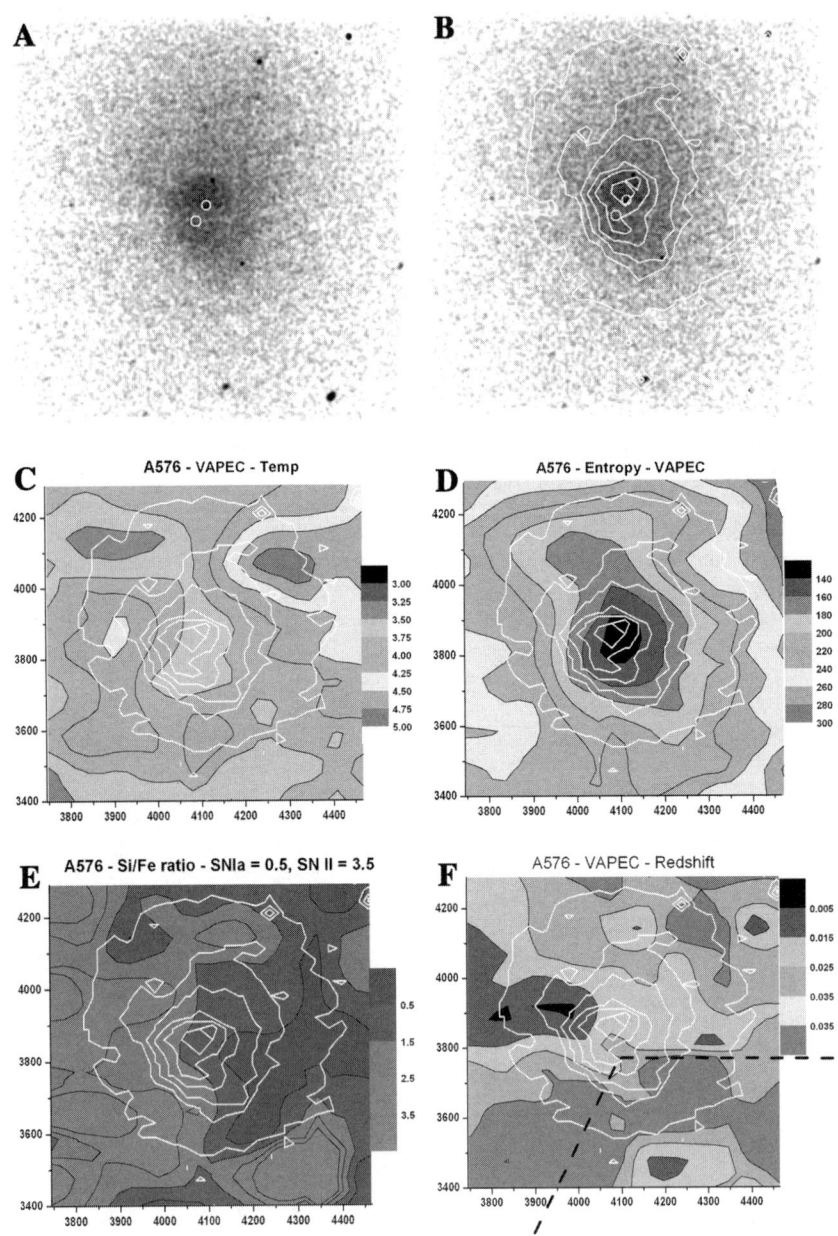

FIGURE 6.

In FIGURE 6 A we show a Chandra X-ray image of A576. Two features stand out: a surface brightness arm from the center to the S–SE forming a near triangular wedge pointing E and also wakes (or fingers) can be seen in the NW region. These are expected in supersonic galaxy motions.[34] FIGURE 6B shows the same as FIGURE 6A but with surface brightness contours overlaid. FIGURE 6C–F show colored contour maps of the distribution of the interesting parameters obtained through spectral fittings using an absorbed VAPEC thermal emission model.

The best fit results for individual points in the map are a result of adaptive smoothing with more than 5,000 cnts/cell and a minimum smoothing radii of 70 pix (35″). The color steps were chosen so as to approximate the 1σ errors. FIGURE 6C–F show the distribution of gas temperatures (keV), entropy ($cm^{2} \cdot K$), Si/Fe ratio, and redshifts. These plots suggest a merger incoming from the NW direction. The X-ray arm is associated with pressure enhancement toward the East with the projected temperature rising by 30% and with marginal indications of a temperature reduction after the shock front (toward the Eastern CCD border). There are two D galaxies near the core of the cluster and they are plotted as circles in FIGURE 6A. Both galaxies are offset from the X-ray isophotal center. The extended low velocity region from the center toward East shows that the merger is also happening near the line of sight going toward the observer. The velocity gradient is significant at the 90% confidence level and shows a difference of $(6 \pm 3.2) \times 10^{3}$ km/sec, in agreement a previous ASCA velocity analysis of similar regions.[30]

The region of high velocities found with ASCA is the sector indicated by the dashed lines in FIGURE 6F. An XMM pointing of similar (but smaller due to CCD gaps) regions also detected a velocity gradient of $(2.8 \pm 1.7) \times 10^{3}$ km/sec.[34] The precision of velocity measurements in that observation was limited by the loss of exposure due to extended flare periods. Although each of the cited individual measurements is not statistically conclusive, *per se*, the fact that different instruments with different systematics are showing velocity differences in the same regions suggests that the presence of velocity gradients is robust.

We show the projected distribution of the Si/Fe ratios in FIGURE 6E. The importance of this ratio is that it allows us discriminate of SN type enrichment, serving as a fingerprint for the gas enrichment history. This is because SNe Ia and II produce different amounts (yields) of different elements. Therefore, we can use elemental abundance ratios to determine the contribution from SN Ia and II to the X-ray emitting plasma.[35] For example, classical SN Ia models predict that the ratio of O to Fe abundances (by number normalized to Solar) should be about 0.04, whereas SN II models predict the same ratio to be about 3.8. Thus, if the measurement of the O/Fe ratio in the IGM is 0.5, it would mean that the SN Ia iron mass fraction is 88%, and the rest (about 12%) would have been produced by SN II. The enrichment history depends on the previous internal characteristics of the merging systems, such as amount of gas stripped by galaxies, SN Ia winds from the central galaxy, strength of protogalactic SN II dominated winds, and so forth. If the edges mentioned above were all generated by some internal phenomena, not connected to a merger, we should see no asymmetry in the distribution of Si/Fe ratio. However, what we see is exactly the opposite. The cluster core sits on the border of a significant separation of media with different enrichment histories. Toward the E the ratio goes from 1.5 near the X-ray arm to 3.5 at 3′ from the center corresponding to 0–65% SN Ia contamination. Toward the W the ratio decreases from 1.0 near the cluster core to 0 at 1.5′ away. This corresponds to 85–100% SN Ia contamination. The difference is significant at more than 90% confidence level and provides direct evidence that the X-ray arm is a surface discontinuity between two different media. The SN Ia enriched "bullet" is trying to push its way through a SN II enriched medium.

The projected temperatures derived for this cluster seem unusually low for such a violent merger. However, two factors must be considered. The first is that the thermalization of the kinetic component could be altered by a pre-existent velocity gradient (gas circulation) from a previous merger. The second is that we are determining projected quantities and since the merging is happening near the line of sight, the effects of projection become more important and we are likely to underestimate the derived gas temperatures. By looking at the temperature map, we can see that the shocked region (E–SE) extends to the South and has a peak directly E of the sharp edge. Selecting the region with the maximum temperature and using a 2-T model we obtain a best-fit hot component $T_{high} = 6.5^{+2.8}_{-1.2}$ keV. The cool component is $T_{low} \sim 2$ keV.

Such high bulk velocities lead to an underestimate of the X-ray derived total mass under the assumption of hydrostatic equilibrium, since there is an extra non-thermal pressure not taken into account. This underestimate of the X-ray derived mass when compared to independent mass estimators can be used as a rough indicator of velocity gradients, if systematics are taken into account. One popular independent method to determine cluster masses is gravitational lensing. Typically, lens modeling methods rely on the detection of as many lensed background galaxies as possible, and the inversion of the lensing effect of the cluster to recover its mass distribution. Lensing techniques are used to determine the projected mass and are sensitive to uncertainties in the three-dimensional configuration, such as elongations on the line-of-sight.

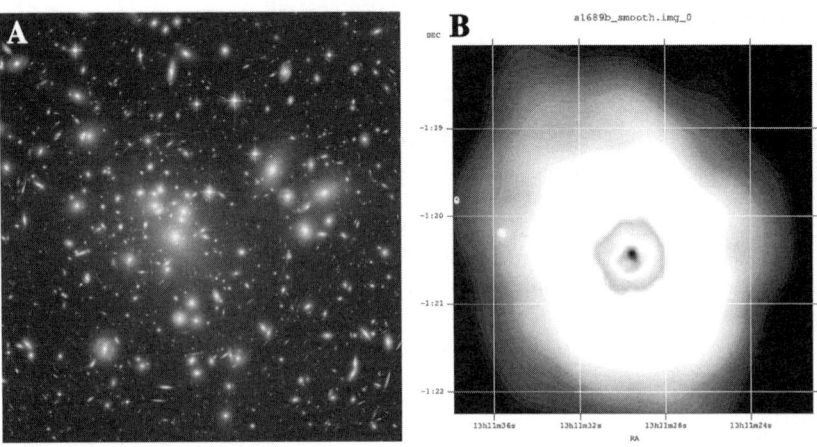

FIGURE 7. (**A**) ACS-WFC image of the lensing cluster Abell 1689 ($z \sim 0.18$). The image is 200″ on one side. (Reproduced from Ref. 36 with permission.) (**B**) Chandra image (10ksec) of A1689, roughly on the same scale.

X-ray and lensing masses do not always agree; often the mass determined through X-ray analyses fall short when compared to those determined through lensing (hereafter X-ray-lens discrepancy). Interestingly, one of the most persistent discrepancies happens to be in the most powerful lens known, A1689[38,39] (see FIGURE 7) Given its optimal physical characteristics, A1689 is a natural laboratory for testing the consistency of different techniques for mass measurements and their limitations. Miralda-Escude and Babul[40] were the first to notice a discrepancy by a factor of two in the cluster masses determined through lensing and X-ray. The authors suggested several explanations for the lower X-ray mass, among them multitemperature gas and nonthermal pressure support, such as intracluster gas bulk motions in the center. Allen[41] was able to reproduce the mass derived from lensing in most of the clusters of his sample, using a combination of ASCA and ROSAT HRI observations and elaborated techniques to compensate for the overestimation of the X-ray core radius sizes. This, however, still could not reconcile the discrepancy in A1689. Andersson and Madejski[39] analyzed A1689 with XMM and confirmed the factor of two discrepancy. However, they performed a velocity analysis, similar to those described above and found a strong velocity gradient of about 5,000 km/sec near the clusters center, indicating that the cluster is in the initial stages of a major merging in the line-of-sight.

The frequency of high velocity gradients is higher than expected theoretically, given the small number of clusters observed. Pawl, Evrard, and Dupke[42] have used the virtual cluster exploratory database of mock clusters from a ΛCDM cosmological simulation to analyze the frequency and magnitude of velocity gradients in the intracluster gas. They found that the likelihood to find velocity gradients greater than Δv decreases as $(\Delta v)^{-4}$, and that only 6% of clusters showed velocity gradients larger than half of the sound speed (see FIGURE 8). Given the current observation difficulties involved in the analysis of intracluster velocities it is likely that the sample of clusters analyzed for velocity gradients so far is biased toward the richest, brightest,

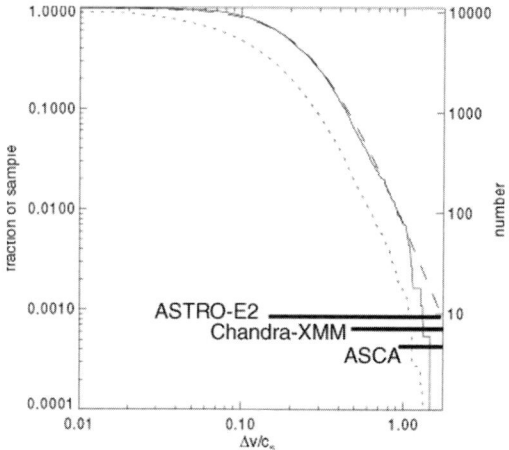

FIGURE 8. Cumulative likelihood of velocity gradients normalized by the ICM sound speed for the maximum velocity difference in a cluster (*solid line*) and for all regions analyzed in all clusters (*dotted line*). (Reproduced from Ref. 41 with permission.)

and most likely to be undergoing strong dynamical processes. In the next few years the study of velocity gradients in intracluster gas is likely to become a major source of information about the nature of cluster formation and the systematics involved in using them as cosmological probes. Larger archival searches with Chandra and XMM will be able to provide better observation constraints to the high velocity tail of the expected frequency of velocity gradients. ASTRO-E2 impressive energy resolution (sensitive to velocity uncertainties of about 100 km/sec) will provide unprecedented bulk velocity measurements as well as turbulent motions through extrathermal line broadening.[43] However, ASTRO-E2 performance will benefit greatly from new Chandra and XMM observations specifically tailored for velocity studies. This is because ASTRO-E2 lifetime is short (two years) and the X-ray calorimeter has a small field of view (2.9′ × 2.9′) and collecting area and, without prior knowledge of the overall velocity distribution, it will need a large number of observations to map a single cluster.

Cold Fronts

Another indication of dynamic activity in clusters is given by the frequent presence of cold fronts. Cold fronts are sharp surface brightness discontinuities characterized by a jump in gas temperature, accompanied by a decline in gas density, such that the gas pressure remains continuous across the front. These characteristics make cold fronts different from bow shocks and the origin for their formation remains unclear. The initial explanation for cold fronts is associated to subsonic (transonic) motions of accreted substructures,[24,25,44] such as gas clumps or small galaxy groups with suppressed thermal conduction. In FIGURE 9A we show schematics for this model from Markevitch *et al.*[24] The top image shows the previous merger epoch. The bottom one shows the stage where cold fronts are seen. Shock fronts have propagated to the cluster outer regions, and the dense core keeps moving through the shocked gas without mixing.

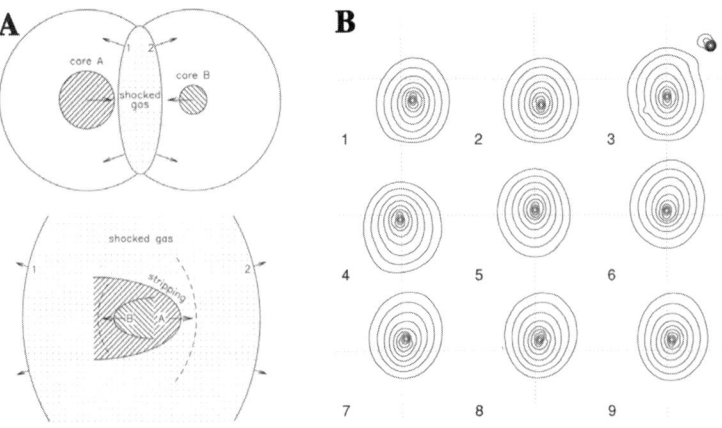

FIGURE 9.

Models that suggest merger core remnant, such as that described above (external models), are theoretically justified[45,46] and hold relatively well for clusters that have clear signs of merging, such as 1E0657-56[26] and A3667.[44] However, they do not work well for the multiple cold fronts discovered in clusters that do not show other strong merging signs, such as A496[7] and A1795.[25] This inconsistency has led to the creation of alternative models for cold fronts that involve internal mechanisms, such as oscillation of the cD plus low entropy gas around the bottom of the potential well[7,47,48] and gravitational field oscillation due to scattering of smaller systems.[49,50] We show a plot from Tittley and Henriksen[49] (FIG. 9B) that illustrates the creation of multiple cold fronts due to the passage and deflection of another system, for illustration. The incoming group approaches from the left in the first two frames (each frame is 50 Myr apart and is 200 kpc in size) and is scattered downwards continuing out of view. In frame 3 closest approach happens (about 100 kpc). The resulting compression of the isodensity contours due to the field oscillation (dark matter and baryonic matter become detached) happens several frames after the deflection, when the deflected group is far away from the core. External models such as this have the advantage of allowing for the formation of multiple coexisting cold fronts.

As pointed out by Dupke and White,[7] internal and external models for cold front formation can be discriminated through the analysis of SN Ia/II contamination (or relative metal enrichment) across the front. If the cold front generation mechanism is external we should expect that the front was also be accompanied by a discontinuity of metal enrichment histories. The gas enrichment history can be determined through the measurement of an ensemble of metal abundance ratios as explained in the previous section. It is unlikely that both metal abundances and abundance ratios would conspire to fake a smooth distribution during merging. Also, if the accreted cluster is the core of a "non-cooling flow" cluster its metal abundance would be that of the average value for clusters (0.3 solar) making it easily distinguishable from the metal enriched central region of the main cluster. If, on the other hand, the cold front is generated by internal mechanisms one would not expect strong chemical gradients at the cold front, except maybe for those associated with the cluster itself, such as radial abundance gradients.

The analysis of heavy element distribution in cold fronts is a powerful tool to understand their nature and currently has to be done with (or in addition to) *Chandra*, given its combination of very high angular and spectral resolutions. Dupke and White[7,47] performed a chemical analysis of the cold front in A496. This is an apparently well behaved nearby bright cluster at a redshift of 0.033. It does not show any obvious signs of merging. At least two sharp edges are seen in the X-ray surface brightness and projected images (see FIGURE 10A). Both edges can also be seen easily in the temperature maps (FIG. 10C). The North edge is characterized by a jump in temperature that can be as high as 1.6 times, accompanied by a density reduction of about 1.8, with a continuous radial decline in pressure (nkT). The gas temperature continues to grow radially in that direction achieving a maximum of about 6 keV within the region analyzed. The overall characteristics are typical of cold fronts. The temperature distribution in the southern edge is anisotropic. The temperature jumps by about 1.3 times to at least 4.0 keV toward the S–SE and the density drops by about 1.7 times over the same spatial scale, consistent with other cold fronts. To the SW

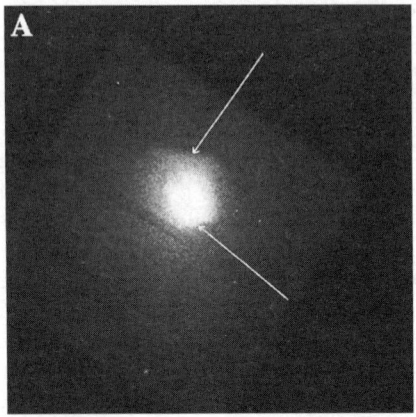
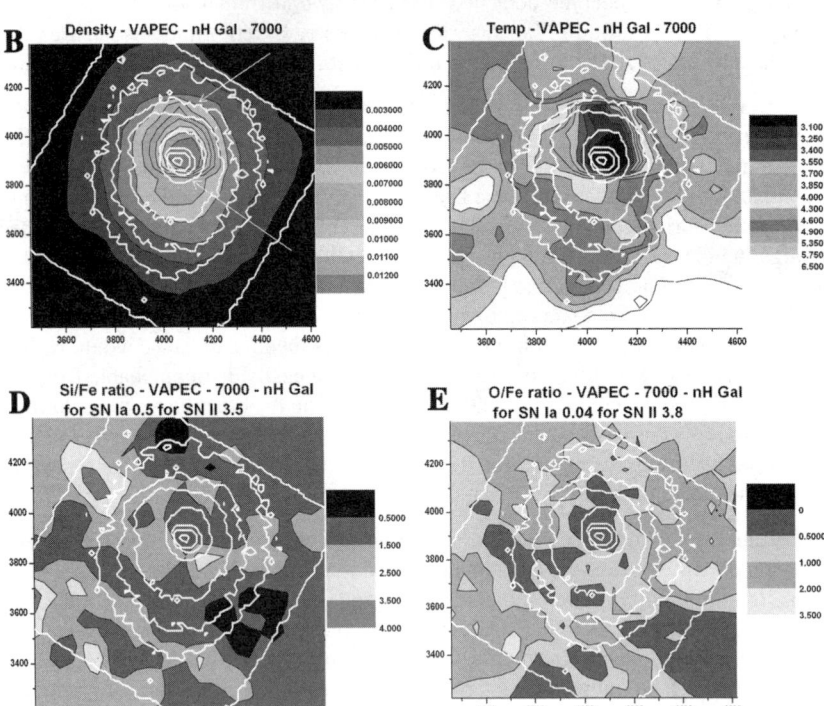

FIGURE 10. Contour maps of the distribution of the interesting parameters obtained through spectral fittings using an absorbed VAPEC thermal emission model. The best fit results for individual points in the map are a result of adaptive smoothing with more than 7,000 cnts/cell and a minimum smoothing radii of 20 pix (10″). The color steps were chosen in a way as to approximate the 1σ errors. **B–D** show the distribution of gas density (cm^{-3}), temperatures (keV), Si/Fe, O/Fe, and redshifts. The outermost square contour shows the CCD border and measurements outside this should be ignored.

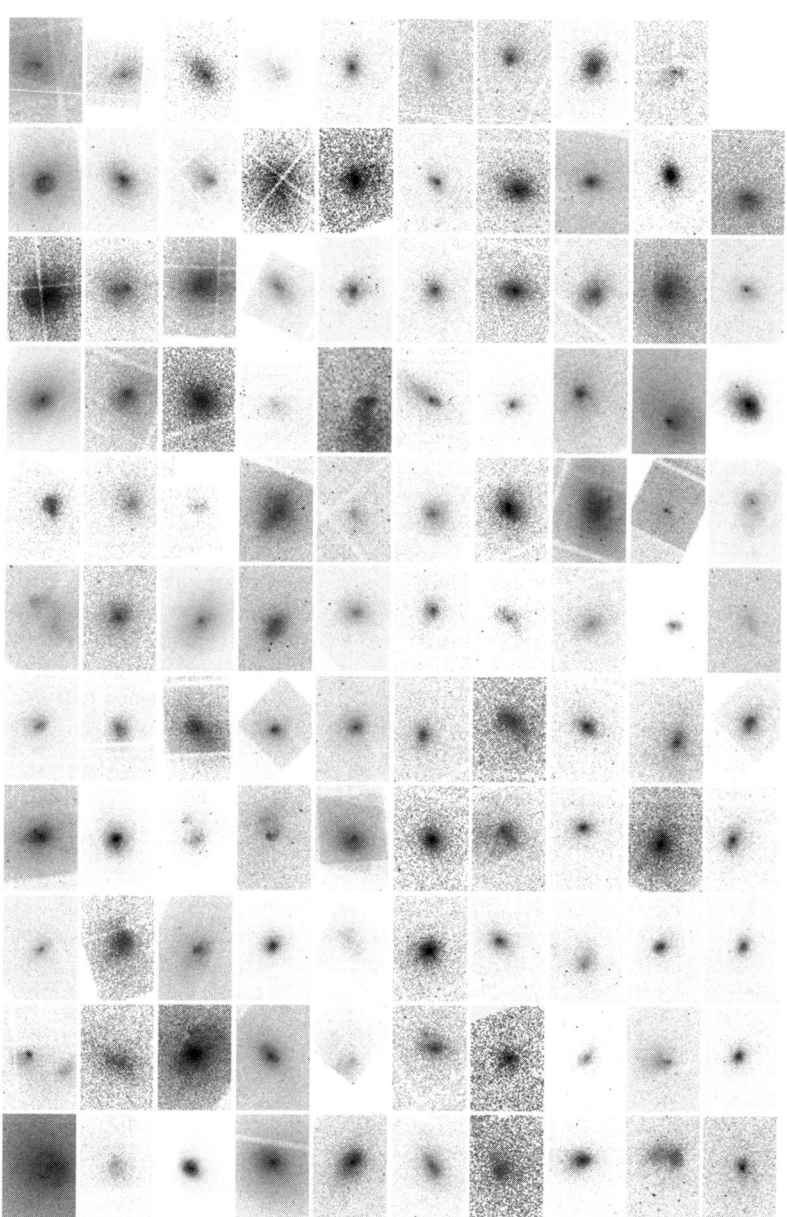

FIGURE 11. Clusters in the archive with easily noticed core edges.

the temperature has a stronger gradient, reaching 4.6 keV (a near 1.6 enhancement) for the same density decline, creating an isobaric tail.

The distribution of Fe, Si, and O abundances showed overall radial gradients, but there were no clear discontinuities related to the cold fronts. The only discontinuities across the cold fronts were consistent with general radial gradients. Both O/Fe and Si/Fe are consistent with a central dominance of SN Ia ejecta (FIG. 10D and E). Within the errors, there was no major change in abundance ratio distribution across the sharp boundaries of the cold front. This is compatible with an *internal* mechanism to generate cold fronts, for example, the cD is oscillating around the cluster potential well. In this model, the cD would be dragging/smearing SN Ia iron enriched cold gas within the spatial oscillation length (about 80 kpc), which defines the distance to cold front.

The velocity map in the regions where it can be well constrained (near the core) suggests that the region of the high temperature tail (S–SW) sits on a lower-than-average velocity (by more than 300 km/sec at 1σ). This suggests that the temperature tail represents stripped material with residual velocities along the line-of-sight towards the observer and also North (up) (Z–Y directions). The moving core, in this scenario, turned around (N) and is currently moving away from the observer.

There are now a large number of clusters observed with Chandra and a significant fraction of them show signs of sharp edges near the core (see FIGURE 11). If most of these edges are real cold fronts it seems unlikely that merging of subclumps is the only way to make them. This is because of their high frequency (near 50%) and the lack of intermediate stages in the accretion process (e.g., before core passage). However, all the mechanisms proposed involve some dynamic activity, either with mergers, scattering of subclumps or sloshing of cold gas and/or cD in the bottom of the potential well. Combined with temperature, density, pressure, and entropy distriubtions, a systematic SN Type enrichment study of cold fronts will provide fundamental clues as to their origin and impact on the dynamics of the intracluster gas.

ACKNOWLEDGMENT

I thank Steve Gottesman for the excellent work in organizing the conference on which this volume is based and for allowing Henry's friends and collaborators to honor him one more time. I also thank Joel Bregman, John Arabadjis, Narciso Benitez, and Nestor Mirabal for helpful discussions. This work was partially supported by NASA Grants 603-4162X, NNG046H856, AR4-5013X, and G04-5145X.

REFERENCES

1. WHITE, S.D., J.F. NAVARRO, A.E. EVRARD & C.S. FRENK. 1993. Nature **366**: 429.
2. SCHRAMM, D. & M. TURNER. 1998. Rev. Mod. Phys. **70**: 303.
3. BAHCALL, N.A., J.P. OSTRIKER, S. PERLMUTTER & P.J. STEINHARDT. 1999. Science **284**: 148.
4. ALLEN, S., R. SCHMIDT & A. FABIAN. 2002. Mon. Not. Roy. Astr. Soc. **334**: L11.
5. RICKER, P.M. 1998. Astrophys. J. **496**: 670.
6. SARAZIN, C. 2004. astro-ph/0406181.
7. DUPKE, R. & R. WHITE. 2003. Astrophys. J. **583**: 13.

8. EVRARD, A.E. 1990. Astrophys. J. **363**: 349.
9. KATZ, N. & S.D.M. WHITE. 1993. Astrophys. J. **412**: 455.
10. ROETTIGER, K., J. BURNS & C. LOKEN. 1993. Astrophys. J. **407**: 53.
11. ROETTIGER, K., J. BURNS & C. LOKEN. 1996. Astrophys. J. **473**: 651.
12. SCHINDLER, S. & E. MULLER. 1993. Astron. Astrophys. **272**: 137.
13. PEARCE, F.R., P.A. THOMAS & H.M.P. COUCHMAN. 1994. Mon. Not. Roy. Astr. Soc. **268**: 953.
14. TAKIZAWA, M. & S. MINESHIGE. 1998. Astrophys. J. **499**: 82.
15. TAKIZAWA, M. 1999. Astrophys. J. **520**: 514.
16. TAKIZAWA, M. 2000. Astrophys. J. **532**: 183.
17. NAVARRO, J.F., C.S. FRENK & S.D.M. WHITE. 1995. Mon. Not. Roy. Astr. Soc., **275**: 720.
18. EVRARD, A.E., C.A. METZLER & J.F. NAVARRO. 1996. Astrophys. J. **469**: 494.
19. ROETTIGER, K., C. LOKEN & J.O. BURNS. 1997. Astrophys. J. **109**: 307.
20. GÓMEZ, P.L., C. LOKEN, K. ROETTIGER & J.O. BURNS. 2002. Astrophys. J. **569**: 122.
21. MOTL, P.M., J.O. BURNS, C. LOKEN, et al. 2004. Astrophys. J. **606**: 635.
22. ROETTIGER, K. & FLORES. 2000. Astrophys. J. **538**: 92.
23. MACHACEK, M., M.W. BAUTZ, C. CANIZARES & G.P. GARMIRE. 2002. Astrophys. J. **567**: 188.
24. MARKEVITCH, M., et al. 2000. Astrophys. J. **541**: 542.
25. MARKEVITCH, M., A. VIKHLININ & P. MAZZOTTA. 2001. Astrophys. J. **562**: L153.
26. MARKEVITCH, M., A.H. GONZALEZ, L. DAVID, et al. 2002. Astrophys. J. **567**: 27.
27. DUPKE, R.A. & J.N. BREGMAN. 2001. Astrophys. J. **547**: 705.
28. DUPKE, R.A. & J.N. BREGMAN. 2001. Astrophys. J. **562**: 266.
29. DUPKE, R.A. & J.N. BREGMAN. 2005. Astrophys. J. Submitted.
30. DUPKE, R.A. & J.N. BREGMAN. 2005. Astrophys. J. Submitted.
31. RINES, K., M.J. GELLER, A. DIAFERIO, et al. 2000. Astrophys. J. **120**: 2338.
32. MOHR, J., M.J. GELLER, D.G. FABRICANT, et al. 1996. Astrophys. J. **470**: 724.
33. KEMPNER, J. & L. DAVID. 2004. Astrophys. J. **607**: 220.
34. STEVENS, I., D. ACREMAN & T. PONMAN. 1999. Mon. Not. Roy. Astr. Soc. **310**: 663.
35. DUPKE, R.A., N. MIRABAL & J.N. BREGMAN. 2005. In preparation.
36. LOEWENSTEIN, M. & R. MUSHOTZKY. 1996. Astrophys. J. **466**: 695.
37. BROADHURST, T., et al. 2005. Astrophys. J. **621**: 53.
38. XUE, S. & X. WU. 2002. Astrophys. J. **576**: 152.
39. ANDERSSON, K. & G. MADEJSKI. 2004. Astrophys. J. **607**: 190.
40. MIRALDA-ESCUDE, J. & A. BABUL. 1995. Astrophys. J. **449**: 18.
41. ALLEN, S. 1998. Mon. Not. Roy. Astr. Soc. **296**: 392.
42. PAWL, A., A. EVRARD & R. DUPKE. 2005. Astrophys. J. Submitted.
43. SUNYAEV, R.A., M.L. NORMAN & G.L. BRYAN. 2003. Astron. Lett. **29**: 783.
44. VIKHLININ, A., M. MARKEVITCH & S.S. MURRAY. 2001. Astrophys. J. **551**: 160.
45. BIALEK, J. & A. EVRARD. 2002. Astrophys. J. **578**: 9.
46. NAGAI, D. & A. KRAVTSOV. 2003. Astrophys. J. **587**: 514.
47. LUFKIN, E., S.A. BALBUS & J.F. HAWLEY. 1995. Astrophys. J. **446**: 529.
48. DUPKE, R. & R. WHITE. 2005. In preparation.
49. TITTLEY, E. & M. HENRIKSEN. 2005. Astrophys. J. **618**: 227.
50. ASCASIBAR, Y. & M. MARKEVITCH. 2005. In preparation.

Resonance Bands and Binary-Star Formation

NORMAN R. LEBOVITZ

Mathematics Department, University of Chicago, Chicago, Illinois, USA

ABSTRACT: Numerical computations on the evolution of realistically stratified, asymmetric, self-gravitating masses reveal the onset of an instability that is suggestive of, but not decisive for, the formation of a binary star. On the other hand, analysis of the evolution and stability of idealized models consisting of uniform-density ellipsoids has been qualitatively accurate in predicting certain features of the behavior of the more realistically stratified figures. This idealized theory is, therefore, reconsidered in an attempt to isolate additional qualitative features that may be common both to the idealized and to realistic figures, and therefore, serve not only as a guide in formulating issues to be addressed in numerical computations but also as a means of interpreting the outcomes of such calculations. Several such features are isolated in this paper, the most striking of which is the existence and importance of resonance bands of instability encountered by the evolutionary trajectories.

KEYWORDS: binary star; resonance; bifurcation

INTRODUCTION

The numerical computations of Cazes[1] and Cazes and Tohline[2] focus on the secular evolution of a rotating, compressible mass beyond the point where a rotationally symmetric figure loses its stability. The object of their investigations is to test, in the framework of realistically stratified masses, an evolutionary picture outlined originally in the framework of ellipsoidal figures of uniform density. This picture[3,4] outlines a possible route to binary fission that is free from a number of objections that had been advanced against the fission theory. Cazes[1] and Cazes and Tohline[2] indeed find an evolution very similar to that found by Lebovitz,[3] where the context was that of secularly evolving Riemann ellipsoids. Moreover, they also find an instability possessing the symmetry characteristics predicted by consideration of these classical ellipsoids, as discussed by Lebovitz.[4] However, the instability that they found did not result in a binary pair by the time they terminated their integrations, but rather seemed to be undergoing large-amplitude oscillations. These and other features of their calculations are incompletely understood. The computations are arduous and expensive: additional guidance in exploring the dynamical phase space would be helpful for further computations.

A number of qualitative (and even semiquantitative) features of realistically stratified masses have been predicted by the corresponding behavior for idealized models in the form of ellipsoids of uniform density, and have provided reliable guidance for

---

Address for correspondence: Norman R. Lebovitz, Mathematics Department, University of Chicago, 5734 S. University Avenue, Chicago, IL 60637, USA. Voice: 773-702-7329; fax: 773-702-9787.

norman@math.uchicago.edu

the numerical computations in the context of the more realistic models. In particular, the existence (and even the approximate location) of the point of dynamical instability of axisymmetric figures, the nature of the perturbation implicated in the instability, and the tendency for the perturbed figure to assume a triaxial shape are all features of the idealized theory that are found also in the numerical computations of stratified figures.[5,6] We therefore seek clues to the evolution and stability of realistically stratified figures via the corresponding evolution and stability of the idealized figures. Another motivation for this investigation is the following.

It is now known that the stability properties of the Riemann ellipsoids are more intricate than had previously been appreciated.[7,8] It appears that the gravitational–rotational instabilities understood and investigated since the nineteenth century are inextricably mixed with *elliptical instability*, studied in laboratory and terrestrial settings beginning in the 1970s (see, for example, Kerswell[9] for an overview of elliptical instability). However, in the discussions of the stability of these figures by Lebovitz[3,4] and in the standard reference on the Riemann ellipsoids by Chandrasekhar,[10] the stability theory is incomplete in that the elliptical instabilities went undetected. In view of the role the Riemann ellipsoids have thus far played in guiding and elucidating our understanding of the evolution of realistically stratified masses, it seems essential to unravel these intricacies in the stability patterns of the Riemann ellipsoids and their role in the secular evolution. That is the main goal of this paper. Particular emphasis is placed on features that can be described qualitatively and that may, therefore, be of common dynamical origin, both in the context of the idealized models discussed here and in the more realistic models discussed by Cazes and Tohline.[2] These are addressed more fully in the DISCUSSION section, but we list the principal conclusions here:

1. A special evolutionary trajectory is found, toward which other evolutionary trajectories tend, almost independently of initial data.
2. These trajectories encounter numerous narrow bands of instability that can be disruptive for the evolution only if $\varepsilon \ll h^2$, where ε is the ratio of dynamical to contraction timescale and h is the width of the band.
3. Many (although not all) of the plausible trajectories encounter a disruptive instability to a perturbation possessing the kind symmetry favorable to the conclusion of fission.

The plan of the paper is as follows. In the next section we present background relating to Riemann ellipsoids and the classical version of the fission theory of binary star formation needed for understanding the remainder of the paper. This is necessarily in shortened form, and references are given to more extensive discussions. We then discuss the stability mechanisms in the context of the evolutionary problem at hand. The principal calculations are presented and the final section is devoted to an interpretation of the results of the calculations as they relate to the modified version of the fission theory proposed.

BACKGROUND

The Riemann ellipsoids represent an idealized family of self-gravitating masses possessing rotation, as well as internal motions of uniform vorticity, and characterized by a density that is spatially constant. Here we confine attention to a subclass referred to by Chandraskhar[10] as *S-type* ellipsoids. For S-type ellipsoids, the angular velocity and the vorticity are both in a single direction that agrees with one of the principal axes of the ellipsoid (we take this to be the z axis). Rectangular axes are chosen in the rotating frame to lie along the principal axes of the ellipsoid. For those S-type ellipsoids that are in a steady state, the velocity field V relative to the rotating frame is

$$V = \lambda\left(\frac{ay}{b}, -\frac{bx}{a}, 0\right), \tag{1}$$

where a, b, and c are the semiaxes of the ellipsoid along the x, y, and z directions, respectively. It was pointed out some years ago[3,4] that a plausible generalization of the classical fission theory of binary-star formation could be formulated within the framework of the steady-state S-type ellipsoids. The steady-state figures among the S-type ellipsoids can be represented in the plane of c/a versus b/a, as in FIGURE 1. As noted in the caption to that figure, each point represents two steady-state solutions. For one of these $|\omega| > |\lambda|$ and for the other this inequality is reversed. In this paper we restrict consideration to the first of these: for this case, rotation dominates internal motion and we regard this, at least provisionally, as the more physically plausible of the two.

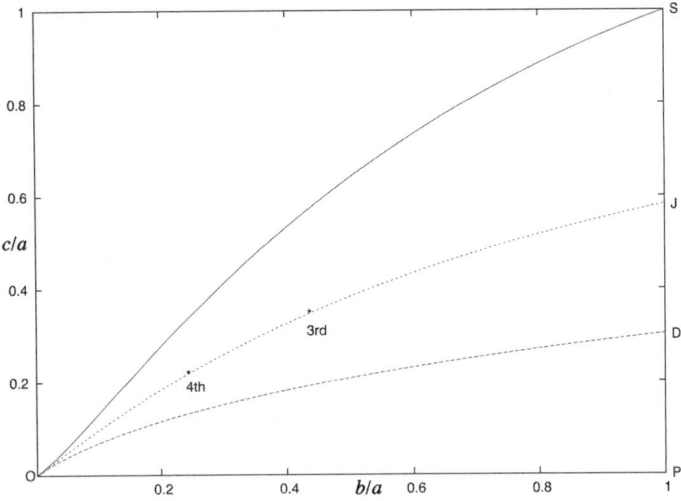

FIGURE 1. The abscissa and ordinate are b/a and c/a, respectively, and in this diagram they lie in the interval [0, 1]. Without loss of generality, it is assumed that $b/a \leq 1$, and it then follows that, for any steady-state S-type Riemann ellipsoid, $c/a \leq 1$ as well. The domain of occupancy of these steady-state S-type ellipsoids is bounded by the curves SDOS. Each point represents two physically distinct figures that can be obtained from one another by interchanging the angular-velocity and vorticity parameters ω and λ. The *direct* figures are (by definition) those for which $-\omega < \lambda < \omega$, and the *adjoint* figures are those for which ω and λ are interchanged in these inequalities. The boundaries SO and DO represent the self-adjoint families, the Maclaurin spheroids lie on the line SP, and the curve JO represents the Jacobi family. In the classical fission theory, the evolution is supposed to take place along the arcs SJ and JO.

The classical fission theory adumbrated by Kelvin (see Thomson and Tait[11]) and fleshed out by Poincaré, Lyapunov, and others at the end of the nineteenth and beginning of the twentieth centuries was formulated for rigidly rotating fluid masses. The rigidly rotating fluids among those depicted in FIGURE 1 are the Maclaurin spheroids (lying on the vertical line SP) and the Jacobi ellipsoids (lying on the curve JO). The restriction to the Maclaurin and Jacobi figures was consistent with the belief held at that time that the fluid viscosity was high enough to enforce rigid-body rotation. The theory envisioned a slow evolution along the Maclaurin curve SP until an instability, stimulated by viscosity at the point J, caused further evolution to take place along the Jacobi curve, which is stable near the point J. Then—the theory ran—a subsequent instability, also stimulated by viscosity, along the curve JO resulted in the gradual breakup of the figure into a pair of detached masses orbiting each other.

The physics underlying Kelvin's outline was based on the energetics of contracting masses; it is briefly recalled below. However, the mathematical realization of fission along the lines outlined by Kelvin required the solution of a series of difficult mathematical problems. Poincaré[12] formulated these problems and carried them a certain distance, identifying the point marked "3rd" in FIGURE 1 as the critical point where the Jacobi family underwent a bifurcation and the transition toward a double star commenced. Concerning the additional mathematical issues that needed to be settled, Poincaré made a series of conjectures that he left to others to resolve. These mathematical issues were indeed settled during the next four decades with the result that Poincaré's conjectures were disproved; this discredited the theory. The monograph by Lyttleton[13] was in fact intended to dispose of this theory once and for all. Although this disposal effort never fully succeeded, the introduction to the Lyttleton monograph is a succinct summary of the situation as it looked at that time.

THE PHYSICAL INSTABILITIES

The role of stability theory in the slow evolution is envisioned as follows. The evolution begins at some initial instant in time with a particular steady state configuration. If this initial configuration is axially symmetric and no perturbations leading to departures from axial symmetry are allowed, the evolution will continue through a succession of axially symmetric figures. If perturbations from axial symmetry *are* allowed and the evolutionary family reaches a figure that is unstable to the perturbation in question, further evolution will then proceed along a new family of figures that is no longer axially symmetric. We envision the perturbation in question as arising from uncertainty in the initial data. In other words, the perturbation is present from the outset but remains small until the underlying axially symmetric family reaches that member of it that is unstable to the perturbation in question. Similarly, if the initial configuration is ellipsoidal in shape (a Riemann ellipsoid), it then remains a Riemann ellipsoid until the evolving family takes on the form of a Riemann ellipsoid that is unstable to some perturbation. We again envision this perturbation as arising from uncertainty in the initial data.

This view is not the only viable view of the role of stability theory. One could regard the evolving figure as subject to essentially random perturbations throughout the course of its evolution. However, in view of the actual uncertainties in the initial data, and in order to fix ideas, we implicitly adopt the view that the perturbations are present from the initial instant.

There are two physically distinct instability mechanisms at work in this problem: (1) a gravitational–rotational instability known and studied since the nineteenth century and (2) elliptical instabilities, first isolated about thirty years ago and explored since that time.

Gravitational–Rotational Instabilities

In a mass of uniform density, any motion that leaves the boundary undisturbed also leaves the gravitational potential undisturbed, so the latter can have no influence on the motion of the fluid elements of the mass. Consequently, in order to find the effect of the perturbation of the gravitational potential on the body, the motion of the free surface must be taken into account. In the case of a spherical figure, the motion of the free surface is conveniently described with the aid of spherical harmonics. In the case of an ellipsoidal figure, it is common to use ellipsoidal harmonics.

Kelvin's picture of fission began with a rotationally symmetric mass of uniform density, rotating like a rigid body (a Maclaurin spheroid). He imagined the mass cooling slowly, that is, radiating energy and contracting, thereby lowering its gravitational potential energy. However, the evolution takes place at constant angular momentum J. Since its moment of inertia I decreases as it contracts, while its angular momentum remains constant, its rotational kinetic energy $J^2/2I$ increases. This partially offsets the energy decrease due to lowering of the gravitational energy. If the initial rotational kinetic energy is small, this effect is small at first. However, at a certain point during the evolution, the two effects become comparable and the fluid mass can no longer lower its energy as effectively by an axisymmetric contraction. At this point, a change in the shape of the figure from rotationally symmetric to ellipsoidal (or bar-shaped) is more effective: it somewhat increases the gravitational potential energy but it decreases the rotational kinetic energy by a greater amount, by increasing the moment of inertia. At this point in the evolution, Kelvin reasoned, the Maclaurin spheroids would become unstable to a bar-shaped deformation, provided viscosity, or some other form of energy dissipation, is present. This proviso is needed since the Maclaurin spheroids are dynamically stable at the point J of FIGURE 1, that is, stable in the absence of viscosity. That they indeed become unstable at J in the presence of viscosity was only proved in 1963 by Roberts and Stewartson.[14]

The Elliptical Instability

Imagine a fluid—water, for example—filling a container that is rotationally symmetric about a vertical axis. A rigid-body rotation is a steady state for such a system, and is known to be stable under laboratory conditions. Now suppose the cross-section of the container is perturbed slightly, into the shape of an ellipse. There is a corresponding steady state in which the fluid particles flow on ellipses similar to the boundary ellipse; that flow is given exactly by Equation (**1**). Even if the departure from rotational symmetry is very small—so that both the boundary and the particle trajectories differ only slightly from circles—this flow may be unstable. Whether it is stable or unstable depends on a second, vertical aspect ratio giving the ratio of the vertical extent of the container to one of the horizontal semiaxes, or to the mean radius of cross-section. The reason for this is that the onset of the linear instability requires a one-to-one frequency resonance, and the vertical aspect ratio serves as a tuning parameter for these frequencies. These elliptical instabilities have been investigated both theoretically and experimentally since the 1970s. Each instability is associated with a certain critical value of the vertical aspect ratio. Several reviews of these instabilities are available.[9,15]

The fluid perturbations in the laboratory problem are constrained to preserve the shape of the figure. The case of the Riemann ellipsoids is somewhat different from the laboratory case in that the boundary of the "container" is free rather than fixed, but these elliptical instabilities play a role for them as well. Since the vertical aspect ratio (taken here to be c/a) goes through large changes during a secular evolution, many of the elliptical instabilities must be triggered as an evolutionary trajectory passes through critical values.

The trajectories pass through bands of instability, but this by itself does not mean that anything dramatic happens to these figures. Clearly, in the limit when the thickness of the band is zero, there is no effect at all. We need a criterion for the instability to set in in a manner that can seriously affect—let us say disrupt—the evolutionary family of quasisteady states.

Criterion for a Disruptive Instability

Suppose an evolutionary trajectory passes through a resonant band of instability at a point where the thickness of the band is h. Measured on the dynamical timescale, the rate of passage through the band is ε, so we may estimate the time interval that the trajectory spends in this band as $t_{band} = h/\varepsilon$. In order for this visit to the unstable band to be even minimally effective, this time interval must equal t_{efold}, the e-folding time at the center of the band. Therefore, we need an estimate for this.

An example of a mechanical system possessing such a band of instability is furnished by the simple, gyroscopic system

$$\ddot{u} - 2\dot{v} + au = 0, \qquad \ddot{v} + 2\dot{u} + bv = 0. \tag{2}$$

The time dependence $e^{i\sigma t}$ is characterized by exponents σ such that

$$\sigma^2 = \frac{1}{2}(4 - a - b \pm \sqrt{(4-a-b)^2 - 4ab}).$$

Choosing a fixed value for $h > 0$ and setting $a = h/2 - t$ and $b = -h/2 - t$ one easily finds that, for the choice of the minus sign in the expression above for σ^2, this quantity takes on negative values for values of the parameter t in the interval $(-h/2, h/2)$. At the center of this instability band, $t = 0$, the value of σ^2 is

$$2 - \sqrt{4 + h^2} \approx -\frac{h^2}{4},$$

so the growth rate at that point is approximately $h/2$, comparable with the width of the interval, which is h as measured by the parameter t. An exact analysis shows that the maximal growth rate occurs not at $t = 0$ but at a nearby value $t \approx -h^2/16$; however, the growth rate at that point is the same as that at $t = 0$ to second order in h.

By constructing simple mechanical models, like that above, of conservative systems possessing such a band structure, one finds that the maximum growth rate at the center of a band is on the order of h, the width of the band, so we estimate

$$t_{\text{efold}} = \frac{1}{h}. \tag{3}$$

A graph of growth rates in an unstable band for the Riemann ellipsoids is reproduced in FIGURE 2, and is seen to be in rough agreement with this choice. It follows then, that for the instability to be effective, $t_{\text{band}} > t_{\text{efold}}$ or $\varepsilon < h^2$. This would appear to be a bare minimum. Not only is it based on the maximum growth rate, but also it provides only for growth by a factor of ε. If $\varepsilon = h^2$ and the perturbation is very small at the point where the trajectory enters the band, it is about a factor of e larger at the point where it exits, the effect being to change a very small perturbation into one that is still rather small. It seems natural, therefore, to expect that for the instability to be capable of producing a serious, nonlinear perturbation, that is, capable of disrupting the original quasisteady figure and transforming it into something substantially different, we need the criterion

$$\varepsilon \ll h^2. \tag{4}$$

In the absence of a satisfactory mathematical theory of the slow passage through instability bands, we adopt this as the criterion.

This criterion represents a necessary condition for a dramatic change in the shape of the figure to take place, but not a sufficient one. It is possible that the nonlinear outcome of the instability is to change the velocity field dramatically while changing the shape rather little.

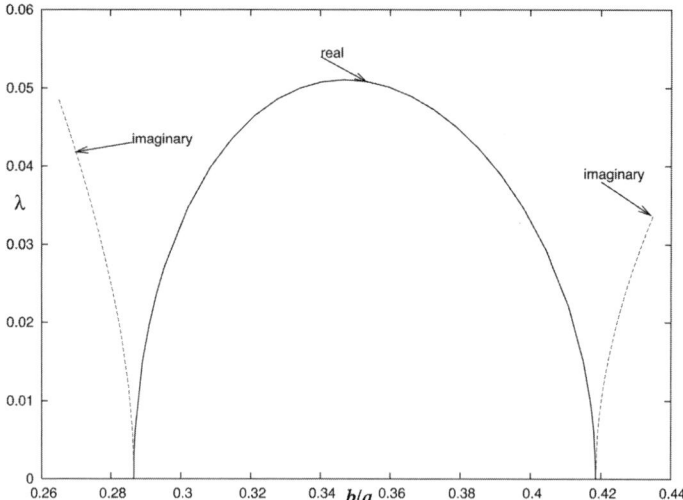

FIGURE 2. An example of the graph of a growth rate across a band of instability. In this case, the abscissa (b/a) represents a portion of the LSA, that part of it near which most evolutionary trajectories must pass (see FIG. 8). The ordinate measures the size of the eigenvalue in the unit $\sqrt{\pi G \rho}$. The eigenvalue is purely imaginary to the right and to the left of the instability band, and purely real inside the band. The width of the band is about 0.14 and its maximum height is about 0.05, in rough agreement with the considerations in the text.

The Full Stability Theory

Although we can distinguish between the two instability mechanisms described above, the Riemann ellipsoids in general mix them up: any normal mode is a combination of a *shape-changing* perturbation and a *shape-preserving* perturbation. Exceptions are the uniformly rotating figures, the Maclaurin spheroids and the Jacobi ellipsoids, for which the two kinds of modes indeed separate (and there is in fact no elliptical instability). However, as emphasized in previous investigations (see for example Lebovitz[16]) we may classify the normal modes according to other criteria. All perturbations can be classified by their polynomial degree. Those for which the velocity perturbation is of degree two in the Cartesian coordinates *include* first- and third-harmonics, which change the shape, but also include the shape-preserving perturbations associated with the elliptical instability. Those of degree three *include* second- and fourth-ellipsoidal harmonics, but also include the shape-preserving perturbations. In the subsequent discussion, we identify the modes by the order of the ellipsoidal harmonics that they include, but this is merely a shorthand and traditional way of referring to modes that include the more general perturbations, which are more properly identified by their polynomial degree.

CALCULATIONS

Trajectories of secular evolution may be obtained by allowing the density to increase while holding fixed the known constants of the motion: mass, angular momentum, and circulation. This we have done for five different sets of initial data, as shown in FIGURE 3. All of the trajectories shown ultimately become indistinguishable, on this scale, from the lower bounding curve marked LSA (for lower self-adjoint family). This tendency for the evolving figures to approach very near to the LSA was observed in similar calculations done many years ago.[3,17]

The steady-state figures are subject to instabilities under perturbations causing the velocity field and the shape of the figure to depart from those of the steady-state figure, and the evolutionary trajectories, which are families of such steady-state figures, encounter regions of instability in the b/a, c/a plane that are not indicated in FIGURE 1. We now proceed to indicate some of these regions.

Third-Harmonics Instabilities

We indicate in FIGURE 4 regions of the ellipsoidal domain in which the steady-state figures are unstable to perturbations that include deformations of the surface described by third ellipsoidal harmonics.

The trajectory A of FIGURE 4 intersects the tongues of instability at points where their thicknesses are about 3×10^{-5} and 10^{-4}; the trajectory E intersects them where their thicknesses are about 3×10^{-4} and 5×10^{-3}. Since the value of ε is about 10^{-6}, only the last of these remains in the tongue long enough for the instability to be effective, by Criterion (**4**).

Consider the possible outcome of the instability for trajectory E as it enters the second of the tongues and remains there for many *e*-folding times. There is, unfortunately, no solid mathematical information concerning the nonlinear outcome of such an instability. There are only educated guesses of the kind that Kelvin employed well over a century ago. In the present case we can speculate based on the origin of the instability. It is an example of an elliptical instability (this is made clear in Ref. 7), which in the laboratory is an instability of the flow rather than one of the shape of the figure. It is possible that the outcome of this instability is a significant change in the flow pattern without a significant change in the shape. In that case we can continue to follow the original trajectory, bearing in mind that flow in the interior is significantly different from that of Equation (1). The free surface must also be altered somewhat, but may not be radically different from the ellipsoidal shape. This is plausible for Riemann ellipsoids that are close to either the Maclaurin or Jacobi lines, since on these lines there is a complete separation between the purely hydrodynamic modes on the one hand and the purely shape-changing modes on the other. It is, therefore, plausible that nearby, the normal modes are mostly of one of these kinds with very little admixture with the other.

However, FIGURE 4 fails to show *all* instabilities associated with the third harmonics. There is a segment of the LSA where there are third-harmonics instabilities, and

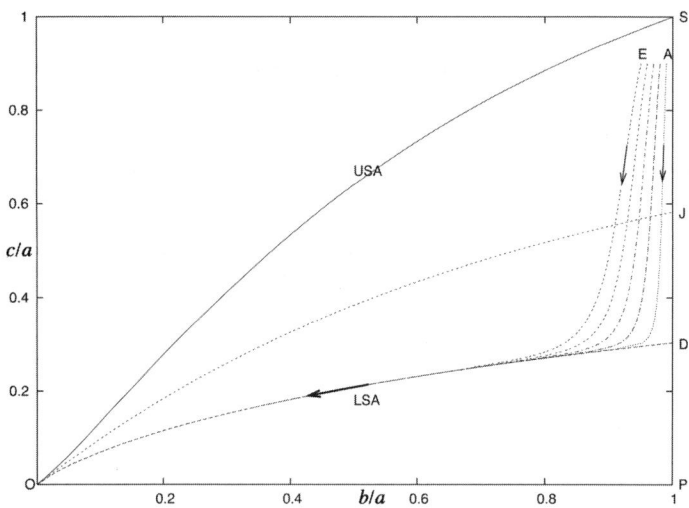

FIGURE 3. The same region of steady-state configurations as that of FIGURE 1, but now five inviscid, evolutionary trajectories are shown. These are quasisteady, that is, they are dynamically steady-state figures for which the density secularly increases while the constants of the motion (mass, angular momentum, and circulation) are held fixed. They are labeled A to E. In each case the figure starts with $c/a = 0.9$, but with five different initial values of b/a: 0.95 (*curve E*), 0.96, 0.97, 0.98, and 0.99 (*curve A*). In each case the density is allowed to increase by a factor of 625. *Arrows* denote the direction of evolution with increasing density. The evolutionary paths depart only slightly from rotational symmetry until the point D of dynamical instability is approached, and then fall very nearly onto the lower self-adjoint family DO (within a vertical distance of 10^{-5} by the time $a_2/a_1 = 1/2$). The role played by the arcs SJ and JO in the classical fission theory are played instead by the arcs SD and DO in the inviscid (modified) theory. The Jacobi family plays no role in this evolutionary scenario.

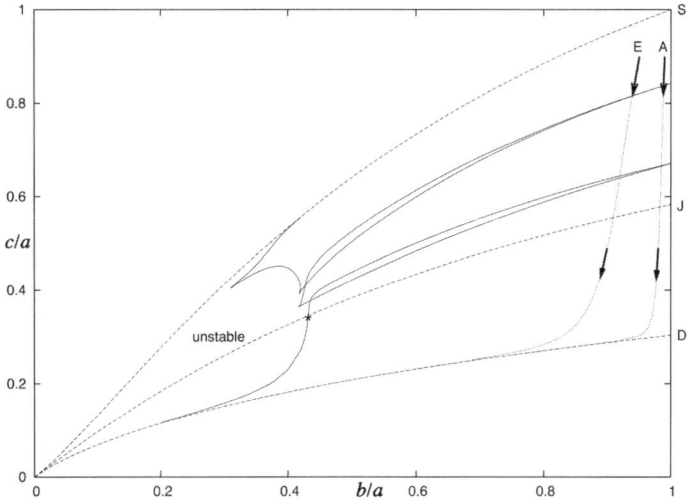

FIGURE 4. As FIGURE 3 with two exceptions: regions of instability to third harmonics are shown and only two of the evolutionary trajectories are shown, those marked A and E above. Note the narrow tongues of instability emanating from points on the Maclaurin line $b/a = 1$. Because the evolving trajectories spend very little time in these tongues, they may have only a slight influence on the subsequent nature of the figures and on their evolution.

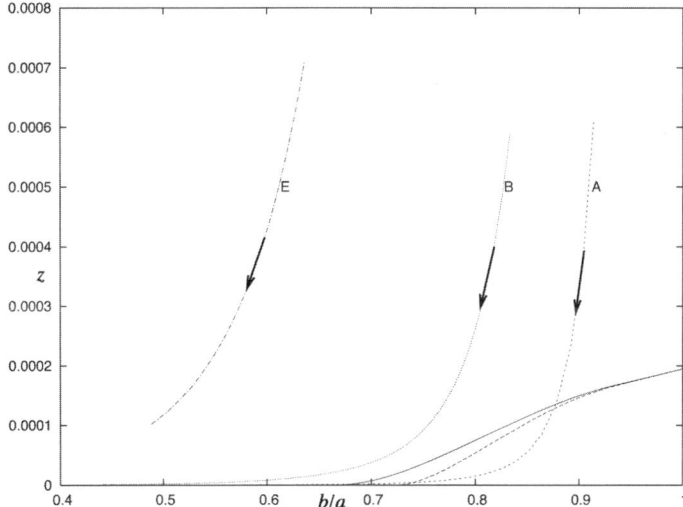

FIGURE 5. Here the LSA has been straightened out, that is, the variable z is not c/a but rather the difference $c/a - (c/a)_{lsa}$, the height above the LSA sequence. This shows the instability tongue for the third-harmonics instability, connecting the band along the LSA with a point on the Maclaurin line less than 2×10^{-4} above the LSA. Of the five trajectories considered, only that marked A is influenced by this tongue.

an extension of this segment into the region of steady-state solutions. Some evolutionary trajectories must encounter this band, or its extension above the LSA. That extension, and some of the trajectories are shown in FIGURE 5. It can be seen that, of the five trajectories calculated, only one of them (marked A), encounters the third-harmonics instability in the region where it is wide enough to disrupt the evolution; the others pass over it and encounter instead a fourth-harmonics instability band (described below). Note how exquisitely close to the LSA this band lies: the vertical axis has been amplified by a factor of 10^4. A similar result was obtained by Lebovitz,[17] although without the guiding knowledge of the existence of the elliptical instability.

We observe that if a figure evolves exactly along the LSA the first instability that it encounters is a third-harmonics instability. This is analogous to the classical situation of Kelvin and Poincaré and may be subject to the same adverse conclusion from the standpoint of the fission theory. However, for such an exact tracking of the LSA to take place, the original figure would have to be exactly axisymmetric. For figures that begin, like that of evolutionary path A, with sufficiently small departure from rotational symmetry, the conclusion is similar. However, figures that begin with slightly larger, but still quite small, departures from rotational symmetry pass over the third-harmonics instability and must meet a different fate.

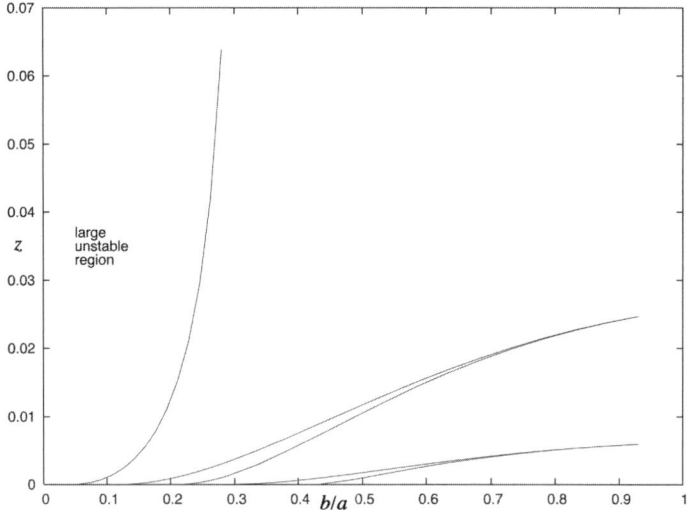

FIGURE 6. The vertical scale is amplified by a factor of 100. The region marked *unstable* is a large domain of instability to the perturbations described by polynomials of degree three (corresponding to ellipsoidal harmonics of order four). The narrow tongues emanating from the abscissa—here representing the LSA—are likewise unstable to the same class of perturbations (fourth-order harmonics). They extend to the Maclaurin line ($b/a = 1$) but become extremely narrow near that line. For example, at $b/a = 0.93$, the thickness of the lower tongue is less than 2×10^{-7}.

Fourth-Harmonics Instabilities

The domain of occupancy of steady-state figures is home not only to third-harmonics instabilities but also to those of the fourth, fifth, and so forth. Here we describe those of the fourth harmonics (i.e., solutions for the velocity perturbation that are polynomials in the cartesian coordinates of degree three).

FIGURE 6 shows the region near the LSA (lower self-adjoint sequence) with the variable z measured from that sequence, and indicates regions of stability and instability to perturbations associated with second and fourth harmonics (polynomials of degree three). The narrow tongues emanating from the LSA in fact extend to the Maclaurin line but become extremely narrow and fail to be picked up by the numerics.

The evolutionary trajectories shown in FIGURE 3 encounter the bands of instability shown in FIGURE 6. Examples of these encounters are shown in FIGURES 7 and 8. It appears from these figures that many, perhaps most, evolutionary trajectories suffer their first disruptive instability to perturbations belonging to the fourth harmonics.

DISCUSSION

This section is separated into two parts: first, the principal qualitative conclusions of interest for pursuing the secular evolution of asymmetric masses in the context of realistically stratified masses and, second, a more general discussion of the strengths and weaknesses of the present formulation.

Qualitative Conclusions

The following remarks represent a distillation of the results presented here, augmented with related remarks:

1. The lower self-adjoint family of S-type Riemann ellipsoids (indicated by LSA in FIG. 3, by the curve OD in FIGS. 1 and 4, and by the abscissa in the remaining figures), represents a kind of attractor for evolutionary trajectories: for a variety of initial data, the evolutionary paths approach this special trajectory. It, therefore, plays a role for an inviscid, secular evolution analogous to that played by the Jacobi family for a viscous, secular evolution—a conclusion that is by no means new to this paper.[3]

2. The regions of instability in the plane of occupancy of the S-type Riemann ellipsoids include narrow, resonant bands, some of which are confined to an extremely narrow region abutting the LSA. The latter are of particular importance since they occur in precisely the region of phase space to which the evolutionary trajectories are attracted.

3. When an evolutionary trajectory encounters a resonant band, the criterion for that encounter to be disruptive is that $\varepsilon < h^2$, where ε is the ratio of dynamical to contraction timescale and h is the width of the band.

4. Many, perhaps most, evolutionary trajectories encounter their first disruptive instability for a fourth-harmonics perturbation, which has a symmetry property that is consistent with fission.[4]

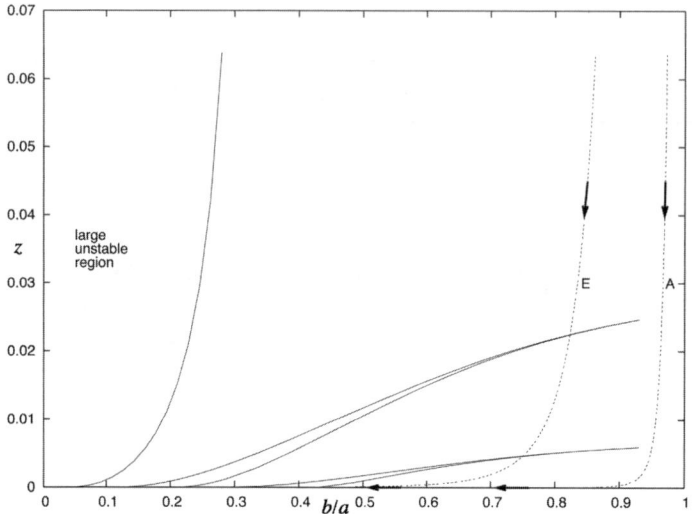

FIGURE 7. The vertical axis has been amplified by a factor of 100. Two of the five evolutionary paths are indicated. They cross the narrow tongues at points that are insignificant for the stability of the paths, but then recross the lower of these essentially on the LSA. Here the instability gap is not at all narrow (it lies roughly in the interval (0.28, 0.42)), and serious consequences for the paths must ensue.

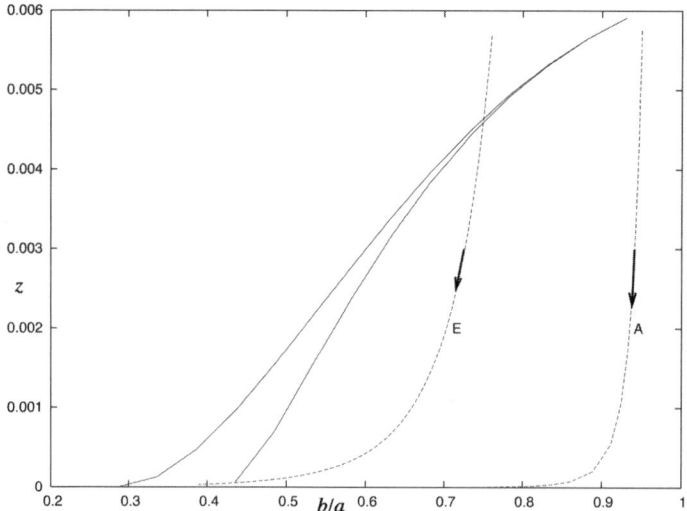

FIGURE 8. A blown-up version of FIGURE 7 (there the vertical scale was amplified by a factor of 100, here by a factor of 1,000) showing only the lower of the two instability tongues associated with fourth-harmonic perturbations. This is the more "dangerous" of the two.

To these conclusions we add observations that refer to the work on numerical computation of secular evolution as described by Cazes[1] and Cazes and Tohline.[2] For the numerical computations of secular evolution of spherical or of axially symmetric families of figures, there are known methods for filtering out the motions on the dynamical timescale in order to follow the secular evolution, that is, the evolution on a longer timescale. For the numerical computations of asymmetric figures, such as those considered by Cazes and Tohline, there is as yet no known, systematic way to filter the dynamical motions so as to be able to choose timesteps based on the contraction timescale. For this reason, the computations are based on the dynamical timescale and the value of ε is greatly increased (i.e., the dynamical timescale is increased or the contraction timescale is decreased) from its actual value (about 10^{-6}) in order to reduce the dynamical timescale to reasonable computational proportions. In Reference 2, the smallest value of ε used is 1/35. This appears to conflict with Criterion (**4**). If the trajectory should encounter a resonant band of width 0.1 (say), it would require a value of ε considerably less than 1/100 to disrupt the quasisteady state and allow the nonlinear development of the instability. This is an example of how the current investigation may serve as a useful guide to the numerical computations of realistically stratified figures.

A second observation relates to the onset of oscillations in the numerical computations. There may be numerical reasons for this, or it could in principle be related to actual oscillations—for example, one might inquire if the instability encountered is an instability to periodic solutions (sometimes called *overstable* oscillations). This latter possibility is *not* borne out by the calculations for the Riemann ellipsoids. The instability occurring in the "dangerous" zone indicated in FIGURES 2 and 8 has zero frequency at the onset of instability and appears to indicate a bifurcation to steady figures.

Strengths and Weaknesses

The curves shown in the diagrams of this paper are of two kinds: (1) the evolutionary trajectories and (2) the borderlines separating regions in parameter space for which the figures are stable to a certain class of perturbation from regions in which the figures are unstable to this same class of perturbation. To keep the diagrams from becoming cluttered, we restrict them to certain classes of perturbations. As discussed elsewhere,[8] the linearized stability equations possess solutions that are represented by polynomials in the Cartesian coordinates. The classes considered there are those belonging to polynomials of degrees two, three, and four; in this paper we go only to degree three, which contains surface deformations given by ellipsoidal harmonics up to order four. It is important to recall, when viewing any *one* of the diagrams, that the fact that a point $(b/a, c/a)$ lies in a stable region of that diagram does not guarantee that the figure it represents is stable, since the same point may lie in an unstable region for a different diagram referring to a different class of perturbations.

Indeed, since in the present paper we go no farther than degree three, we effectively limit the perturbations to those of fairly large wavelength, and the full stability picture is in fact not presented. We know that in the limit of perturbations of asymptotically *small* wavelength, almost all of the parameter space is unstable.[7] However, this last remark ignores *all* effects of viscous dissipation, which must ultimately set in for sufficiently small wavelength, where it suppresses the instability (see, for example, Saffman[18]). Although it may not be sufficient to ignore perturbations of polynomial degree greater than three, it does appear reasonable to make some choice of maximal polynomial degree (and, therefore, minimal wavelength).

A weakness of the picture presented here is that it only presents the nonlinear outcome of the fourth-harmonic instability speculatively, based on the symmetry-preserving character of the perturbation. A full bifurcation calculation, such as that carried out by Lyapunof, Darwin, and others for the Jacobi figure, has not yet been carried out for the point of instability along the LSA. This elaboration is a natural extension of the work presented here.

REFERENCES

1. CAZES, J.E. 1999. The Formation of Short Period Binary Star Systems From Stable, Self-Gravitating, Gaseous Bars. Dissertation submitted to Louisiana State University.
2. CAZES, J.E. & J.E. TOHLINE. 2000. Self-gravitating gaseous bars. I. Compressible analogs of Riemann ellipsoids with supersonic internal flow. Astroph. J. **532:** 1051–1068.
3. LEBOVITZ, N. 1972. On the fission theory of binary stars. Astroph. J. **175:** 171-193.
4. LEBOVITZ, N. 1987. Binary fission via inviscid trajectories. Geoph. Astroph. Fl. Dyn. **38:** 15–24.
5. TASSOUL, J.-L. & J. OSTRIKER. 1968. On the oscillations and stability of rotating stellar models. I. Mathematical techniques. Astroph. J. **154:** 613–626.
6. DURISEN, R.H., R.A. GINGOLD, J.E. TOHLINE & A.P. BOSS. 1986. The binary fission hypothesis: a comparison of results from finite difference and smoothed particle hydrodynamics codes. Astroph. J. **305:** 281–308.
7. LEBOVITZ, N. & A. LIFSCHITZ. 1996. Short-wavelength instabilities of Riemann ellipsoids. Phil. Trans. R. Soc. Lond. A **354:** 927–950.
8. LEBOVITZ, N. & A. LIFSCHITZ. 1996. New global instabilities of the Riemann ellipsoids. Astroph. J. **458:** 699–713.
9. KERSWELL, R. 2002. Elliptical instability. Annu. Rev. Fluid Mech. **34:** 83–113.
10. CHANDRASEKHAR, S. 1969. Ellipsoidal Figures of Equilibrium. Yale University Press, New Haven.
11. THOMSON, W. & P.G. TAIT. 1912. Treatise on Natural Philosphy. Cambridge University Press, Cambridge.
12. POINCARÉ, H. 1885. Sur l'équilibre d'une masse fluide animée d'un mouvement de rotation. Acta Math. **7:** 259–380.
13. LYTTLETON, R.A. 1953. The Stability of Rotating Liquid Masses. Cambridge University Press, Cambridge.
14. ROBERTS, P. & K. STEWARTSON. 1963. On the stability of a Maclaurin spheroid of small viscosity. Astroph. J. **137:** 777–790.
15. GLEDZER, E.B. & V.M. PONOMAREV. 1992. Instability of bounded flows with elliptical streamlines. J. Fluid Mech. **240:** 1–30.
16. LEBOVITZ, N. 1989. Mathematical status of the fission theory. Highlights Astron. **8:** 129–131.
17. LEBOVITZ, N. 1974. On the fission theory of binary stars. II. Stability to third-harmonics disturbances. Astroph. J. **190:** 121–130.
18. SAFFMAN, P.G. 1992. Vortex Dynamics. Cambridge University Press, Cambridge.

The Symplectic Group and Classical Mechanics

ALEX J. DRAGT

*Center for Theoretical Physics, University of Maryland,
College Park, Maryland, USA*

ABSTRACT: The symplectic group is the underlying symmetry group for Hamiltonian dynamics. Yet relatively little is commonly known about its properties including its Lie structure and representations. This paper describes and summarizes some of these properties; and, as a first application of symplectic group theory, provides a symplectic classification of all first-order differential equations in an even number of variables.

KEYWORDS: symplectic group; classical mechanics

INTRODUCTION

Historically there are several mathematical groups that have been studied in detail because of their relevance to our understanding of the physical world. A detailed knowledge of the three-dimensional rotation group is of great use in many areas, including condensed matter physics, chemistry, atomic physics, nuclear physics, and elementary particle physics. Knowledge of the rotation–translation group leads to a classification of crystals and quasicrystals. Knowledge of the unitary group is helpful for quantum mechanics and is essential to understanding quantum information and quantum computing. Knowledge of the Lorentz group leads to the construction of spinors, four-vectors, general tensors, and classical fields. Knowledge of the Poincaré group (the Lorentz group plus translations in space and time) leads to a classification of elementary particles and the construction of quantum fields. Knowledge of various other groups facilitates many-body theory calculations. Finally, there are the various "internal" and/or gauge symmetry groups that play an important role in our current understanding of elementary particles and the fundamental forces.

The symplectic group is the underlying group in classical mechanics for Hamiltonian systems. Yet, in contrast to the groups just mentioned, almost nothing is commonly known or readily available about the symplectic group. For example, many are familiar with aspects of the rotation group, including spin (irreducible representations and how they are labeled) and how spins couple and combine (the Clebsch–Gordan series and coefficients for the rotation group). Yet few have heard or read about representations of the symplectic group, knowledge of its Clebsch–Gordan series is not widespread, and little is known in detail about its Clebsch–Gordan coefficients.

The purpose of this paper is to describe aspects of the finite-dimensional representations of the symplectic group with the hope that this knowledge, like that for

---

Address for correspondence: Alex J. Dragt, Center for Theoretical Physics, University of Maryland, College Park, MD 20742, USA.
dragt@physics.umd.edu

the well-studied groups, will also ultimately prove useful. We first define the symplectic group and its Lie algebra and relate them to Hamiltonian dynamics. We then define Lie operators and general vector fields and present the Cartan form for the symplectic Lie algebra (in the six-variable case). Subsequently, we describe representations of this Lie algebra and, as a first consequence of the Lie-algebraic effort, we find a symplectic classification of all analytic vector fields (in an even number of variables), including those for systems with dissipation.

THE SYMPLECTIC GROUP AND LIE ALGEBRA

Definitions and Theorems

Consider $2n$-dimensional phase space with canonical coordinates $q_1, ..., q_n$ and $p_1, ..., p_n$. Collect them together into $2n$ coordinates $z_1, ..., z_{2n}$ by writing

$$z = (q_1, ..., q_n; p_1, ..., p_n). \tag{1}$$

Recall that if f and g are any two functions of z, their *Poisson bracket* $[f, g]$ is defined by the rule

$$[f, g] = \sum_i \left(\frac{\partial f}{\partial q_i}\right)\left(\frac{\partial g}{\partial p_i}\right) - \left(\frac{\partial f}{\partial p_i}\right)\left(\frac{\partial g}{\partial q_i}\right). \tag{2}$$

Suppose that f, g, and h are any three functions of z. Then it can be shown that the Poisson bracket satisfies the remarkable *Jacobi identity* or condition

$$[f, [g, h]] + [g, [h, f]] + [h, [f, g]] = 0. \tag{3}$$

The fundamental Poisson bracket results follow from definition (2):

$$[q_i, q_j] = [p_i, p_j] = 0, \tag{4}$$

$$[q_i, p_j] = \delta_{ij}. \tag{5}$$

Define a $2n \times 2n$ matrix J by the rule

$$J_{ab} = [z_a, z_b]. \tag{6}$$

According to (1), (4), and (5) J takes the form

$$J = \begin{bmatrix} 0 & I \\ -I & 0 \end{bmatrix}, \tag{7}$$

where all entries are $n \times n$ blocks. We note that, with the aid of J, the Poisson bracket (2) can also be written in the form

$$[f, g] = \sum_{ab} \left(\frac{\partial f}{\partial z_a}\right) J_{ab} \left(\frac{\partial g}{\partial z_b}\right). \tag{8}$$

A $2n \times 2n$ matrix M is called *symplectic* if it satisfies the relation

$$M^T J M = J, \tag{9}$$

where M^T denotes the transpose of M. It is easily verified that if M is symplectic, its inverse M^{-1} exists and is symplectic. Moreover, the identity matrix I is symplectic. Finally, if M and M' are any two symplectic matrices, simple calculation shows that their product MM' is also a symplectic matrix. We say that symplectic matrices form a group. This group is called the symplectic group and is denoted by $Sp(2n)$.

Let S be any $2n \times 2n$ *symmetric* matrix. Matrices of the form JS are called *Hamiltonian* matrices. Define M to be the *exponential* of JS,

$$M = \sum_{m=0}^{\infty} \frac{(JS)^m}{m!} = \exp(JS). \tag{10}$$

Then it is a theorem that M is symplectic. Conversely, if M is symplectic and sufficiently near I, then it can be written uniquely in the form $M = \exp(JS)$.

It is easily checked that the sum (and any linear combination) of two Hamiltonian matrices is again a Hamiltonian matrix. Consequently, Hamiltonian matrices form a *linear vector space*. Let $\{JS, JS'\}$ be the *commutator* of any two Hamiltonian matrices,

$$\{JS, JS'\} = (JS)(JS') - (JS')(JS). \tag{11}$$

Then it can be shown that the commutator of any two Hamiltonian matrices is again a Hamiltonian matrix. That is, given S and S', there is a symmetric matrix S'' such that

$$\{JS, JS'\} = JS''. \tag{12}$$

We say that Hamiltonian matrices form a *Lie algebra* with the commutator playing the role of a *Lie product*. The Lie algebra formed by Hamiltonian matrices is called the symplectic Lie algebra and denoted by $sp(2n)$. As in (10), it can be shown that in general Lie groups and Lie algebras are related by exponentiation.

Since a Lie algebra is a linear vector space, it has a dimension. In the case of $sp(2n)$, this dimension can easily be verified to be given by the relation

$$\dim sp(2n) = n(2n+1). \tag{13}$$

For the most part we are interested in the case $n = 3$, which describes the motion of a particle in three-dimensional (configuration) space. For this case we have the result

$$\dim sp(6) = 21. \tag{14}$$

Dynamical Significance

Consider any dynamical system whose evolution is governed by a set of first-order differential equations of motion. Suppose z^{in} is a set of *initial* conditions that, under this evolution, is sent to a set of *final* conditions z^{fin}. We write this relation symbolically in the form

$$z^{\text{fin}} = \mathcal{M} z^{\text{in}} \tag{15}$$

and call $\mathcal{M}$ a transfer map, see FIGURE 1.

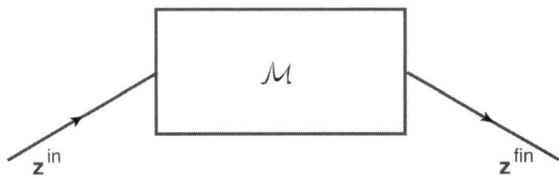

FIGURE 1. The transfer map $\mathcal{M}$ obtained by integrating a set of first-order differential equations sends the initial conditions z^{in} to the final conditions z^{fin}.

The matrix M, defined by

$$M(z^{\text{in}})_{ab} = \frac{\partial(z_a^{\text{fin}})}{\partial(z_b^{\text{in}})}, \tag{16}$$

is called the *Jacobian* matrix associated with the map $\mathcal{M}$. It describes the small changes produced in the final conditions z^{fin} when small changes are made in the initial conditions z^{in}. Suppose the Jacobian matrix $M(z^{\text{in}})$ is symplectic. Then, we say that $\mathcal{M}$ is a symplectic map. Note that relation (9) is *nonlinear* and must hold for *all* z^{in}. Therefore, sympletic maps are severely restricted. Suppose the equations of motion that produce $\mathcal{M}$ can be written in *Hamiltonian* form. Then, remarkably, the associated transfer map $\mathcal{M}$ is symplectic, and conversely.

At this point we remark that the commutator $\{*, *'\}$ employed in (11) obeys the same Jacobi identity (3) as the Poisson bracket $[*, *']$ defined in (2). Indeed, this property (along with antisymmetry under the interchange of $*$ and $*'$) is the defining characteristic of a Lie product. We also observe that the set of all phase-space functions forms a linear vector space. It follows that the set of all phase-space functions forms a Lie algebra, with the Poisson bracket playing the role of a Lie product. Since the set of all phase-space functions is infinitely dimensioned, this Lie algebra is infinitely dimensioned.

LIE OPERATORS AND GENERAL VECTOR FIELDS

Let $f(z)$ be any function of the phase-space variables z. Associated with each f is a *Lie operator* that we denote by $:f:$. The Lie operator $:f:$ is a *differential* operator defined by

$$:f: \equiv \sum_i \left(\frac{\partial f}{\partial q_i}\right)\left(\frac{\partial}{\partial p_i}\right) - \left(\frac{\partial f}{\partial p_i}\right)\left(\frac{\partial}{\partial q_i}\right). \tag{17}$$

In particular, if $:f:$ acts on any phase-space function g, one obtains the result

$$:f:g = \sum_i \left(\frac{\partial f}{\partial q_i}\right)\left(\frac{\partial g}{\partial p_i}\right) - \left(\frac{\partial f}{\partial p_i}\right)\left(\frac{\partial g}{\partial q_i}\right) = [f, g]. \tag{18}$$

Thus, one may heuristically view a Lie operator as a Poisson bracket waiting to happen. Note that, in view of (8), the defining relation (17) can also be written in the form

$$:f: = \sum_{a,b} \left(\frac{\partial f}{\partial z_a}\right) J_{ab} \left(\frac{\partial}{\partial z_b}\right). \tag{19}$$

We also remark that in the mathematics literature the Lie operator $:f:$ is sometimes referred to as $ad(f)$ where ad is shorthand for *adjoint*. We use the notation $:f:$ instead of $ad(f)$ because it facilitates the writing of complicated expressions.

It is easily shown that the set of Lie operators forms a linear vector space. Moreover, the commutator of two Lie operators is again a Lie operator. Indeed, as a consequence of the Jacobi identity (3), there is the relation

$$\{:f:, :g:\} = :f::g: - :g::f: = :[f, g]:. \tag{20}$$

Thus, the set of Lie operators forms a Lie algebra. Note that there are now two Lie algebras that need to be kept in mind. First, there is the Lie algebra of functions of z

with the Lie product defined to be the Poisson bracket. Second, there is the Lie algebra of Lie operators with the Lie product defined to be the commutator. Relation (20) indicates that they are essentially the same up to an additive constant. It can be shown that the set of all symplectic maps forms a Lie group and its associated Lie algebra is the Lie algebra of all Lie operators.

There is a Lie-algebraic correspondence between *quadratic* polynomials in z and Hamiltonian matrices. Suppose that f and g are any two such polynomials. They can be written in the form

$$f = \frac{1}{2}\sum_{a,b} S^f_{ab} z_a z_b, \qquad (21)$$

$$g = \frac{1}{2}\sum_{a,b} S^g_{ab} z_a z_b, \qquad (22)$$

where S^f and S^g are real *symmetric* matrices. Evidently, there is a one-to-one correspondence between homogeneous second-degree polynomials and symmetric matrices. We indicate a one-to-one correspondence by the symbol $\leftrightarrow$. Since J is invertible, there is also an associated one-to-one correspondence between homogeneous second degree polynomials and matrices of the form JS. Indeed, relations (21) and (22) can also be written in the form

$$f \leftrightarrow JS^f \Leftrightarrow f = \frac{1}{2}(Jz, JS^f z), \qquad (23)$$

$$g \leftrightarrow JS^g \Leftrightarrow f = \frac{1}{2}(Jz, JS^g z). \qquad (24)$$

The symbol $\Leftrightarrow$ denotes a logical bijection.

It is easily verified that the set of all quadratic polynomials forms a Lie algebra under the Poisson bracket operation. Indeed, suppose the second-degree polynomial h is defined by the relation

$$h = [f, g]. \qquad (25)$$

Then it can be shown that h is given by

$$g = \frac{1}{2}\sum_{a,b} S^h_{ab} z_a z_b, \qquad (26)$$

where S^h is defined by the equation

$$JS^h = \{JS^f, JS^g\}. \qquad (27)$$

Observe that (25) is a Lie-algebraic relation in the Poisson bracket Lie algebra of second-degree polynomials, and (27) is a Lie-algebraic relation in $sp(2n)$. Thus, we have the logical equivalence

$$h = [f, g] \Leftrightarrow JS^h = \{JS^f, JS^g\}. \qquad (28)$$

What we have just shown is that these two Lie algebras are isomorphic under the one-to-one correspondence given by (23), (24), and (26).

We close this section by noting that what we have called a Lie operator is actually a special case of a more general object. Let x denote a collection of N variables x_1, x_2, ..., x_N. Also, let $\boldsymbol{g} = (g_1, g_2, ..., g_N)$ be a collection of N functions of x. The Lie operator $\mathcal{L}_{\boldsymbol{g}}$ associated with the collection of functions $g_b(x)$ is defined to be the differential operator given by

$$\mathcal{L}_g = \sum_{b=1}^{N} g_b(x)\left(\frac{\partial}{\partial x_b}\right). \tag{29}$$

Relation (29) is the general definition of a Lie operator. The general Lie operator $\mathcal{L}_g$ is also sometimes called a vector field. It is easily verified that if $\mathcal{L}_f$ and $\mathcal{L}_g$ are any two vector fields, then so is their commutator. That is, there is a vector field $\mathcal{L}_h$ such that

$$\{\mathcal{L}_f, \mathcal{L}_g\} = \mathcal{L}_f \mathcal{L}_g - \mathcal{L}_g \mathcal{L}_f = \mathcal{L}_h. \tag{30}$$

Therefore, general vector fields also form a Lie algebra. Indeed, it can be shown that they constitute the Lie algebra of the group of all *diffeomorphic* maps.

There is an intimate connection between vector fields and ordinary differential equations. Consider the set of first-order differential equations

$$\dot{x}_a = g_a(x). \tag{31}$$

Then, using (29), this set can also be written in the form

$$\dot{x}_a = \mathcal{L}_g x_a. \tag{32}$$

Furthermore, let h be any function of x. Then, by the chain rule, the time derivative of h along a trajectory is given by

$$\dot{h} = \sum_b \left(\frac{\partial h}{\partial x_b}\right)\dot{x}_b = \sum_b g_b\left(\frac{\partial h}{\partial x_b}\right) = \mathcal{L}_g h. \tag{33}$$

Comparing (29) with (19), we see that we have assumed $N = 2n$ and

$$g_b(z) = \sum_a \left(\frac{\partial f}{\partial z_a}\right) J_{ab}. \tag{34}$$

We conclude that, in the case of interest for Hamiltonian systems, the collection of functions g_b arises from a *single* function f according to (34). Thus, to be more precise, what we have called and will continue to call a Lie operator, could better be called a *Hamiltonian* Lie operator or a *Hamiltonian* vector field.

CARTAN FORM FOR $sp(6)$ LIE ALGEBRA

Killing and Cartan classified and studied all the so-called *simple* Lie algebras, and found both *regular families* and the *exceptional* Lie algebras. The symplectic Lie algebras $sp(2n)$ form one such regular family. They discovered that all simple Lie algebras could be described in a beautiful geometric form by a proper choice of basis called the Cartan basis. In this basis there are elements of two types: those of the first type form a subalgebra of elements, all of which *commute* with each other, called the *Cartan subalgebra*; the remaining elements are *ladder* elements, whose properties are analogous to those of the *raising* and *lowering* operators that are familiar from the quantum theory of angular momentum.

In the case of $sp(6)$ the Cartan subalgebra is three dimensional. These mutually *commuting* elements, call them c_j, can be conveniently taken to be the monomials

$$c_j = -q_j p_j \quad \text{for } j = 1, 2, 3, \tag{35}$$

where we use the fact (as illustrated in the previous section) that the Poisson bracket Lie algebra of the quadratic polynomials in the quantities z_a (for $a = 1$ to 6) is a realization of $sp(6)$.

Corresponding to the fact that the Cartan subalgebra for $sp(6)$ is three dimensional, the ladder elements are labeled by *vectors* $\boldsymbol{\mu}$ in a three-dimensional Euclidean space. These vectors are called *root* vectors. (We are describing the situation for $sp(6)$. Killing and Cartan discovered that all the simple Lie algebras could be described by root vector structures, and that the length of and angles between these root vectors can only take on certain well-defined values as a result of the Jacobi identity.) Since $sp(6)$ is 21 dimensional, and the Cartan subalgebra is three dimensional, there must be 18 ladder elements. They are labelled by 18 three-component vectors consisting of nine vectors and their negatives. For convenience, we call these nine vectors $\boldsymbol{\alpha}^j$, $\boldsymbol{\beta}^j$, and $\boldsymbol{\gamma}^j$, where j ranges from 1 to 3. They are given in terms of three orthogonal unit vectors e^1, e^2, and e^3 by the relations

$$\boldsymbol{\alpha}^1 = 2e^1 \tag{36}$$

$$\boldsymbol{\alpha}^2 = e^1 + e^2 \tag{37}$$

$$\boldsymbol{\alpha}^3 = e^1 - e^2 \tag{38}$$

$$\boldsymbol{\beta}^1 = 2e^2 \tag{39}$$

$$\boldsymbol{\beta}^2 = e^2 + e^3 \tag{40}$$

$$\boldsymbol{\beta}^3 = e^2 - e^3 \tag{41}$$

$$\boldsymbol{\gamma}^1 = 2e^3 \tag{42}$$

$$\boldsymbol{\gamma}^2 = e^3 + e^1 \tag{43}$$

$$\boldsymbol{\gamma}^3 = e^3 - e^1. \tag{44}$$

The 18 $sp(6)$ root vectors are shown in FIGURE 2. Note they all have the form $(\pm e^i, \pm e^j)$ with the signs taken independently and the zero vector omitted.

The ladder elements $r(\boldsymbol{\mu})$ labelled by these *root* vectors are monomials given by the relations

$$r(\boldsymbol{\alpha}^1) = \frac{1}{\sqrt{2}} q_1^2 \tag{45}$$

$$r(-\boldsymbol{\alpha}^1) = -\frac{1}{\sqrt{2}} p_1^2 \tag{46}$$

$$r(\boldsymbol{\alpha}^2) = q_1 q_2 \tag{47}$$

$$r(-\boldsymbol{\alpha}^2) = -p_1 p_2 \tag{48}$$

$$r(\boldsymbol{\alpha}^3) = q_1 p_2 \tag{49}$$

$$r(-\boldsymbol{\alpha}^3) = q_2 p_1 \tag{50}$$

$$r(\boldsymbol{\beta}^1) = \frac{1}{\sqrt{2}} q_2^2 \tag{51}$$

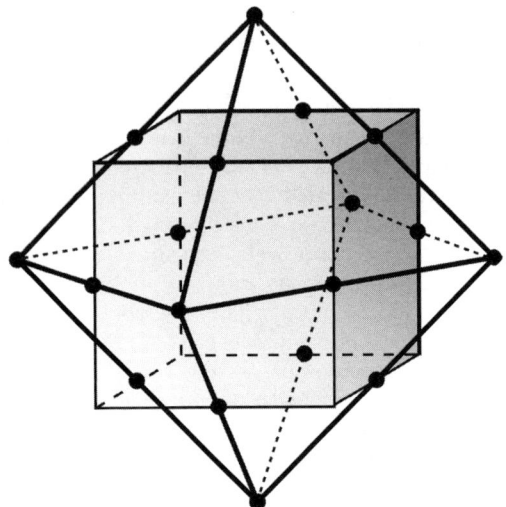

FIGURE 2. Root diagram showing the root vectors for $sp(6)$. The six tips of the long root vectors $\pm\alpha^1$, $\pm\beta^1$, and $\pm\gamma^1$ form the vertices of a regular octahedron. These root vectors have length 2. The remaining 12 short root vectors have length $\pm\sqrt{2}$ and their tips lie at the midpoints of the 12 edges of the unit cube (the cube with edge 2).

$$r(-\beta^1) = -\frac{1}{\sqrt{2}}p_2^2 \tag{52}$$

$$r(\beta^2) = q_2 q_3 \tag{53}$$

$$r(-\beta^2) = -p_2 p_3 \tag{54}$$

$$r(\beta^3) = q_2 p_3 \tag{55}$$

$$r(-\beta^3) = q_3 p_2 \tag{56}$$

$$r(\gamma^1) = \frac{1}{\sqrt{2}}q_3^2 \tag{57}$$

$$r(-\gamma^1) = -\frac{1}{\sqrt{2}}p_3^2 \tag{58}$$

$$r(\gamma^2) = q_3 q_1 \tag{59}$$

$$r(-\gamma^2) = -p_3 p_1 \tag{60}$$

$$r(\gamma^3) = q_3 p_1 \tag{61}$$

$$r(-\gamma^3) = q_1 p_3. \tag{62}$$

The virtue of the Cartan basis is that the commutation rules take a particularly illuminating form. As desired, the elements of the Cartan subalgebra commute,

$$[c^j, c^k] = 0. \tag{63}$$

The Lie product of a Cartan element c^j with a ladder element $r(\mu)$ reproduces $r(\mu)$ multiplied by the jth component of μ,

$$[c^j, r(\mu)] = (e^j \cdot \mu) r(\mu). \tag{64}$$

The Lie product of a ladder element $r(\mu)$ and the corresponding element $r(-\mu)$ is a sum of Cartan elements with coefficients given by the components of μ,

$$[r(\mu), r(-\mu)] = \sum_j (e^j \cdot \mu) c^j. \tag{65}$$

Finally, the Lie product of two ladder elements $r(\mu)$ and $r(\nu)$, with $\mu \neq -\nu$, is proportional to the ladder element $r(\mu + \nu)$ if the sum($\mu + \nu$) is itself a root vector,

$$[r(\mu), r(\nu)] = N(\mu, \nu) r(\mu + \nu). \tag{66}$$

All other Lie products vanish. For the case of $sp(6)$, the $N(\mu, \nu)$ have values $\pm \sqrt{2}$. The reader is invited to verify, by direct computation, relations (63)–(66) using definitions (35) and (45)–(62).

REPRESENTATIONS OF $sp(6)$

A (linear) *representation* of a Lie algebra consists of a vector space and a set of linear operators that act on this vector space. The linear operators form a Lie algebra with the commutator serving as the Lie product. This linear-operator Lie algebra is required to be the same as the Lie algebra being represented. In particular, suppose the operators C^j and $R(\mu)$ are *any* set of linear operators that obey the rules (63)–(66), with the Poisson bracket replaced by a commutator. Suppose further that a scalar product can be set up in such a way that the C^j are Hermitian. Let $|w\rangle = |w_1 w_2 w_3\rangle$ denote an eigenvector of the C^j with the property

$$C^j |w_1 w_2 w_3\rangle = w_j |w_1 w_2 w_3\rangle \quad \text{or} \quad C^j |w\rangle = (e^j \cdot w) |w\rangle. \tag{67}$$

Since the C^j are Hermitian, the w_j are real. It is convenient, as shown, to treat them together as the components of a single vector w called a *weight*.

Consider the vector $R(\mu)|w\rangle$. From the commutation rules (64) we obtain the relation

$$\begin{aligned} C^j R(\mu)|w\rangle &= R(\mu) C^j |w\rangle + (e^j \cdot \mu) R(\mu)|w\rangle \\ &= R(\mu) w_j |w\rangle + (e^j \cdot \mu) R(\mu)|w\rangle \\ &= [e^j \cdot (w + \mu)] R(\mu)|w\rangle. \end{aligned} \tag{68}$$

It follows that if $R(\mu)|w\rangle$ is different from zero, then it is an eigenvector of the C^j with weight $w + \mu$. Consequently, from a single weight we can produce a whole set of weights. The set of weights can be *ordered* by means of the following definitions:

1. A vector is *positive* if its first nonvanishing component is positive.
2. A vector w is *higher* than the vector w' if $w - w'$ is positive.

We now state the fundamental theorems of Cartan concerning representations:

1. In any irreducible representation, there is an eigenvector with highest weight, and this eigenvector is unique, that is, non-degenerate.

2. Two irreducible representations are equivalent if they have the same highest weight.
3. Every highest weight w^h is a linear combination, with non-negative integer coefficients, of what are called *fundamental* weights. For a rank ℓ Lie algebra there are ℓ such fundamental weights. (The *rank* of a Lie algebra is the dimension of its Cartan subalgebra.)

In the case of $sp(6)$, which has rank 3, the three fundamental weights ϕ_1, ϕ_2, and ϕ_3, are given by

$$\phi^1 = e^1 \tag{69}$$

$$\phi^2 = e^1 + e^2 \tag{70}$$

$$\phi^3 = e^1 + e^2 + e^3. \tag{71}$$

These fundamental weights are shown in FIGURE 3 together with the $sp(6)$ root vectors. Thus, for $sp(6)$, every highest weight w^h has the form

$$w^h = \ell\phi^1 + m\phi^2 + n\phi^3 = (\ell + m + n)e^1 + (m + n)e^2 + ne^3, \tag{72}$$

where ℓ, m, and n are arbitrary nonnegative integers.

Taken together, the theorems of Cartan show that an irreducible representation of $sp(6)$ is completely characterized by the three integers ℓ, m, and n. We denote this representation by $\Gamma(\ell, m, n)$. It can be shown that the dimension of $\Gamma(\ell, m, n)$ is given by

$$\Gamma(\ell, m, n) = \frac{1}{720}(\ell + 2m + 2n + 5)(\ell + m + 2n + 4)(\ell + m + n + 3) \tag{73}$$
$$\times (\ell + m + 2)(m + 2n + 3)(m + n + 2)(\ell + 1)(m + 1)(n + 1).$$

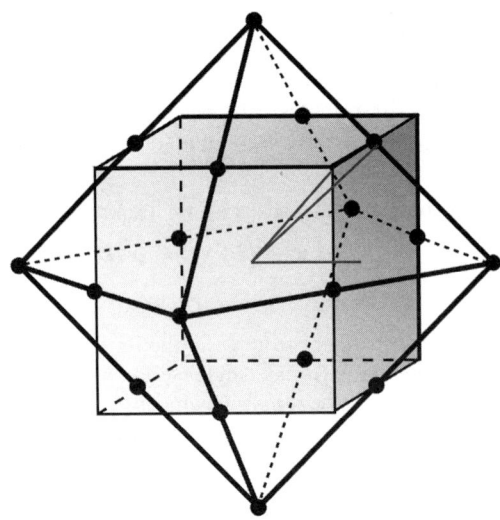

FIGURE 3. Fundamental weights ϕ^1, ϕ^2, and ϕ^3 for $sp(6)$. The root vectors are also shown.

TABLE 1. Dimensions of representations of $sp(6)$

ℓ	m	n	$\dim \Gamma(\ell,m,n)$	ℓ	m	n	$\dim \Gamma(\ell,m,n)$
0	0	0	1	3	0	0	56
1	0	0	6	2	1	0	189
0	1	0	14	2	0	1	216
0	0	1	14	1	2	0	350
2	0	0	21	1	1	1	512
1	1	0	64	1	0	2	378
1	0	1	70	0	3	0	385
0	2	0	90	0	2	1	616
0	1	1	126	0	1	2	594
0	0	2	84	0	0	3	330

For quick reference the dimensions of the first few representations are listed in TABLE 1. Where there is no possibility of confusion, we sometimes refer to a representation by its dimension.

Consider the representation $\Gamma(0,0,0)$. According to **(72)** its highest weight is the vector 0, and according to **(73)** this representation is one dimensional. Thus, $\Gamma(0,0,0)$ has only one weight vector. FIGURE 4 displays this vector in what is called a *weight diagram*. Since $\Gamma(0,0,0)$ is one dimensional, it is often referred to by its dimension, 1.

Consider the representation $\Gamma(1,0,0)$. The highest weight w^h for this representation is shown in FIGURE 5. Also shown are all other weights obtained from w^h by adding and subtracting various integer multiples of the root vectors α^j, β^j, and γ^j. Observe that there are six different weights in accord with **(73)** evaluated for $\ell = 1$ and $m = n = 0$. This representation is called the *fundamental* representation because it is the representation given by 6×6 matrices of the form JS.

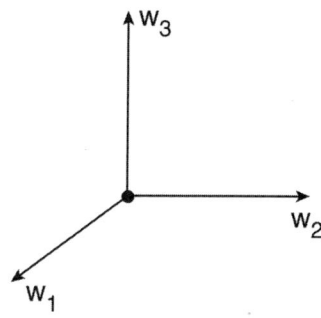

FIGURE 4. Weight diagram for the representation $1 = \Gamma(0,0,0)$.

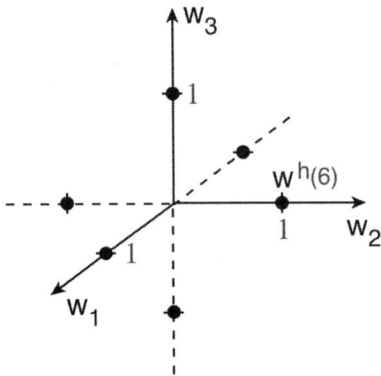

FIGURE 5. Weight diagram for the fundamental representation $6 = \Gamma(1,0,0)$.

FIGURES 6 and 7 show the highest weights and weight diagrams for the representations $\Gamma(0,1,0)$ and $\Gamma(2,0,0)$. It can be shown that the eigenvectors $|w\rangle$ on the boundaries of the diagram are nondegenerate. However, there are two linearly independent eigenvectors corresponding to the weight at the origin in FIGURE 6, and three linearly independent eigenvectors corresponding to the weight at the origin in FIGURE 7. The representative $\Gamma(2,0,0)$ is called the *adjoint* representation because it is the representation obtained by letting the Lie algebra act on itself by way of the Lie product.

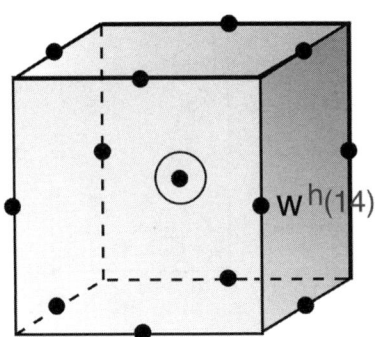

FIGURE 6. Weight diagram for the representation $14 = \Gamma(0,1,0)$. The *circled* weight at the origin has multiplicity two. Observe from FIGURE 2 that the 12 other weights are located at the tips of the root vectors having length $\sqrt{2}$.

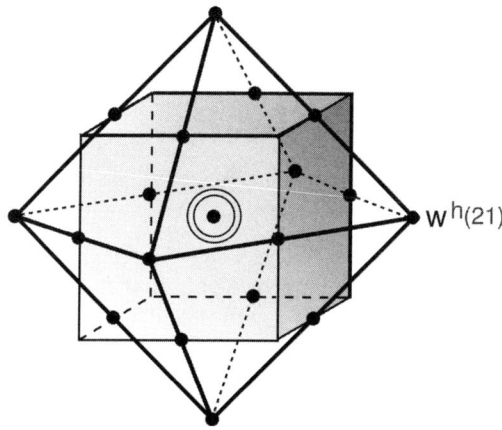

FIGURE 7. Weight diagram for the adjoint representation $21 = \Gamma(2,0,0)$. The *doubly circled* weight at the origin has multiplicity three. The 18 other weights are located at the tips of the $sp(6)$ root vectors.

SYMPLECTIC CLASSIFICATION OF ANALYTIC VECTOR FIELDS IN SIX VARIABLES

Let $\mathcal{L}_f$ be a general vector field in $N = 2n$ variables where f denotes the collection of $2n$ functions $(f_1, f_2, \ldots, f_{2n})$. Assume that the functions f_1 through f_{2n} are analytic at some common point $(z_1^0, z_2^0, \ldots, z_{2n}^0)$. Without loss of generality we may take this point to be the origin. (If not, make a linear change of variables that sends $(z_1^0, z_2^0, \ldots, z_{2n}^0)$ to the origin.) Then, we may decompose the Taylor expansions of the components of f into sums of homogeneous polynomials and consider, individually, vector fields of the form $\mathcal{L}_{f^m}$, where the components f^m are *homogeneous* polynomials of degree m. Thus, we have the decomposition,

$$\mathcal{L}_f = \sum_{m=0}^{\infty} \mathcal{L}_{f^m}. \tag{74}$$

The purpose of this section is to show that, in the case of six variables ($N = 6$), each $\mathcal{L}_{f^m}$ can be *uniquely* decomposed into vector fields that carry well-defined representations of $sp(6)$. A consequence of this result is that each $\mathcal{L}_{f^m}$, and hence, any analytic $\mathcal{L}_f$, can be uniquely decomposed into Hamiltonian and non-Hamiltonian parts. (Analogous results also hold for the case $N = 2n$ and representations of $sp(2n)$. The results for $n = 3$ are sufficiently general that one can infer from them the results for any n.)

Suppose $\mathcal{L}_f$ is any vector field. Then we associate with $\mathcal{L}_f$ what we call a *super* Lie operator and denote it by $\#\mathcal{L}_f\#$. It acts on general vector fields $\mathcal{L}_g$ by the rule

$$\#\mathcal{L}_f\#\mathcal{L}_g = \{\mathcal{L}_f, \mathcal{L}_g\}. \tag{75}$$

Note that according to (**30**), super Lie operators act on vector fields to produce other vectors fields. Also, observe the resemblance of (**75**) to (**18**). Finally, in the case of

Hamiltonian vector fields, we use the abbreviated rotation #*f*# to denote the super Lie operator #:*f*:#.

Let Σ denote the vector field

$$\Sigma = \sum_a z_a \left(\frac{\partial}{\partial z_a}\right). \tag{76}$$

Then, by Euler's relation for homogeneous functions, we obtain the result

$$\#\Sigma\#\mathcal{L}_{f^n} = \{\Sigma, \mathcal{L}_{f^n}\} = (n-1)\mathcal{L}_{f^n}. \tag{77}$$

We say that $\mathcal{L}_{f^n}$ is homogeneous of degree $(n-1)$. Next, suppose f_n is a homogeneous polynomial of degree n. Then, in view of **(34)**, in the special case of a Hamiltonian vector field of the form $:f_n:$ there is the result

$$\#\Sigma\#:f: = \{\Sigma, :f:\} = (n-2):f:. \tag{78}$$

Thus, the vector field $:f_n:$ is homogeneous of degree $(n-2)$. Finally, it is easily verified that there is a *grading* relation of the form

$$\{\mathcal{L}_{f^\ell}, \mathcal{L}_{g^m}\} = \mathcal{L}_{h^n} \quad \text{with} \quad n = \ell + m - 1. \tag{79}$$

Let f_2 be a quadratic polynomial in z. We saw previously that such polynomials are associated with the Lie algebra $sp(2n)$. We can let $sp(2n)$ act on vector fields $\mathcal{L}_{g^m}$ with the aid of super operators of the form #f_2#. Indeed, according to **(79)** we have a relation of the form

$$\#f_2\#\mathcal{L}_{g^m} = \{:f_2:, \mathcal{L}_{g^m}\} = \mathcal{L}_{h^m}. \tag{80}$$

We draw the important conclusion that the set of homogeneous vector fields $\mathcal{L}_{g^m}$ transforms under, and forms a representation of, $sp(2n)$.

What irreducible representations occur? Consider first the case of Hamiltonian vector fields. In this case for a homogeneous polynomial g_m of degree m there is the result

$$\#f_2\#:g_m: = \{:f_2:, :g_m:\} = :[f_2, g_m]: = :(:f_2:g_m):. \tag{81}$$

Note that the quantity $:f_2:g_m$ is itself a homogeneous polynomial of degree m. It follows that Hamiltonian vector fields of the form $:g_m:$ are transformed into each other under the action of $sp(2n)$.

At this point let us specialize our discussion to the case of $sp(6)$. Consider the homogeneous of degree m monomial q_1^m. From **(35)** we see that it satisfies the relations

$$:c_1:q_1^m = mq_1^m, \tag{82}$$

$$:c_2:q_1^m = 0, \tag{83}$$

$$:c_3:q_1^m = 0. \tag{84}$$

Consequently, this monomial has the weight vector w given by

$$w = (m, 0, 0). \tag{85}$$

From relations **(45)**–**(62)** we see that it also has the properties

$$:r(\mu):q_1^m = 0 \tag{86}$$

unless $\mu = -\alpha^1, -\alpha^2, -\alpha^3, -\gamma^2,$ or γ^3. Inspection of relations **(36)**–**(44)** shows that none of these μ is positive. We conclude that w as given by **(85)** is the highest weight for all monomials of the form $:r(\mu):q_1^m$. Moreover, it is easily verified that all monomials of degree m can be obtained by repeated application of various $:r(\mathbf{v}):$ on q_1^m. It follows that the set of homogeneous monomials of degree m carries the representation $\Gamma(m,0,0)$ of $sp(6)$, and hence, according to (**81**), the set of all Hamiltonian vector fields of the form $:g_m:$ also carries the representation $\Gamma(m,0,0)$ of $sp(6)$.

Next consider the vector fields $\mathcal{L}_{g^0}$. They are linear combinations of the vector fields $\partial/\partial z_a$. From (**9**) we have the result

$$:z_c: = \sum_b J_{cb}\left(\frac{\partial}{\partial z_b}\right), \tag{87}$$

which can be solved for the $\partial/\partial z_a$ to give the relation

$$\frac{\partial}{\partial z_a} = -\sum_c J_{ac}:z_a:. \tag{88}$$

It follows that the $\partial/\partial z_a$, and hence, all vector fields of the form $\mathcal{L}_{g^0}$, are linear combinations of vector fields of the form $:g_1:$. We already know that they are Hamiltonian and carry the representation $\Gamma(1,0,0)$. Thus, the vector fields $\mathcal{L}_{g^0}$ are Hamiltonian and carry the representation $\Gamma(1,0,0)$.

We now turn to the general case $\mathcal{L}_{g^\ell}$ with $\ell \geq 1$. Any such vector field (for $N = 6$) can be written in the form

$$\mathcal{L}_{g^\ell} = \sum_{a=1}^{6} g_a^\ell\left(\frac{\partial}{\partial z_a}\right). \tag{89}$$

We know that the g_a^ℓ carry the representation $\Gamma(\ell,0,0)$ and the $(\partial/\partial z_a)$ carry the representation $\Gamma(1,0,0)$. It follows that the $\mathcal{L}_{g^\ell}$ must carry the direct product representation $\Gamma(\ell,0,0) \otimes \Gamma(1,0,0)$. It can be shown that for $sp(6)$ there is the Clebsch-Gordan result

$$\Gamma(\ell, 0, 0) \otimes \Gamma(1, 0, 0) = \Gamma(\ell + 1, 0, 0) \oplus \Gamma(\ell - 1, 1, 0) \oplus \Gamma(\ell - 1, 0, 0). \tag{90}$$

Consequently, any $\mathcal{L}_{g^\ell}$ with $\ell \geq 1$ has the unique decomposition

$$\mathcal{L}_{g^\ell} = \mathcal{H}^{\ell+1,0,0} + \mathcal{G}^{\ell-1,1,0} + \mathcal{G}^{\ell-1,0,0}. \tag{91}$$

It can be shown that $\mathcal{H}^{\ell+1,0,0}$ is a Hamiltonian vector field that carries the representation $\Gamma(\ell+1,0,0)$ and has the form $:h_{\ell+1}:$. The quantities $\mathcal{G}^{\ell-1,1,0}$ and $\mathcal{G}^{\ell-1,0,0}$ are what we call *non-Hamiltonian* vector fields. They carry the representations $\Gamma(\ell-1,1,0)$ and $\Gamma(\ell-1,0,0)$, respectively. We show that the $\mathcal{G}^{\ell-1,0,0}$ have the form

$$\mathcal{G}^{\ell-1,0,0} = f_{\ell-1}\Sigma, \tag{92}$$

where $f_{\ell-1}$ is any homogeneous polynomial of degree $(\ell - 1)$. The construction of the vector fields that span $\mathcal{G}^{\ell-1,1,0}$ requires a special effort that is beyond the scope of this paper.

Let us work out some details of the simplest case $\ell = 1$, for which we have the result

$$\mathcal{L}_{g^1} = \mathcal{H}^{2,0,0} + \mathcal{G}^{0,1,0} + \mathcal{G}^{0,0,0}. \tag{93}$$

Since $\mathcal{L}_{g^1}$ is spanned by the vector fields $z_a(\partial/\partial z_b)$ with $a,b = 1,2,...,6$, it has dimension 36. We know that $\mathcal{H}^{2,0,0}$ has the form $:h_2:$ and, therefore, has dimension 21, and $\mathcal{G}^{0,0,0}$ has dimension 1. It follows that $\mathcal{G}^{0,1,0}$ has dimension $(36 - 21 - 1) = 14$, which we know is the dimension of $\Gamma(1,0,0)$; see FIGURES 4, 6, and 7. It is easy to identify $\mathcal{G}^{0,0,0}$. For the case $n = 2$, relation (**94**) can be rewritten in the form

$$\#f_2\#\Sigma = \{:f_2:,\Sigma\} = -\{\Sigma,:f_2:\} = 0. \tag{94}$$

Evidently Σ belongs to $\mathcal{L}_{g^1}$, and (94) shows that it carries the representation $\Gamma(0,0,0)$. Therefore, $\mathcal{G}^{0,0,0}$ must be a multiple of Σ. Finally, we remark that since $f_{\ell-1}$ carries the representation $\Gamma(\ell-1,0,0)$ and (as we have seen) Σ carries the representation $\Gamma(0,0,0)$, it follows that, in general, $\mathcal{G}^{\ell-1,0,0}$ must have the form (92).

CONCLUDING COMMENTS

The symplectic family of Lie algebras $sp(2n)$ has been defined and specific results have been summarized for the physically interesting case of $sp(6)$. Analogous results are also known for the simpler cases of $sp(2)$ and $sp(4)$. The Lie algebraic structure of all $sp(2n)$, for example root vectors and fundamental weight vectors, is also known. In particular, for $sp(2n)$, a representation is characterized by n non-negative integers $k_1, k_2, \ldots, k_n$ and may be denoted by $\Gamma(k_1,k_2,\ldots,k_n)$. Homogeneous polynomials of degree ℓ in the $2n$ components of z again carry representations of $sp(2n)$, and for these representations there is the result

$$k_1 = \ell \tag{95}$$

$$k_j = 0 \text{ for } j = 2, 3, \ldots, n. \tag{96}$$

There is also an analogous Clebsch–Gordon result of the form (90), where all entries in $\Gamma(k_1,k_2,\ldots,k_n)$ are zero save for the first two,

$$\begin{aligned}&\Gamma(\ell, 0, 0, \ldots) \otimes \Gamma(1, 0, 0, \ldots) \\&= \Gamma(\ell + 1, 0, 0, \ldots) \oplus \Gamma(\ell - 1, 1, 0, \ldots) \oplus \Gamma(\ell - 1, 0, 0, \ldots).\end{aligned} \tag{97}$$

Thus, the symplectic classification of all analytic vector fields in any (even) dimension is in principle known.

Related results can also be obtained for the associated Lie groups. For example, it can be shown that any analytic map has a unique formal Lie factorization of the form

$$\begin{aligned}\mathcal{M} = \;&\exp(\mathcal{G}_2)\exp(\mathcal{G}_3)\exp(\mathcal{G}_4)\ldots \\&\times \exp(:f_1:)\exp(:f_2:)\exp(:f_3:)\exp(:f_4:)\ldots,\end{aligned} \tag{98}$$

where the f_m are homogeneous polynomials of degree m in the components of z and the product $\exp(:f_1:)\exp(:f_2:)\exp(:f_3:)\ldots$ describes the symplectic part of $\mathcal{M}$. The $\mathcal{G}_m$ are non-Hamiltonian vector fields (homogeneous of degree $m-2$) and the product $\exp(\mathcal{G}_2)\exp(\mathcal{G}_3)\ldots$ describes the nonsymplectic part of $\mathcal{M}$.

Another application of the symplectic group is the construction of *moment invariants*. Suppose $\rho(z)$ is a density that describes a particle distribution in phase space and that we form moments $\langle z_a z_b z_c z_d \ldots \rangle$ using the rule

$$\langle z_a z_b z_c z_d \ldots \rangle = \int \rho(z) d^6z (z_a z_b z_c z_d \ldots). \tag{99}$$

Then it can be shown the moments of degree ℓ transform under $sp(6)$ according to the representation of $\Gamma(\ell,0,0)$.

It can also be shown that for $sp(6)$ there is the Clebsch-Gordan series

$$\begin{aligned}\Gamma(2, 0, 0) \otimes \Gamma(2, 0, 0) = \;&\Gamma(0, 0, 0) \oplus \Gamma(0, 1, 0) \oplus \Gamma(0, 2, 0) \\&\oplus \Gamma(2, 0, 0) \oplus \Gamma(2, 1, 0) \oplus \Gamma(4, 0, 0).\end{aligned} \tag{100}$$

Note that $\Gamma(0,0,0)$, the *identity* representation, occurs. Therefore, by taking suitable linear combinations of products of second-order moments, it should be possible to construct a symplectic invariant. This is indeed the case. Doing so provides a generalization of the familiar two-dimensional phase space *emittance* invariant

$$I_2 = (\langle q^2 \rangle)(\langle p^2 \rangle) - (\langle qp \rangle)^2 \qquad (101)$$

to fully (linearly) coupled six-dimensional phase space.

By a continuation of this line of reasoning, one finds that there are two more invariants that can be constructed from linear combinations of products of second-order moments. Together they comprise three *eigenemittances* that remain unchanged under arbitrary (but linear) symplectic transport.

In similar fashion, one can also prove the existence of, and construct invariants out of, linear combinations of products of higher order moments.

END REMARKS

A more detailed exposition of the results summarized in this paper, including results for $sp(2)$ and $sp(4)$, may be found elsewhere.[1] Expositions of the general theory of Lie algebras may be found in the books by Fulton and Harris[2] or by Georgi.[3]

ACKNOWLEDGMENT

This work was supported by DOE Grant DE-FG02-96ER40949.

REFERENCES

1. DRAGT, A. 2005. Lie methods for nonlinear dynamics with applications to accelerator physics, Chapter 17. Physics Department Technical Report, University of Maryland.
2. FULTON, W. & J. HARRIS. 1991. Representation Theory, A First Course. Springer–Verlag.
3. GEORGI, H. 1999. Lie Algebras in Particle Physics. Perseus Books.

Chaos and Quantum Mechanics

SALMAN HABIB,[a] TANMOY BHATTACHARYA,[a] BENJAMIN GREENBAUM,[b] KURT JACOBS,[a,c] KOSUKE SHIZUME,[d] AND BALA SUNDARAM[e]

[a]*Theoretical Division, The University of California, Los Alamos National Laboratory, Los Alamos, New Mexico, USA*

[b]*Columbia University, New York, New York, USA*

[c]*Griffith University, Queensland, Australia*

[d]*Tsukuba University, Tsukuba, Japan*

[e]*City University of New York, New York, New York, USA*

ABSTRACT: The relationship between chaos and quantum mechanics has been somewhat uneasy—even stormy, in the minds of some people. However, much of the confusion may stem from inappropriate comparisons using formal analyses. In contrast, our starting point here is that a complete dynamical description requires a full understanding of the evolution of *measured systems*, necessary to explain actual experimental results. This is of course true, both classically and quantum mechanically. Because the evolution of the physical state is now conditioned on measurement results, the dynamics of such systems is intrinsically nonlinear even at the level of distribution functions. Due to this feature, the physically more complete treatment reveals the existence of dynamical regimes—such as chaos—that have no direct counterpart in the linear (unobserved) case. Moreover, this treatment allows for understanding how an effective classical behavior can result from the dynamics of an observed quantum system, both at the level of trajectories as well as distribution functions. Finally, we have the striking prediction that time-series from measured quantum systems can be chaotic far from the classical regime, with Lyapunov exponents differing from their classical values. These predictions can be tested in next-generation experiments.

KEYWORDS: chaos; quantum mechanics

PROLOGUE

I met Henry Kandrup as a graduate student at Maryland in 1985, having recently decided to switch from experiment to theory. My first interaction with postdoctoral scholars—at the time an intimidatingly higher form of life—occurred when Henry suggested that he and another postdoctoral scholar, Ping Yip, and I take up the question of Landau damping and stability of star clusters. Although I was happy to work with Henry and Ping, most of the time I was struggling to understand cryptic conversations laced with mathematical jargon—functions of compact support, consider the following inner product, and so on. Since I was not following too much of this,

Address for correspondence: Salman Habib, MS B285, Theoretical Division, The University of California, Los Alamos National Laboratory, Los Alamos, NM 87545, USA.

habib@lanl.gov

I decided it was better to go away and catch up by reading every paper that was even vaguely related to the topic. This turned out to be much easier than expected, and one night I came up with a simple way of combining Henry's previous work on stability with conventional Landau damping theory from plasma physics. Coming in to the department late in the morning I showed the first set of notes to Henry. He looked at them, did not say much—which was unusual—and went back home. Next day, as I entered my office, I was stunned to find, slipped under the door, a complete preprint of a paper, all equations written in by hand in Henry's beautiful copperplate. He had gone home, generalized my notes to the problem at hand, worked through the entire thing, come back late at night, and typed the preprint on an electric typewriter (this was just before the advent of word processing), finishing as the Sun came up. After this incident, I was really afraid of postdoctoral scholars!

My early interactions with Henry were very wide-ranging; we discussed all sorts of topics, from classical statistical mechanics to quantum gravity, and on all of them he was very well-informed and entertainingly opinionated. The years went by quickly, Henry moved on to other places and so did I. Although we argued and collaborated now and then as of old, in my memory the early years have a certain luminescence. My favorite remembrance of Henry is that after he had demolished somebody's hapless piece of research in one of our discussions, he would look up, smile in a disarming way, and say, "True?" It usually was.

One of the topics Henry and I discussed at considerable length and depth was the nature of chaos in multiparticle systems and its role in controlling aspects of the dynamical behavior of statistical averages. Although we did not always agree, these discussions certainly attuned my thinking about the problem. In this contribution, I present a discussion of how to think about chaos in a physical way, from the point of view of realistic experiments. The basis of the arguments applies to both classical and quantum systems and serves to bring together these two great dynamical traditions that are seemingly at such odds with each other. The work reported here is the result of several collaborations between subsets of the authors. I do not know what Henry's opinions would have been on this subject, however, I am sure he would not have been quiet!

INTRODUCTION

In classical theory—unlike in quantum mechanics—the status of dynamical chaos is apparently clear: chaos exists observationally and is well-described theoretically by Newton's equations. (Nevertheless, even here, a deeper look at the physical meaning of chaos is certainly helpful; we return to this presently.) It is in the context of quantum theory, however, that the notion of chaos appears so puzzling and mysterious. Because of the Kosloff–Rice theorem[1,2] and related results,[3,4] it is clear that quantum evolution of the wave function or the density matrix is integrable; hence, chaos cannot exist in quantum mechanics in the canonical sense. This is the basic stumbling block to defining a quantum notion of nonintegrability.

One may argue that real quantum systems are always coupled to an environment and hence their evolution —"for all practical purposes," (FAPP), in Bell's famous phrase[5]—should be described by unitarity-breaking master equations rather than the unitary evolution assumed by the Kosloff–Rice theorem. Perhaps this way out, although not fundamentally satisfying to the purist, is enough by itself, but it is easy to see what is wrong with the argument. Fundamentally, any fully quantum dynamical description must arise from a Hamiltonian describing the system, its environment, and their coupling. The master equation represents the evolution of the reduced density matrix for the system that arises from tracing over the environment variables in the full (system plus environment) density matrix. Since the full evolution must satisfy Kosloff–Rice, the evolution of the reduced density matrix cannot be nonintegrable.

Thus, the fundamental problem we are faced with is this: we are familiar with chaos in the real world, but our fundamental theory of dynamics—which passes every experimental test beautifully—seemingly does not have a natural place within it to tolerate even the existence of the concept. This should not come as a surprise; after all, the trajectories of classical mechanics are apparently "real" and effortless to contemplate, but they too, have no natural place in quantum mechanics. Now it is true that quantum mechanics is an intrinsically probabilistic theory, but that, in itself, is not the real issue. Classical theory can be easily cast as fundamentally probabilistic as well, via the classical Liouville equation describing the evolution of a classical probability in phase space. (For an attempt at an even closer analogy, see Ref. 6 and the discussion in Ref. 7.) As discussed below, the key point is rather that, unlike special relativity, where $v/c \to 0$ smoothly transitions between Einstein and Newton, the limit $\hbar \to 0$ is singular. The symmetries underlying quantum and classical dynamics—unitarity and symplecticity, respectively—are fundamentally incompatible with the opposing theory's notion of a physical state: quantum-mechanically, a positive semidefinite density matrix; classically, a positive phase-space distribution function.

In the rest of this article, we expose the singular nature of the $\hbar \to 0$ limit and discuss a physical point of view—applicable to both classical and quantum systems—that will enable us to explain how trajectories and chaos appear in real experiments.

At this point, it should be clear that the questions taken up in this contribution are not those usually considered under the research area called "quantum chaos." There, one is primarily interested in the quantum behavior of a system with a classically chaotic Hamiltonian, what might happen to the validity of certain approximations (e.g., semiclassical approaches to calculating the quantum propagator) and whether classical trajectories and phase space structures can provide insight into the nature of quantum wavefunctions. However, one does not actually study quantum chaos.

We distinguish between *isolated* evolution, where the system state evolves without any coupling to the external world, *unconditioned open* evolution, where the system evolves coupled to an external environment but where no information regarding the system is extracted from the environment, and *conditioned open* evolution, where such information *is* extracted. In the third case, the evolution of the physical state is driven by the system evolution, the coupling to the external world, and by the fact that observational information relating to the state has been obtained. This last aspect—system evolution conditioned on the measurement results via Bayesian inference—leads to an intrinsically nonlinear evolution for the system state, and distinguishes it from unconditioned evolution. Although the concept of conditioned evolution of the system state is familiar to engineers and mathematicians, especially systems engineers and control theorists,[8,9] it is not yet completely familiar territory to the majority of physicists. Nevertheless, driven by the impressive progress in the experimental state-of-the-art in quantum and atomic optics and in nanoscience,[10,11] these notions are now being employed as everyday tools at least in some fields.

The conditioned evolution provides, in principle, the most realistic possible description of an experiment. To the extent that quantum and classical mechanics are eventually just methodological tools to explain and predict the results of experiments, this is the proper context in which to compare them and discuss the nature of predictions for real experiments. The explicit incorporation of information gained via measurement also provides a structure to address the quantum–classical transition more generally, and to frame the question of where chaos exists within this structure.

The fact that quantum and classical mechanics are fundamentally incompatible in many ways, yet the macroscopic world is well-described by classical dynamics has puzzled physicists ever since the laying of the foundations of quantum theory. It is fair to say that not everyone is satisfied with the state of affairs—including many seasoned practitioners of quantum mechanics.

Of course, the notion of measurement in quantum mechanics—the denial of reality to system properties unless they are measured—is such a revolutionary concept that it engenders much more unease,[12] even today. The problem is that, were quantum mechanics the final theory, it could deny reality to the measurement results themselves unless they were observed by another system and so on, *ad infinitum*. In order to "solve" the "measurement problem," it originally appeared impossible to think of quantum mechanics as a fundamental theory without relying on the existence of a classical world-view within which to embed it.[13] Although we still cannot dispel the unease invoked by the measurement problem, it is important to stress that the quantum–classical transition can be understood independently. This transition should not be confused with the measurement problem.

A partial understanding of the classical limit arises from the idea—familiar from nonequilibrium statistical mechanics—that weak interactions of a system with an environment are universal.[14] These interactions can effectively suppress certain nonclassical terms in the quantum evolution.[15–18] However, at best they only allow for the emergence of a classical probabilistic evolution and it can be shown that the mere existence of such interactions is insufficient to yield classical evolution in all cases.[19] Finally, this picture alone cannot explain the results of actual measurements, where information can be continuously extracted from the environment and used to define operational notions of a trajectory. We now go into these questions in more detail.

ISOLATED AND OPEN EVOLUTION

Suppose we are given an arbitrary system Hamiltonian $H(x, p)$ in terms of the dynamical variables x and p; we will be more specific about the precise meaning of x and p as position and momentum later. The Hamiltonian is the generator of time evolution for the physical system state, provided there is no coupling to an environment or measurement device. In the classical case, we specify the initial state by a positive phase space distribution function $f_{Cl}(x, p)$; in the quantum case, by the (position-representation) positive semidefinite density matrix $\rho(x_1, x_2)$ or, completely equivalently, by the corresponding Wigner distribution function $f_W(x, p)$ (not positive). The Wigner distribution[20–23] is a "half-Fourier" transform of $\rho(x_1, x_2)$, defined by

$$f_W(x, p) = \frac{1}{2\pi\hbar}\int d\Delta \rho\left(x + \frac{1}{2}\Delta, x - \frac{1}{2}\Delta\right)\exp\left(\frac{-ip\Delta}{\hbar}\right), \quad (1)$$

where $x \equiv (x_1 + x_2)/2$ and $\Delta \equiv x_1 - x_2$.

The evolution of an *isolated* system is then given by the classical and quantum Liouville equations for the *fine-grained* distribution functions (i.e., the evolution is entropy-preserving):

$$\partial_t f_{Cl}(x, p) = -\left[\frac{p}{m}\partial_x - \partial_x V(x)\partial_p\right]f_{Cl}(x, p), \quad (2)$$

$$\partial_t f_W(x, p) = -\left[\frac{p}{m}\partial_x - \partial_x V(x)\partial_p\right]f_W(x, p)$$
$$+ \sum_{\lambda=1}^{\infty} \frac{(\hbar/2i)^{2\lambda}}{(2\lambda+1)!}\partial_x^{2\lambda+1} V(x)\partial_p^{2\lambda+1} f_W(x, p), \quad (3)$$

where we have assumed for simplicity that the potential $V(x)$ can be Taylor expanded; this does not alter the nature of any of the following arguments. Note that these evolutions are both linear in the respective distribution functions.

The limiting form $f_{Cl}(x, p) = \delta(x - \bar{x})\delta(p - \bar{p})$ is allowed classically, and, on substitution in Equation (2), yields the expected Newton equations. These may then be interpreted as equations for the particle position and momentum, although we must emphasize that this identification is only formal at this stage. Quantum mechanically, this ultralocal limit is not permitted since $f_W(x, p)$ must be square-integrable, therefore—even formally—no direct particle interpretation can exist. In both cases, if one allows for initially localized distributions but which nevertheless have some finite width, it is easy to see that if $V(x)$ is nonlinear, quite generically the distribution will eventually spread over the allowed phase space and not remain localized.

As alluded to in the INTRODUCTION, the extension to open systems is conceptually trivial, but very difficult to implement in practice. To the original system Hamiltonian, we now add pieces representing the environment and the system–environment coupling. If the environment is in principle unobservable, then a (nonlocal in time) linear master equation for the reduced density matrix of the system is—in theory—derivable by tracing over the environmental variables. In practice, tractable equations are impossible to obtain without drastic simplifying assumptions, such as weak coupling, timescale separations, and simple forms for the environmental and coupling Hamiltonians. In any case, the important point to note is that the act of tracing over the environment does not change the linear nature of the equations. Generally speaking, master equations describing open evolution of *coarse-grained* distributions augment the right-hand-side of Equations (2) and (3) with terms containing dissipation and diffusion kernels connected via generalized fluctuation–dissipation relations.[24–26] Although the classical diffusion term vanishes in the limit of zero temperature for the environment, this is not true quantum mechanically due to the presence of zero-point fluctuations.

CONTINUOUS MEASUREMENT AND CONDITIONED EVOLUTION

In contrast to classical theory, where measurement can be, in principle, a passive process, in quantum theory measurement creates an irreducible disturbance on the observed system (quantum "backaction"). This being so, if our aim is that measurement yield dynamical information—rather than strongly influence dynamics—the desired measurement process must yield a limited amount of information in a finite time. Hence, simple projective (von Neumann) measurements are clearly not appropriate because they yield complete information instantaneously via state projection. Nevertheless, this fundamental notion of measurement can be easily extended[7] to devise schemes that extract information continuously.[27–34] The basic idea is to have the system of interest interact weakly with another (e.g., atom interacting with an electromagnetic field) and make projective measurements on the auxiliary system (e.g., photon counting). Because of the weak interaction, the state of the auxiliary system gathers very little information about the system of interest, and therefore this system, in turn, is only perturbed slightly by the measurement backaction. Only a small component of the information gathered by the projective measurement of the auxiliary system relates to the system of interest, and a continuous limit of the measurement process can be taken.

In the continuous limit, the evolution of the system density matrix is fundamentally different from the equations discussed above for the case of open evolution. The master equation describing the evolution of the reduced density matrix conditioned on the results of the measurements contains a term that reflects the gain in information arising from the measurement record ("innovation" in the language of control theory). This term, arising from applying a continuous analog of Bayes' theorem, is intrinsically nonlinear in the distribution function. The coupling to an external probe (and the associated environment) will also cause effects very similar to the open evolution considered earlier, and there can once again be dissipation and diffusion terms in the evolution equations. The primary differences between the classical and quantum treatments, aside from the kinematic constraints on the distribution functions, are the following: (1) the (nonlocal in p) quantum evolution term in Equation (**3**), and (2) an irreducible diffusion contribution due to quantum backaction reflecting the *active* nature of quantum measurements.

We now consider a simple model of position measurement to provide a measure of concreteness. In this model, we assume that there are no environmental channels aside from those associated with the measurement. Suppose we have a single quantum degree of freedom, position in this case, undergoing a weak, ideal continuous measurement.[27–34] Here "ideal" refers to no loss of information during the measurement, that is, a fine-grained evolution with no increase in entropy. Then, we have two coupled equations, one for the measurement record $y(t)$,

$$dy = \langle x \rangle dt + \frac{dW}{\sqrt{8k}}, \qquad (4)$$

where dy is the infinitesimal change in the output of the measurement device in time dt, the parameter k characterizes the rate at which the measurement extracts information about the observable, that is, the *strength* of the measurement,[35] and dW is the Wiener increment describing driving by Gaussian white noise,[36] the difference between the actually observed value and that expected. The other equation—the nonlinear stochastic master equation (SME)—specifies the resulting conditioned evolution of the system density matrix, given in the Wigner representation,

$$f_W(x, p, t + dt) = \left[1 + dt\left[-\frac{p}{m}\partial_x + \partial_x V(x)\partial_p + D_{BA}\partial_p^2\right]\right.$$

$$\left. + dt \sum_{\lambda = 1}^{\infty} \frac{(\hbar/2i)^{2\lambda}}{(2\lambda + 1)!}\partial_x^{2\lambda + 1}V(x, t)\partial_p^{2\lambda + 1}\right]f_W(x, p, t) \quad (5)$$

$$+ dt\sqrt{8k}(x - \langle x \rangle)f_W(x, p, t)dW,$$

where $D_{BA} = \hbar^2 k$ is the diffusion coefficient arising from quantum backaction and the last (nonlinear) term represents the conditioning due to the measurement. In principle, there is also a (generalized) damping term,[37] but if the measurement coupling is weak enough, it can be neglected. If we choose to average over all the measurement results, which is the same as ignoring them, then the conditioning term vanishes, but *not* the diffusion from the measurement backaction. Thus, the resulting linear evolution of the coarse-grained quantum distribution is not the same as the linear fine-grained evolution (3), but yields a conventional open-system master equation. Moreover, for a given (coarse-grained) master equation, different underlying fine-grained SMEs may exist, specifying different measurement possibilities.

In contrast to the quantum case, the corresponding (ideal) classical conditioned master equation (set $\hbar = 0$ in (5), holding k fixed),

$$f_{Cl}(x, p, t + dt) = \left[1 - dt\left[\frac{p}{m}\partial_x - \partial_x V(x)\partial_p\right]\right]f_{Cl}(x, p, t)$$

$$+ dt\sqrt{8k}(x - \langle x \rangle)f_{Cl}(x, p, t)dW, \quad (6)$$

does not have the backaction term as these classical measurements are *passive*: averaging over all measurements simply gives back the Liouville equation (2), and there is no difference between the fine-grained and coarse-grained evolutions in this special case. (In general, classical diffusion terms from ordinary open evolution can also coexist, as in the more general *a posteriori* evolution specified by the Kushner–Stratonovich equation.[38]) As a final point, we delay our discussion of how the classical trajectory limit is incorporated in Equation (6), that is, the precise sense in which the "the position of a particle is what a position-detector detects", to the next section.

QCT: THE QUANTUM–CLASSICAL TRANSITION

If quantum mechanics is really the fundamental theory of our world, then an effectively classical description of macroscopic systems must emerge from it—the so-called quantum–classical transition (QCT). It turns out that this issue is inextricably connected with the question of the physical meaning of dynamical nonlinearity discussed above. Having written down the relevant evolution equations, we now analyze two notions of the QCT and how they emerge from the equations.

Quantum mechanics is intrinsically probabilistic, but classical theory—as shown above by the existence of the delta-function limit for the classical distribution function—is not. Since the Newton equations provide an excellent description of observed classical systems, including chaotic systems, it is crucial to establish how such a localized, or trajectory, description can arise quantum mechanically. We call this the *strong* form of the QCT. Of course, in many situations, only a statistical description is possible even classically, and here we demand only the agreement of quantum and classical distributions and the associated dynamical averages. This defines the *weak* form of the QCT.

It is clear that if the strong form of the QCT holds, then, via trivial coarsegraining, the weak form follows automatically. The reverse is not true, however: results from a coarse-grained analysis cannot be applied to the finegrained situation. Moreover, the violation of the conditions necessary to establish the strong form of the QCT need not prevent the existence of a weak QCT. We now discuss and establish the conditions under which these transitions occur. Since the strong form of the QCT requires treating the localized limit, a cumulant expansion for the distribution function immediately suggests itself, whereas, for the more nonlocal issues relevant to the weak form of the QCT, a semiclassical analysis turns out to be natural.

Strong Form of the QCT: Chaos in the Classical Limit

It is easy to see that the strong form of the QCT is impossible to obtain from either the isolated or open evolution equations for the density matrix orWigner function. As mentioned already, for a generic dynamical system, a localized initial distribution tends to distribute itself over phase space—and then continue to evolve—either in complicated ways (isolated system) or asymptote to an equilibrium state (open system), whether classically or quantum mechanically. In the case of conditioned evolution, however, the distribution can be localized due to the information gained from the measurement, and evolve in a quite different manner. In order to quantify how this happens, let us first apply a cumulant expansion to the (fine-grained) conditioned classical evolution (6). This results in the following equations for the centroids ($\bar{x} \equiv \langle x \rangle$, $\bar{p} \equiv \langle p \rangle$),

$$d\bar{x} = \frac{\bar{p}}{m}dt + \sqrt{8k}C_{xx}dW,$$
$$d\bar{p} = \langle F(x) \rangle dt + \sqrt{8k}C_{xp}dW, \tag{7}$$

where

$$F(x) = -\partial_x V(x),$$
$$C_{AB} = \frac{\langle AB \rangle + \langle BA \rangle - 2\langle A \rangle \langle B \rangle}{2}, \tag{8}$$

along with a hierarchy of coupled equations for the time-evolution of the higher cumulants. These equations are the continuous measurement, real-world, analog of the formal ultralocal Newtonian limit of the distribution function in the classical Liouville equation (2). Whereas Equations (7) always apply, our aim is to determine the conditions under which the cumulant expansion effectively truncates and brings their solution very close to that of the Newton equations. This will be true provided the noise terms are small (in an average sense) and the force term is localized, that is, $\langle F(x) \rangle = F(\bar{x}) + \ldots$, the corrections being small. The required analysis involves higher cumulants and was carried out elsewhere.[39–41] It turns out that the distribution is localized provided

$$8k \gg \sqrt{\frac{(\partial_x^2 F)^2 |\partial_x F|}{2mF^2}} \qquad (9)$$

and the motion of the centroid effectively defines a smooth classical trajectory—the low-noise condition—as long as

$$k \gg \frac{2|\partial_x F|}{S}, \qquad (10)$$

where S is the action scale of the system. Note that this condition does not bound the measurement strength: classically we can always extract as much information as needed—at least in principle—to gain the trajectory limit. This, then, is a "realistic" derivation of the Newton equations.

We now turn to the quantum version of these results. In this case, the analogous cumulant expansion gives exactly the same equations for the centroids as above, whereas the equations for the higher cumulants are different. (The evolution of classical and quantum averages is the same to Gaussian order, with the first differences arising at the next order.[42]) We can again investigate whether a trajectory limit exists. Localization holds in the weakly nonlinear case if the classical condition above is satisfied. In the case of strong nonlinearity, the inequality becomes[39–41]

$$8k \gg \frac{(\partial_x^2 F)^2 \hbar}{4mF^2}. \qquad (11)$$

Because of the backaction, the low-noise condition is implemented in the quantum case by a double-sided inequality:

$$\frac{2|\partial_x F|}{s} \ll \hbar k \ll \frac{|\partial_x F|s}{4}, \qquad (12)$$

where the action is measured in units of $\hbar$, s being dimensionless. The left inequality is the same as the classical one discussed above, however the right inequality is essentially quantum mechanical. The measurement strength cannot be made arbitrarily large as the backaction will result in too large a noise in the equations for the centroids. As the action s is made larger, both inequalities are satisfied for an ever wider range of values of k. For sufficiently large s, the actual value of k becomes irrelevant and the dynamics becomes effectively classical.

To recapitulate, for continuously measured quantum systems, trajectories that emerge in the macroscopic limit follow the Newton equations, and hence, can be chaotic as shown elsewhere.[39–41] Thus, as speculated in a prescient paper by Chirikov,[43] measurement indeed provides the missing link between "quantum" and "chaos," at least in the classical limit.

Finally, in experiments one usually considers the measurement record itself rather than the estimated state of the system as we have discussed so far. Because measurement introduces a white noise, it is important to investigate the condition under which the record tracks the estimate faithfully. If Δt is the time over which the continuous measurement is averaged to obtain the record (this averaging being a necessary part of any finite-bandwidth experiment), and we allow ourselves a maximum of Δx as the position noise, it is easy to see that the measurement strength needs to satisfy[39,40]

$$8k > \frac{1}{\Delta t (\Delta x)^2}. \quad (13)$$

To demonstrate these results for a concrete example, we revisit the results of References 39–40 for a driven, Duffing oscillator, with system Hamiltonian

$$H = \frac{p^2}{2m} + Bx^4 - Ax^2 + \Lambda x \cos(\omega t), \quad (14)$$

with $m = 1$, $B = 0.5$, $A = 10$, $\Lambda = 10$, and $\omega = 6.07$. This Hamiltonian has been used before in studies of quantum chaos[44] and quantum decoherence[45] and, in the parameter regime used, a substantial area of the accessible phase space is stochastic.

Numerical calculations at various values of $\hbar$ confirm that as $\hbar$ is reduced, both the steady-state variance, and the resulting noise (for optimal measurement strengths) are reduced, as expected. Because the dynamical time scale of this problem is 1–0.1 (in units of the driving period), the continuous observation record was averaged over a period of 0.01. Similarly, since the range of the motion covers distances of $O(10)$, we demand that the position be tracked to an accuracy of 0.01 to define an effective "trajectory." To satisfy this, we need to have $k \sim O(10^5)$ or larger (compare with (13)). In our example, we choose the energy to be $O(10^2)$, the corresponding typical action turns out to be $O(10)$, and the typical nonlinearity makes the right-side of Equation (9), $O(1)$. We see that a choice of $\hbar = 10^{-5}$ and $k = 10^5$, satisfies all the constraints for a classical motion. In FIGURE 1 we demonstrate that in this regime, localization is maintained along with low levels of trajectory noise. FIGURE 1A shows a typical phase space trajectory, with the position variance during the evolution, $V_x \equiv (\Delta x)^2$, plotted in FIGURE 1B. We find that the width Δx is always bounded by 3.4×10^{-3}. Furthermore, as is immediately evident from the smoothness of the trajectory in FIGURE 1A, the noise is also negligible on these scales. Additionally, one can verify that the quantum trajectory evolution and that given by a classical trajectory with an equivalent noise are essentially identical—and chaotic—yielding a Lyapunov exponent of 0.57. We return to discuss the Lyapunov exponent later.

Chaos and the Weak Form of the QCT

The weak form of the QCT utilizes the *coarse-grained* distribution function (averaging over all measurements), whereas the strong form refers to the *fine-grained* distribution for a single measurement realization. It is important to reiterate that nonexistence of the strong form of the QCT does not influence the existence of the weak form of the QCT: it does not matter if the distribution is too wide, as long as the classical and quantum distributions agree, and, even if the backaction noise is large, the coarse-grained distribution can remain smooth and the weak quantum-classical correspondence still exist. Consequently, this correspondence has to be approached in a different manner. In fact, the weak version is just another way to state the conventional decoherence idea;[15–18] however, as discussed elsewhere,[46] mere suppression of quantum interference does not guarantee the QCT, even in the weak form.

We now focus on a semiclassical analysis of the weak QCT for bounded, classically chaotic open systems.[46] This analysis is best regarded as a *regularization* of the singular $\hbar \to 0$ limit via the environmental interaction. This is distinct from the state *localization* characteristic of the strong form of the QCT. Given a small, but finite, value of $\hbar$, the aim is to establish the existence of a timescale beyond which the dynamics of open quantum and classical systems becomes statistically equivalent if the environmental interaction is sufficiently strong.

It has been demonstrated[46] that, for a bounded open system with a classically chaotic Hamiltonian, the weak form of the QCT is achieved by two parallel processes, both relying essentially on the existence of environmental diffusion. First, the semiclassical approximation for quantum dynamics, which breaks down for classically chaotic systems due to overwhelming nonlocal interference, is recovered as the environmental interaction filters these effects. Second, environmental noise restricts the foliation of the unstable manifold, the set of points that approach a hyperbolic point in reverse time, allowing the semiclassical wavefunction to track this modified classical geometry. In this way, the noise prevents classical chaos from breaking the semiclassical approximation as $\hbar \to 0$, and thus regularizes this limit. Note that this approach explicitly incorporates both the stretching and folding typical of hyperbolic regions, as well as the role of the environment as a filter on a phase-space quantum distribution.

We begin with a simple model of a quantum system weakly coupled to the environment so as to maintain complete positivity for the subsystem density matrix,

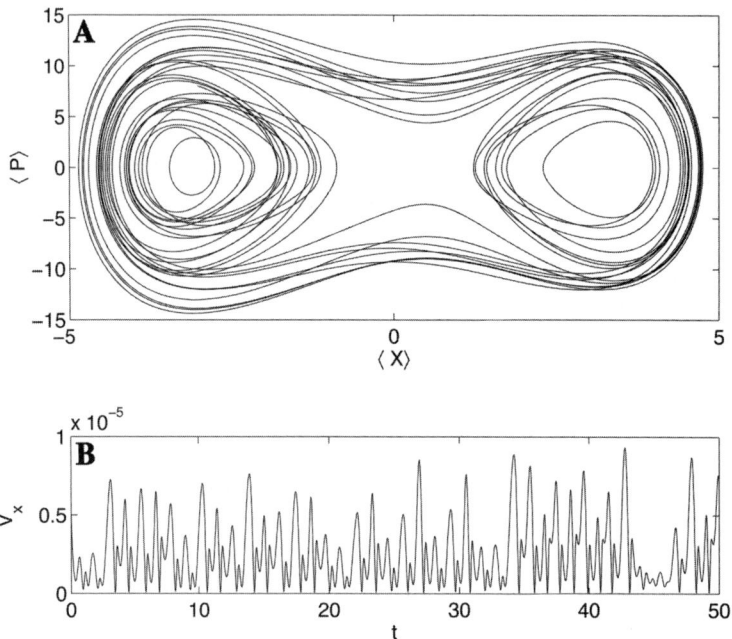

FIGURE 1. (**A**) The quantum trajectory in phase space for a continuously measured Duffing oscillator,[39–41] with $\hbar = 10^{-5}$ and $k = 10^5$. (**B**) The position variance, V_x, as a function of time. Note the smallness of the scale on the y-axis.

$\rho(t)$, while subjecting it to a, time-local, unitarity-breaking interaction. These conditions mathematically constrain the master equation to be of the (so-called) Lindblad form.[47,48] If this environmental interaction couples to the position, as is often the case, the master equation takes the form

$$\frac{\partial f_w}{\partial t} = L_{cl} f_w + L_q f_w + D \frac{\partial^2 f_w}{\partial p^2}, \qquad (15)$$

where L_{cl}, the classical Liouville operator, and L_q, the quantum correction, can be easily identified from Equation (3). We note in passing that whereas the sum of L_{cl} and L_q is clearly unitary, individually the operators are not unitary.[19] In this simple master equation, we have neglected the dissipative environmental channel and retained the diffusive channel for two reasons: (1) the coupling to the environment is always assumed to be weak and the dissipative timescales are, hence, very long, longer than the dynamical timescales of interest, (2) the weak form of the QCT arises only from the diffusive channel, hence, dissipative effects are not of interest here.

When $L_q = 0$, this equation reverts to the classical Fokker–Planck equation. It is important to keep in mind that the specific form of the diffusion coefficient depends strongly on the physical situation envisaged. Thus, if the master equation describes a weakly coupled, high temperature environment, $D = 2m\gamma k_B T$ (γ is the damping coefficient),[49] whereas for a weak, continuous measurement of position, the diffusion due to quantum backaction is $D = \hbar^2 k$.[27-34] The following discussion holds for all of these cases.

Once the QCT occurs, the effects of L_q in the evolution specified by Equation (15) are subdominant. Therefore, to understand how environmental noise acts in this limit, it suffices to consider the behavior of the corresponding classical Fokker–Planck equation. To do this, it is convenient to examine the underlying Langevin equations for noisy trajectories that unravel the evolution of the classical distribution function when $L_q = 0$. These are given by

$$dq = \frac{p}{m} dt \qquad (16)$$

$$dp = f(q) dt + \sqrt{2D} dW.$$

Using weak-noise perturbation theory, one can perform an expansion about a hyperbolic fixed point and in this way obtain the spreading of the position and momentum due to the diffusion. As a trajectory evolves, it simultaneously smooths over a transverse width in phase space of size $\sqrt{Dt/m\lambda}$, where λ is the local Lyapunov exponent.[46]

The smoothing implies a termination in the development of new phase space structures at some finite time t^*, whose scaling behavior can be determined. (Caveat: this need not be true in a non-compact phase space.) The average motion of a trajectory is identical to its deterministic motion, so that at time t, if the initial length in phase space is u_0 (u has units of square-root of phase-space area), its current length will be approximately $u_0 e^{\bar{\lambda} t}$ since its forward time evolution will be dominated by its component in the unstable direction. Here $\bar{\lambda}$ is the time-averaged positive Lyapunov exponent. If the region is bounded within a phase space area A, the typical distance between neighboring folds of the trajectory is given by

$$l(t) \approx \frac{A}{u_0} e^{-\bar{\lambda}t}, \qquad (17)$$

where $l(t)$ still carries the units of the square root of phase space area. However, since phase structures can only be known to within the width specified above, the time at which any new structure will be smoothed over is defined by

$$l(t^*) \approx \sqrt{\frac{Dt^*}{m\bar{\lambda}}}. \qquad (18)$$

These two equations can be used to determine t^*, which only weakly depends on D and the prefactor in Equation (**17**). Due to smoothing, one does not see an ergodic phase space region, but one in which the large, short-time features that develop prior to t^* are pronounced and the small, long-time features that develop later are smoothed over by the averaging process. Therefore, to approximate noisy classical dynamics, a quantum system need not track all of the fine scale structures, but only the larger features that develop before the production of small scale structures terminates.

To establish the conditions under which quantum dynamics can track this modified phase space geometry, a semiclassical analysis can be performed. In the Wigner function formalism, the breakdown of the semiclassical approximation for chaotic systems can be associated with an appealing geometric picture[23,50,51] based on a uniform approximation in phase space—the Berry construction. We now use this construction to understand how quantum interference in phase space is smoothed over by the diffusion associated with environmental coupling.

A general mixed state is an incoherent superposition of pure state Wigner functions, where an individual semiclassical pure state Wigner function can be formed by substituting the Van-Vleck semiclassical wavefunction in Equation (**1**). If we allow q to be perturbed by noise we can rewrite the classical action,[52]

$$S(q, t) \approx S(q_C, t) - \sqrt{2D} \int_0^t dt\, \xi(t) q_C(t).$$

Following Berry,[50] we rewrite the action for the ith solution to the Hamilton–Jacobi equation as

$$\begin{aligned} S_i(q_C, t) &= \int_{q_C(0)}^{q_C(t)} dq'\, p_i(q', t) - \int_0^t dt'\, H(q_C(0), p_i(q_C(0), t')) \\ &\equiv \int_0^t dt'\, \mathcal{H}_i(t'), \end{aligned} \qquad (19)$$

where $p_i(q, t)$ is the ith branch of the momentum curve for a given q. If we average over all noisy realizations, after separating the contributions from identical branches, the following suggestive expression for the noise averaged semiclassical Wigner function obtains:

$$\frac{1}{2\pi\hbar}\int_{-\infty}^{\infty} dX \exp\left(-\frac{DtX^2}{2\hbar^2}\right)\left(\sum_i \mathcal{I}_{ii} \times \exp\left[\frac{i}{\hbar}\left\{\int_{\bar{q}_-}^{\bar{q}_+} dq' p_i(q',t) - pX\right\}\right]\right.$$

$$\left. + 2i\sum_{i<j} \mathcal{I}_{ij} \sin\left[\frac{1}{\hbar}\left\{\int_{q_C(0)}^{\bar{q}_+} dq' p_i(q',t) - \int_{q_C(0)}^{\bar{q}_-} dq' p_j(q',t)\right\}\right.\right. \quad (20)$$

$$\left.\left. -\int_0^t (dt'(\mathcal{H}_i - \mathcal{H}_j) + \phi_i - \phi_j)\right]\right),$$

$$\mathcal{I}_{ij} \equiv \frac{C_i(\bar{q}_+, t) C_j(\bar{q}_-, t)}{\sqrt{|J_i(\bar{q}_+, t)||J_j(\bar{q}_-, t)|}}, \quad (21)$$

for Jacobian determinant $J_i(q, t)$ and transport coefficient $C_i(q, t)$; $\bar{q}_\pm \equiv q \pm X/2$ and $\phi_i = \pi v_i$ where v_i is the ith Maslov index.[53]

The dominant contributions to the integrals can be analyzed in the stationary phase approximation.[54] If $D = 0$, these would contribute phase coherences at values of X that satisfy $p_i(q + X/2, t) + p_i(q - X/2, t) - 2pX = 0$ for the first term in the sum and $p_i(q + X/2, t) + p_j(q - X/2, t) - 2pX = 0$ for the second term, the former being the famous Berry midpoint rule. For a chaotic system, Berry argued that, due to the proliferation of momentum branches, $p_i(q, t)$, arising from the infinite number of foldings of a bounded chaotic curve as $t \to \infty$, a semiclassical approximation would eventually fail, since the interference fringes stemming from a given p_i could not be distinguished after a certain time from those emanating from the many neighboring branches.[54] Although the precise value of this time has since been challenged numerically, the essential nature of this physical argument has remained valid.[55,56]

In the present case, however, the presence of noise acts as a dynamical Gaussian filter, damping contributions for any solutions to the above equation that are greater than $X \approx \hbar/\sqrt{Dt}$. In other words, noise dynamically filters the long "De Broglie" wavelength contributions to the semiclassical integral, the very sort of contributions that generally invalidate such an approximation. If we rescale the above result and combine it with our understanding of how noise affects classical phase space structures, we can qualitatively estimate whether or not a semiclassical picture is a valid approximation to the dynamics. As already discussed, t^* is the time when the formation of new classical structures ceases and $l(t^*)$ is the associated scale over which classical structures are averaged. The key requirement is then that the semiclassical phase filters contributions of size

$$\frac{\sqrt{\bar{\lambda} m \hbar}}{\sqrt{Dt^*}} \lesssim l(t^*). \quad (22)$$

In other words, for a given branch, the phases with associated wavelengths long enough to interfere with contributions from neighboring branches are strongly damped, and the intuitive semiclassical picture of classical phase space distributions decorated by local interference fringes recovered.

The weak form of the QCT is completed when the inequality (22) is satisfied. Substituting the scale of classical smoothing (18) in this inequality, we find

$$Dt^* \gtrsim \bar{\lambda} m \hbar. \quad (23)$$

(Note that the purely classical quantity t^* is first independently determined by solving **(18)** and then compared to the right side of the above equation.) The left hand side of the inequality contains the mutually dependent t^* and D, the right hand side, however, depends only on fixed properties of the system and $\hbar$. This condition, therefore, defines a threshold at which the semiclassical approximation becomes stable and that may be set in terms of either D or t^*. Once the threshold is met, t^* becomes the time beyond which the semiclassical description is valid. The semiclassical nature of this condition becomes more evident on defining $s = l(t^*)^2$, which, given that l^2 is an areal scale in phase space for the diffusion averaged dynamics, has dimensions of action. A physical interpretation is more apparent on rewriting **(23)** as $S = l(t^*)^2 \gtrsim \hbar$, which is readily identified as the usual condition for the validity of a semiclassical analysis.

The weak form of the QCT can also be demonstrated by using the Duffing example.[46] The dynamical evolution of the bounded motion is dominated by the

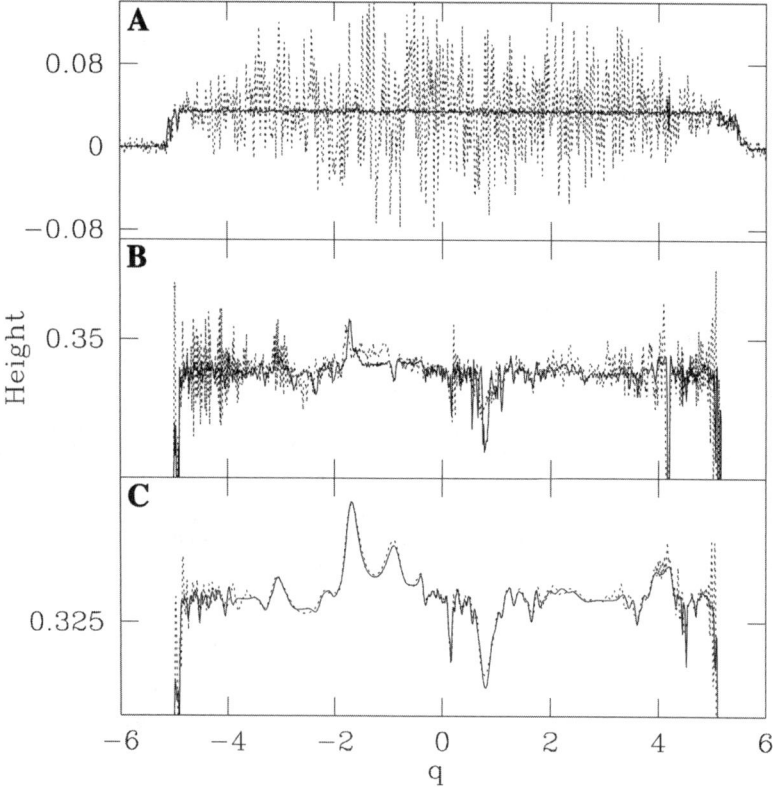

FIGURE 2. The weak form of the QCT for a driven Duffing oscillator:[46] sectional cuts of Wigner functions (*dashed lines*) and classical distributions (*solid lines*), after 149 drive periods, taken at $p = 0$ for diffusion coefficient values **(A)** $D = 10^{-5}$, **(B)** $D = 10^{-3}$, and **(C)** $D = 10^{-2}$. Other parameter values are stated in the text; the height is specified in scaled units.

homoclinic tangle of a single hyperbolic fixed point. As a result, the long-time chaotic evolution can be completely characterized by the unstable manifold associated with that fixed point.[56] The value of $\hbar$ is now set to $\hbar = 0.1$, significantly larger than when studying the strong form of the QCT.

The evolution of the corresponding distributions was numerically calculated for both the classical and quantum master equations. FIGURE 2 shows sectional cuts at $p = 0$ of the quantum and classical phase space distribution functions for three different values of the diffusion coefficient, $D = 10^{-5}$, 10^{-3}, and 10^{-2} after time $T = 149$ evolution periods. As already mentioned, t^* varies slowly with D, and in the three cases shown, t^* ranges only from about 20 to 14 (note that $t^* \ll T$). It is easy to check that the inequality (23) is strongly violated $D = 10^{-5}$, mildly violated for $D = 10^{-3}$, and approximately satisfied for $D = 10^{-2}$. For $D = 10^{-5}$, the classical and quantum sections show no similarities, as expected. The quantum Wigner function also shows large negative regions, reflecting strong quantum interference. On increasing D to 10^{-3} the magnitude of quantum coherence decreases dramatically and the classical and quantum slices have the same average value, as well as specific agreement on some large scale features. The two disagree, as expected, on the small scale structures. This indicates that, although the quantum and classical distributions do not exactly match, the Wigner function has now become sensitive to the larger features of the noise averaged classical distribution function, indicative of the transition to a semiclassical regime. At $D = 10^{-2}$, there is near perfect agreement between classical

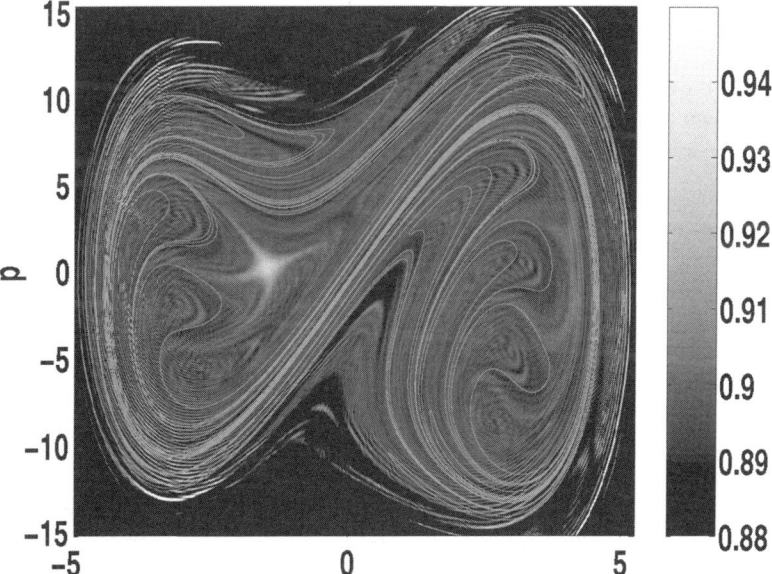

FIGURE 3. Phase space rendering of the Wigner function at time $t = 149$ periods of driving.[40] The early time part of the unstable manifold associated with the noise-free dynamics is shown in gray. The value of $D = 10^{-5}$ is not sufficient to wipe out all the quantum interference, which, as expected, is most prominent near turns in the manifold.

and quantum distribution functions, save on the smallest scales. When D is of order unity, the inequalities enforcing the strong QCT at the level of individual trajectories[39–41] are satisfied and the agreement is essentially exact. However, as indicated by FIGURE 2C, detailed agreement for quantum and classical distribution functions can begin at much smaller values of the diffusion constant.

For more detailed evidence that, at $D = 10^{-3}$, one is entering a semiclassical regime, in FIGURE 3 we superimpose an image of the large scale features of the classical unstable manifold on top of the full quantum Wigner distribution at $D = 10^{-3}$ after 149 drive periods (FIG. 2B). The quantum phase space clearly exhibits local interference fringes around the large lobe-like structures associated with the short-time evolution of the unstable manifold. The appearance of local fringing about classical structures is direct evidence of a semiclassical evolution, where interference effects appear locally around the backbone of a classical evolution. This is in sharp contrast to the global diffraction pattern seen for $D = 0$, where the contributions from individual curves cannot be distinguished, suppressing the appearance of any classical structure.[45]

CHAOS IN QUANTUM MECHANICS

At this point, our analysis of measured quantum dynamical systems may be said to have harmonized quantum and classical mechanics in the sense that the strong and weak forms of the QCT have appeared naturally. Although this is certainly pleasing, we wish to go further and ask whether the formalism can be tested by making predictions that are experimentally verifiable and depend uniquely on the nonlinear nature of the conditioned evolution. One very interesting idea is the real-time control of quantum systems using state-estimation as pioneered by Belavkin[57–59] or direct feedback of the measured classical current.[60] Quantum feedback control applications[57–59] have their own importance, but we now return to the original burning question: is there chaos in quantum mechanics?

In a limiting case, the answer is clearly in the affirmative. We have already shown that quantum distributions, provided certain conditions are met, can evolve while staying localized and be only very weakly perturbed by noise. In the classical limiting case, we recover localized classical trajectories, and these can certainly be chaotic. However, what if these conditions are not satisfied?

This is the question addressed and answered in Reference 38. By defining and computing the Lyapunov exponent for an observed quantum system deep in the quantum regime, we were able show that the system dynamics is chaotic. Furthermore, the Lyapunov exponent is not the same as that of the classical dynamics that emerges in the classical limit. Since the quantum system in the absence of measurement is not chaotic, this chaos must emerge as the strength of the measurement is increased, and we examined the nature of this emergence.

To do this, we must first make certain that we can quantify the existence of chaos in a robust way. The rigorous quantifier of chaos in a dynamical system is the maximal Lyapunov exponent.[66] The exponent yields the (asymptotic) rate of exponential divergence of two trajectories that start from neighboring points in phase space, in the limit in which they evolve to infinity, and the neighboring points stay infinitesimally close. The maximal Lyapunov exponent characterizes the sensitivity of the system evolution to changes in the initial condition: if the exponent is positive, then the system is exponentially sensitive to initial conditions, and is said to be chaotic. We now discuss how this notion can be applied to observation-conditioned evolution of quantum expectation values.

A single quantum mechanical particle is in principle an infinite dimensional system. However, for the purpose of defining an observationally relevant Lyapunov exponent, it is sufficient to use a single projected data stream: let us consider the expectation value of the position, $\langle x(t) \rangle$. The important quantity is the divergence, $\Delta(t) = |\langle x(t) \rangle - \langle x_{\text{fid}}(t) \rangle|$, between a fiducial trajectory and a second trajectory infinitesimally close to it. It is important to keep in mind that the system is driven by noise. Since we wish to examine the sensitivity of the system to changes in the initial conditions, and not to changes in the noise, we must hold the noise realization fixed when calculating the divergence. The Lyapunov exponent is, therefore,

$$\lambda \equiv \lim_{t \to \infty} \lim_{\Delta_s(0) \to 0} \frac{\ln \Delta_s(t)}{t} \equiv \lim_{t \to \infty} \lambda_s(t), \qquad (24)$$

where the subscript s denotes the noise realization. This definition is the obvious generalization of the conventional ODE definition to dynamical averages, where the noise is treated as a drive on the system. Indeed, under the conditions when (noisy) classical motion emerges, and thus when localization holds, it reduces to the conventional definition, and yields the correct classical Lyapunov exponent. To combat slow convergence, we measure the Lyapunov exponent by averaging over an ensemble of finite-time exponents $\lambda_s(t)$ instead of taking the asymptotic long-time limit for a single trajectory.

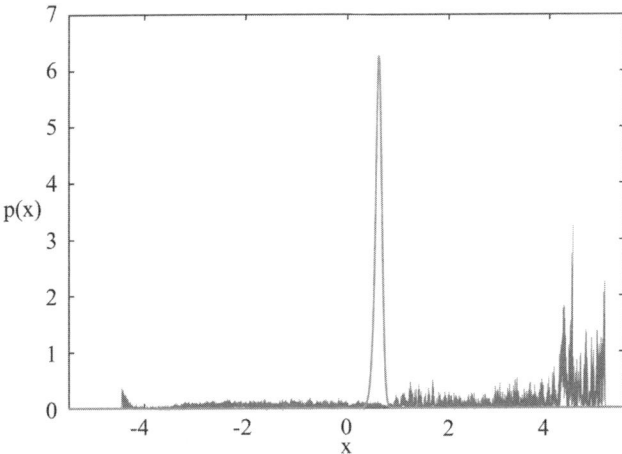

FIGURE 4. Position distribution for the Duffing oscillator with measurement strengths $k = 0.01$ (*dark gray*) and $k = 10$ (*light grey*), demonstrating measurement-induced localization ($k = 10$) as the measurement coupling is increased.[65] The momentum distribution behaves similarly.

A key result now follows: in unobserved, that is, isolated quantum dynamical systems, it is possible to prove, by employing unitarity and the Schwarz inequality, that λ vanishes; the finite-time exponent, $\lambda(t)$, decays as $1/t$.[67] From the Kosloff–Rice theorem we know, of course, that the Lyapunov exponent must be zero, since the overall evolution is integrable, but this result gives us a quantitative statement about the decay of the exponent. It turns out that this particular result applies also to the evolution of averages in isolated classical systems and, in this sense, is more general than Kosloff–Rice. As we emphasized previously, once measurement is included, the evolution becomes nonlinear and the Lyapunov exponent need not vanish classically or quantum mechanically.

As a particular system of interest, we turn once again to the Duffing oscillator, this time with $\hbar = 10^{-2}$, which is small enough so that the system makes a transition to classical dynamics when the measurement is sufficiently strong. As we increase the measurement strength, we can examine the transformation from essentially isolated quantum evolution all the way to the known chaos of the classical Duffing oscillator. To examine the emergence of chaos, we solved[65] for the evolution of the system for $k = 5 \times 10^{-4}$, 10^{-3}, 0.01, 0.1, 1, and 10. When $k \leq 0.01$, the distribution is spread over the entire accessible region, and the Ehrenfest theorem is not satisfied. Conversely, for $k = 10$, the distribution is well-localized (see FIGURE 4), and the Ehrenfest theorem holds throughout the evolution. Since the backaction noise, characterized by the momentum diffusion coefficient, $D = \hbar^2 k$, remains small, at this value of k the motion is that of the classical system, to a very good approximation.

Stroboscopic maps help reveal the global structural transformation in phase space in going from quantum to classical dynamics (see FIGURE 5). The maps consist of points through which the system passes at time intervals separated by the period of the driving force. For very small k, $\langle x \rangle$, and $\langle p \rangle$ are largely confined to a region in the center of phase space. Somewhat remarkably, at $k = 0.01$, although the system is largely delocalized, as shown in FIGURE 4, nontrivial structure appears, with considerable time being spent in certain outer regions. By $k = 1$ the localized regions have formed into narrower and sharper swirling coherent structures. At $k = 10$ the swirls disappear, and we retrieve the uniform chaotic sea of the classical map (the small "holes" are periodic islands). The swirls in fact correspond to the unstable manifolds of the classical motion. Classically, these manifolds are only visible at short times, as continual and repeated folding eventually washes out any structure in the midst of a uniform tangle. In the quantum regime, however, the weakness of the measurement, with its inability to crystallize the fine structure, has allowed them to survive: we emphasize that the maps result from long-time integration, and are therefore essentially time-invariant.

To calculate the Lyapunov exponent we implemented a numerical version of the classical linearization technique,[68] suitably generalized to quantum trajectories. The method was tested on a classical noisy system with comparison against results obtained from solving the exact equations for the Lyapunov exponents.[69] The calculation is numerically intensive, since it involves integrating the stochastic Schrodinger equation equivalent to the SME (5) over thousands of driving periods, and averaging over many noise realizations; parallel supercomputers were invaluable for this task.

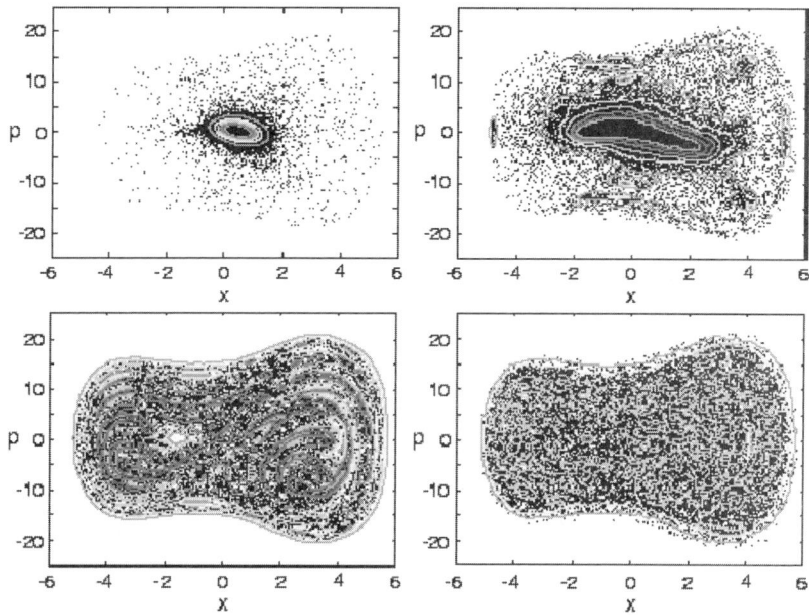

FIGURE 5. Phase space stroboscopic maps[65] for the observed Duffing oscillator for four different measurement strengths, $k = 5 \times 10^{-4}$, 0.01 (*top*), and 1, 10 (*bottom*). Contour lines are superimposed to provide a measure of local point density at relative density levels of 0.05, 0.15, 0.25, 0.35, 0.45, and 0.55.

The computations show that as t is increased, for nonzero k, the value obtained for $\lambda(t)$ falls as $1/t$, following the behavior expected for $k = 0$, until a point at which an asymptotic regime takes over, stabilizing at a finite value of the Lyapunov exponent as $t \to \infty$. This behavior is shown in FIGURE 6 for three values of k. The Lyapunov exponent as a function of k is shown in FIGURE 7. The exponent increases over two orders of magnitude in an approximately power-law fashion as k is varied from 5×10^{-4} to 10, before settling to the classical value, $\lambda_{Cl} = 0.57$. The results in FIGURES 6 and 7 show clearly that chaos emerges in the observed quantum dynamics well before the limit of classical motion is obtained.

We also computed the Lyapunov exponent for the quantum system when its action is sufficiently small that smooth classical dynamics cannot emerge, even for strong measurements.[65] Taking a value of $\hbar = 16$, we find that for $k = 5 \times 10^{-3}$, $\lambda = 0.029 \pm 0.008$, for $k = 0.01$, $\lambda = 0.046 \pm 0.01$, and for $k = 0.02$, $\lambda = 0.077 \pm 0.01$. Thus, the system is once again chaotic and becomes more strongly chaotic the more strongly it is observed. From these striking results, it is clear that there exists a purely *quantum* regime in which an observed system, while behaving in a fashion quite distinct from its classical limit, nevertheless evolves chaotically with a finite Lyapunov exponent, also distinct from the classical value.

It is worth pointing out that an analogous analysis can also be carried out for a continuously observed classical system. As mentioned previously, an *unobserved* probabilistic classical system also has provably zero Lyapunov exponent: the average of x for an ensemble of classical particles does not exhibit chaos, due to the linearity of the Liouville equation.[67] If we consider a noiseless observed chaotic classical system—possible since classical measurements are by definition passive

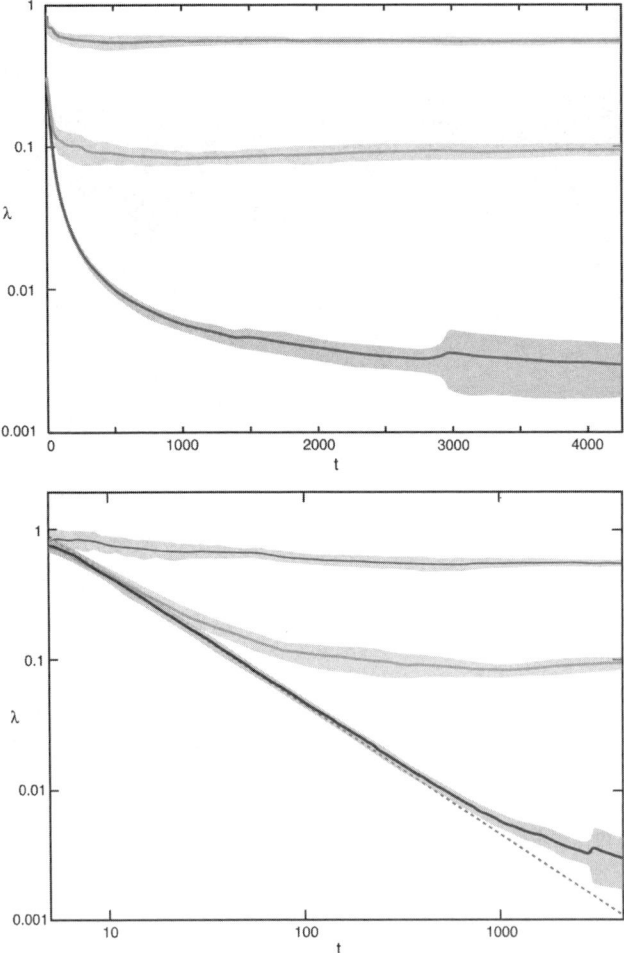

FIGURE 6. Finite-time Lyapunov exponents $\lambda(t)$ for measurement strengths $k = 5 \times 10^{-4}$, 0.01, and 10 averaged over 32 trajectories for each value of k (linear scale in time, **top**, and logarithmic scale, **bottom**; bands indicate the standard deviation over the 32 trajectories).[65] The (analytic) $1/t$ fall-off at small k values, prior to the asymptotic regime, is evident in the **bottom panel**. The unit of time is the driving period.

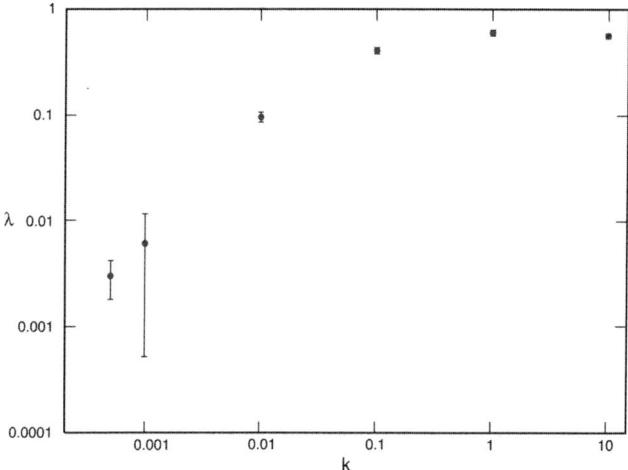

FIGURE 7. The emergence of chaos: [65] the Lyapunov exponent λ as a function of measurement strength k. *Error-bars* follow those of FIGURE 6, taken at the final time.

(no backaction noise)—then even the weakest meaningful measurement will, over time, localize the probability density, generating an effective trajectory limit, and thus, the classical Lyapunov exponent, λ_{Cl}.[67] Noise can always be injected into classical systems as an external drive, nevertheless, in the limit of weak noise, the system will once again possess the noiseless exponent λ_{Cl}. In a classical system the external noise is not connected to the strength of the measurement, so one can simultaneously have strong measurement and weak noise, which, as we have seen, is possible in the quantum theory only under specific conditions.[39–41]

As one way to understand the classical case, we can employ the quantum result as an intermediate step. Consider the quantum Lyapunov exponent at a fixed value of k (where $\lambda < \lambda_{Cl}$) as in FIGURE 7. If the value of $\hbar$ is now reduced, the dynamics of the system must tend to the classical limit as the quantum–classical correspondence inequalities[3] are better satisfied. Thus, the Lyapunov exponent in the classical limit of quantum theory—which, to a very good approximation, is just classical dynamics driven by weak noise—must tend to λ_{Cl}. If, however, the noise is not weak, an observed classical system, like a quantum system outside the classical regime, will also not be localized, and may well have an exponent different from λ_{Cl}. In addition, one may expect the non-localized quantum and classical evolutions to have quite different Lyapunov exponents, especially when $\hbar$ is large on the scale of the phase space, because quantum and classical evolutions generated by a given nonlinear Hamiltonian are essentially different.[19] The nature of the Lyapunov exponent for nonlocalized classical systems, and its relationship to the exponent for quantum systems is a very interesting open question.

CONCLUDING REMARKS

To summarize, we have presented a simple analysis of continuously observed classical and quantum dynamical systems. This analysis is in fact required to deal with next-generation experiments and underlies the nascent field of real-time quantum feedback control. Major results include an intuitive and quantitative understanding of the quantum-classical transition. It is pleasing that both the strong and weak forms of the QCT can eventually be understood as a macroscopic limit of observed-system quantum mechanics, that is, whenever the observed system action $S \gg \hbar$.

Perhaps, most interestingly, we have obtained clear predictions for dynamical chaos in observed quantum systems that are far from the classical regime. We emphasize that the chaos identified here is not merely a formal result—even deep in the quantum regime, the Lyapunov exponent can be obtained from measurements on a real system as in near-future cavity QED and nanomechanics experiments.[10,11] Experimentally, one would use the known measurement record to integrate the SME (5); this provides the time evolution of the mean value of the position. From this fiducial trajectory, given the knowledge of the system Hamiltonian, the Lyapunov exponent can be obtained by following the procedure described here.

ACKNOWLEDGMENTS

SH thanks the organizers of the 16th Florida Workshop in Nonlinear Astronomy and Physics, dedicated to the memory of Henry Kandrup, for their kind invitation to lecture at the meeting. Large-scale parallel computing support from Los Alamos National Laboratory's Institutional Computing Initiative is gratefully acknowledged. This research is supported by the Department of Energy, under contract W-7405-ENG-36.

The authors declare that they have no competing financial interests.

REFERENCES

1. KOSLOFF, R. & S.A. RICE. 1981. J. Chem. Phys. **74:** 1340.
2. MANZ, J. 1989. J. Chem. Phys. **91:** 2190.
3. BOCCHIERI, P. & A. LOINGER. 1957. Phys. Rev. **107:** 337.
4. HOGG, T. & B.A. HUBERMAN. PHYS. 1982. Rev. Lett. **48:** 711.
5. BELL, J.S. 1990. Phys. World **8:** 33.
6. KOOPMAN, B.O. 1931. Proc. Natl. Acad. Sci. USA **17:** 31.
7. PERES, A. 1993. Quantum Theory: Concepts and Methods. Kluwer, Boston.
8. MAYBECK, P.S. 1982. Stochastic Models, Estimation and Control. Academic Press, New York.
9. JACOBS, O.L.R. 1993. Introduction to Control Theory. Oxford University Press, Oxford.
10. MABUCHI, H. & A.C. DOHERTY. 2002. Science **298:** 1372.
11. LAHAYE, M.D., O. BUU, B. CAMAROTA & K.C. SCHWAB. 2004. Science **304:** 74.
12. BELL, J.S. 1988. Speakable and Unspeakable in Quantum Mechanics. Cambridge University Press, New York.
13. LANDAU, L.D. & E.M. LIFSHITZ. 1965. Quantum Mechanics: Non-Relativistic Theory. Pergamon Press, New York.
14. LANDAU, L.D. & E.M. LIFSHITZ. 1980. Statistical Physics. Pergamon Press, New York.
15. HEPP, K. 1972. Helv. Phys. Acta **45:** 237.
16. ZUREK, W.H. 1981. Phys. Rev. D **24:** 1516.
17. ZUREK, W.H. 1981. Phys. Rev. D **26:** 1862.

18. JOOS, E. & H.D. ZEH. 1985 Z. Phys. B **59**: 223.
19. HABIB, S., K. JACOBS, H. MABUCHI, et al. 2002. Phys. Rev. Lett. **88**: 040402.
20. WIGNER, E.P. 1932. Phys. Rev. **40**: 749.
21. TATARSKII, V.I. 1983. Usp. Fiz. Nauk **139**: 587. [Sov. Phys. Uspekhi **26**: 311].
22. HILLERY, M., R.F. O'CONNELL, M.O. SCULLY & E.P.WIGNER. 1984. Phys. Rep. **106**: 121.
23. HABIB, S. 1990. Phys. Rev. D **42**: 2566.
24. KADANOFF, L.P. & G. BAYM. 1989. Quantum Statistical Mechanics. Addison-Wesley, Redwood City.
25. ZWANZIG, R. 2001. Nonequilibrium Statistical Mechanics. Oxford University Press, New York.
26. BLUM, K. 1996. Density Matrix Theory and Applications. Plenum Press, New York.
27. DIOSI, L. 1988. Phys. Lett. **129A**: 419.
28. BELAVKIN, V.P. & P. STASZEWSKI. 1989. Phys. Lett. **140A**: 359,
29. SALAMA, Y. & N. GISIN. 1993. Phys. Lett. **181A**: 269.
30. CAVES, C.M. & G.J. MILBURN. 1987. Phys. Rev. A **36**: 5543.
31. WISEMAN, H.M. & G.J. MILBURN. 1993. Phys. Rev. A **47**: 642.
32. MILBURN, G.J. 1996. Quantum semiclass. Opt. **8**: 269.
33. DOHERTY, A.C. & K. JACOBS. 2003. Phys. Rev. A **60**: 2700.
34. WARSZAWSKI, P. & H.M. WISEMAN. 2003. J. Opt. B **5**: 1.
35. DOHERTY, A.C., K. JACOBS & G. JUNGMAN. 2001. Phys. Rev. A **63**: 062306.
36. GILLESPIE, D.T. 1996. Am. J. Phys. **64**: 225.
37. MOZYRSKY, D. & I. MARTIN. 2002. Phys. Rev. Lett. **89**: 018301.
38. MCGARTY, T.P. 1974. Stochastic Systems and State Estimation. Wiley-Interscience, New York.
39. BHATTACHARYA, T., S. HABIB & K. JACOBS. 2000. Phys. Rev. Lett. **85**: 4852.
40. BHATTACHARYA, T., S. HABIB & K. JACOBS. 2003. Phys. Rev. A **67**: 042103.
41. GHOSE, S., P. ALSING, I. DEUTSCH, et al. 2004. Phys. Rev. A **69**: 052116.
42. HABIB, S. quant-ph/0406011.
43. CHIRIKOV, B.V. 1991. Chaos **1**: 95.
44. LIN, W.A. & L.E. BALLENTINE. 1990. Phys. Rev. Lett. **65**: 2927.
45. HABIB, S., K. SHIZUME & W.H. ZUREK. 1998. Phys. Rev. Lett. **80**: 4361.
46. GREENBAUM, B.D., S. HABIB, K. SHIZUME & B. SUNDARAM. 2005. Chaos. In press. quant-ph/0401174.
47. LINDBLAD, G. 1976. Comm. Math. Phys. **48**: 199.
48. GORINI, V., A. KOSSAKOWSKI & E.C.G. SUDARSHAN. 1976. J. Math. Phys. **17**: 821.
49. CALDEIRA, A.O. & A.J. LEGGETT. 1985. Phys. Rev. A **31**: 1059.
50. BERRY, M.V. 1977. Phil. Trans. Roy. Soc. A **287**: 237.
51. HELLER, E.J. 1977. J. Chem. Phys. **67**: 3339.
52. KOLOVSKY, A.R. 1996. Phys. Rev. Lett. **76**: 340.
53. MASLOV, V.P. & M.V. FEDORIUK. 1981. Semi-Classical Approximation in Quantum Mechanics. Reidel, Holland.
54. BERRY, M.V. & N.L. BALAZS. 1979. J. Phys. A **12**: 625.
55. HELLER, E.J. & S. TOMSOVIC. 1993. Phys. Today **7**: 38.
56. GUCKENHEIMER, J. & P. HOLMES. 1986. Nonlinear Oscillations, Dynamical Systems, and Bifurcations of Vector Fields. Springer-Verlag, Berlin.
57. BELAVKIN, V.P. 1992. Comm. Math. Phys. **146**: 611.
58. BELAVKIN, V.P. 1999. Rep. Math. Phys. **43**: 405.
59. DOHERTY, A.C., S. HABIB, K. JACOBS, et al. 2000. Phys. Rev. A **62**: 012105.
60. WISEMAN, H.M. & G.J. MILBURN. 1993. Phys. Rev. Lett. **70**: 548.
61. WISEMAN, H.M., S. MANCINI & J. WANG. 2002. Phys. Rev. A **66**: 013807.
62. RUSKOV, R. & A.N. KOROTKOV. 2002. Phys. Rev. B **66**: 041401(R).
63. HOPKINS, A., K. JACOBS, S. HABIB & K. SCHWAB. 2003. Phys. Rev. B **68**: 235328.
64. STECK, D.A., K. JACOBS, H. MABUCHI, et al. 2004. Phys. Rev. Lett. **92**: 223004.
65. HABIB, S., K. JACOBS & K. SHIZUME. quant-ph/0412159.
66. ECKMANN, J.-P. & D. RUELLE. 1985. Rev. Mod. Phys. **57**: 617.
67. HABIB, S., K. JACOBS & K. SHIZUME. 2005. In preparation.
68. WOLF, A., J.B. SWIFT, H.L. SWINNEY & J.A. VASTANO. 1985. Physica **16D**: 285.
69. HABIB, S. & R.D. RYNE. 1995. Phys. Rev. Lett. **74**: 70.

Afterword

I hope that this volume of *Annals of the New York Academy of Sciences* and the meeting on which it was based are the memorials that Henry would have liked us to have for him. The meeting involved a lot of science as well as some lighter moments. I cannot possibly summarize all that went on, so I will not try, and please excuse the fact that what I have to say is colored by my own interests. Henry's interests were broad, and the editors and conference organizers have done well in making sure that all are covered to some degree. Those interests included quantum as well as classical mechanics, general relativity, accelerator dynamics, statistical mechanics, in addition to his work on galaxies, which I know best. David Merritt, one of our opening day speakers, gave an excellent account of Henry's ideas concerning how stellar systems relax, and of the significance of Henry's ideas of chaotic mixing. I well remember the fervor with which Henry used to berate galactic theoreticians for ignoring the role of chaos. One could almost envision him as an Old Testament prophet. The reason why there was less of this in later years was not due to any loss of fervor on his part, but simply because his message began to be heeded.

The volume reminds us that Henry's collaborations were likewise broad. Court Bohn, another opening day conference speaker, told us of how he and Henry made contact only as recently as 2001 through their mutual interest in accelerator beams. Their collaboration bloomed impressively since then, and it is good to know that there is now a thriving center of Kandrupian ideas, staffed by former coworkers of Henry and others, on the plains of Illinois, in the Physics Department of Northern Illinois University.

I did not begin to know Henry until 1991, shortly after he moved to the University of Florida. He soon made his presence felt more widely in Florida by organizing an inter-institutional project on collisions and mergers of elliptical galaxies. This included Eddie Qian and myself from FSU, and was funded by the Florida Space Grant Consortium, who helped support the meeting. It was embarrassing to us how much more productive Henry and his UF colleagues Elaine Mahon and Haywood Smith were than the rest of us. It made me realize what a great catch the University of Florida had made with Henry. Another of my early memories of Henry is of a foreign meeting. Unfortunately it was not in such elegant surroundings as the Villa Borghese, of which Katie Freese reminisced, but in the somewhat less prepossessing surroundings of Petrozavodsk in post-Soviet Russia. It was held to mark the 60th and 80th birthdays, respectively, of Vadim Antonov and T.A. Agekian. Henry and I were the only two Western scientists to attend, which made me realize how hardy Florida must make us. Henry was the only speaker at the meeting who was invited to give two talks, and he was clearly an honored guest. That meeting also introduced me to Henry's love of beer. One of our Russian hosts took us on a walking tour of St. Petersburg after the meeting. It was a long walk because Peter the Great built his capital on a grand scale. The day was sunny, and I remember Henry's delight when we finally came upon a street vendor selling beer, and the dispatch with which Henry drank. I was reminded of that event during the memorial meeting for Henry, when Balša Terzić, my former student and Henry's former postdoctoral student, were dining in one of Henry's favorite restaurants. It turns out that Henry had taken Balša there while he was recruiting him, to make the point that Gainesville had some important cultural advantages, such as the availability of Newcastle brown and Guinness on draft.

Henry thrived at the University of Florida. He spent several years as a wandering scholar before finding his home in Gainesville. Anyone who ever heard him speak could see what an inspiring teacher he must have been. Stan Dermott, his Department chair, confirmed that his student course evaluations showed this too. During the meeting we heard that same message and of the concern he had for students, from the former students who have spoken. My most vivid memory, from the memorial service in the Baughman Center in December 2003, was of the aching sense of loss felt and expressed by the many students who had known Henry and had been influenced by him.

Shortly after he died, I learned that Henry had an ambitious plan to extend his teaching to a wider audience. He had proposed to Springer a three-volume book on Hamiltonian galactic dynamics and Springer had received enthusiastic reviews for this work. That is understandable because it would be of an unrivalled breadth. It would cover basic principles and would include nonlinear dynamics, solar system work, plasma physics, and stochastic differential equations within its scope. It would also "translate important results from papers too mathematical for many galactic astronomers to follow easily." This refers to results that have been obtained by a group associated with Professor Jürgen Batt of the Ludwig Maximilian University in Munich, who have made important advances and obtained rigorous results for the collisionless Boltzmann equation. Gerhard Rein, another of the opening day speakers, hails from that group. He and Yan Guo have obtained results on nonlinear stability of stellar systems that are rigorous, as well as surpassing anything previously known. I use this example to illustrate Henry's knowledge of and appreciation of mathematics, which is probably less well known to this audience than his physics and astrophysics. It is not clear that there is anyone else who has Henry's capabilities for writing this book and since it sounds as though Henry himself expected that it would take him another two years to complete, we may have lost much that would have been valuable.

Meetings are not the same without one of Henry's talks. Although not one to push himself forward, his eyes would light up when his time to speak came and he had the stage. He clearly relished his chance to teach us something. He had a big influence on his field, his colleagues, and his students. His life made a difference. It was encouraging to learn that the Department of Astronomy has now recognized the need for new appointments in theoretical astrophysics and intends to make them; something for which Henry long campaigned. We have done the best that we can to honor Henry's memory. The meeting was a success because of all those who knew Henry, and who came because of that to tell us about new and exciting developments, and to explain how Henry's work continues to have influence.

CHRIS HUNTER

Index of Contributors

Athanassoula, E., 168–192

Baskaran, A., 93–102
Bernal, S., 45–54
Bhattacharya, T., 308–332
Bohn, C.L., 12–33, 34–44
Boonyasait, C., 203–224
Buchler, J.-R., ix–x

Contopoulos, G., 139–167

Dragt, A.J., 291–307
Dufty, J.W., 93–102
Dupke, R., 260–275

Gottesman, S.T., ix–x, 203–224
Greenbaum, B., 308–332

Haber, I., 45–54
Habib, S., 308–332
Harris, J.R., 45–54
Harsoula, M., 139–167
Hueckstaedt, R.M., 246–259
Hunter, C., 120–138, 333–334
Hunter, Jr., J.H., 246–259
Huo, Y., 45–54

Ipser, J.R., 1

Jacobs, K., 308–332

Kandrup, H.E., 12–33, 68–78
Kishek, R.A., 12–33, 45–54

Lebovitz, N.R., 276–290
Li, H., 45–54
Lovelace, R.V.E., 246–259

Mahon, M.E., ix–x
Merritt, D., 3–11, 225–231
Mikkola, S., 225–231

O'Shea, P.G., 12–33, 45–54

Patsis, P.A., 203–224
Pfenniger, D., 193–202
Pogorelov, I.V., 55–67, 68–78

Rein, G., 103–119
Reiser, M., 12–33, 45–54
Revaz, Y., 193–202

Shizume, K., 308–332
Sideris, I.V., 12–33, 79–92
Sundaram, B., 308–332
Szell, A., 225–231

Terzić, B., 55–67

Williams, R.K., 232–245